T0238205

Lecture Notes in Computer Science 805

Edited by G. Goos and J. Hartmanis

Advisory Board: W. Brauer D. Gries J. Stoer

Michel Cosnard Afonso Ferreira
Joseph Peters (Eds.)

Parallel and
Distributed Computing

Theory and Practice

First Canada-France Conference
Montréal, Canada, May 19-21, 1994
Proceedings

Springer-Verlag
Berlin Heidelberg New York
London Paris Tokyo
Hong Kong Barcelona
Budapest

Michel Cosnard   Afonso Ferreira
Joseph Peters (Eds.)

Parallel and Distributed Computing

Theory and Practice

First Canada-France Conference
Montréal, Canada, May 19-21, 1994
Proceedings

Centre Jacques Cartier

CENTRE NATIONAL
DE LA RECHERCHE
SCIENTIFIQUE

Université de Montréal

Concordia University

CRIM - Montréal

Springer-Verlag

Berlin Heidelberg New York
London Paris Tokyo
Hong Kong Barcelona
Budapest

Series Editors

Gerhard Goos
Universität Karlsruhe
Postfach 69 80
Vincenz-Priessnitz-Straße 1
D-76131 Karlsruhe, Germany

Juris Hartmanis
Cornell University
Department of Computer Science
4130 Upson Hall
Ithaca, NY 14853, USA

Volume Editors

Michel Cosnard
Afonso Ferreira
CNRS - Laboratoire de l'Informatique du Parallélisme
Ecole Normale Supérieure de Lyon
46, Allée d'Italie, F-69364 Lyon Cédex 07, France

Joseph Peters
School of Computing Science, Simon Fraser University
Burnaby, B.C. V5A 1S6 Canada

CR Subject Classification (1991): D.4.2-4, D.4.7, C.2.1, C.2.4, F.1.2, G.1, I.3.5

ISBN 3-540-58078-6 Springer-Verlag Berlin Heidelberg New York
ISBN 0-387-58078-6 Springer-Verlag New York Berlin Heidelberg

CIP data applied for

© Springer-Verlag Berlin Heidelberg 1994
Printed in Germany

Typesetting: Camera-ready by author
SPIN: 10131120   45/3140-543210 - Printed on acid-free paper

Preface

With the arrival of the 90's, a successful series of small annual workshops have been organized in Vancouver (Canada) in the summers of 1990 and 1992 and in Marseille (France) in the summers of 1991 and 1993. The main subject of such workshops was information dissemination in interconnection networks for distributed memory parallel computers.

The first Canada-France Conference on Parallel Computing grew out of such workshops, along with the well established and very productive collaboration between Canadian and French researchers, aiming to foster collaboration between theoreticians who study the design and analysis of parallel and distributed algorithms and networks, and practitioners who apply, adapt, and extend the theoretical results to solve real-world problems.

This Conference, that is open to all researchers from the international community working in the area, is a complement to the annual workshops. Its scientific program consists of four sessions composed by one invited speaker followed by five contributed talks each. Their themes are parallel algorithms and complexity, interconnection networks and distributed computing, algorithms for unstructured problems, and structured communications. The contributed papers in this proceedings were selected by the Program Committee on the basis of referee reports. Each paper appearing in these proceedings was reviewed by at least two referees who judged the papers for originality, quality, and consistency with the themes of the conference. Nick Pippenger's invited paper was also refereed. We wish to thank all of the authors who responded to the call for papers, our invited speakers, and all of the referees and Program Committee members who reviewed papers.

We are grateful to our partner, the Centre Jacques Cartier, and to the PRC C3 of the French CNRS for financial support, to Centre de Recherche en Informatique de Montréal, to Département d'Informatique et de Recherche Opérationnelle, Université de Montréal, and Laboratoire de l'Informatique du Parallélisme, Ecole Normale Supérieure de Lyon for secretarial support, and to Concordia University for providing the facilities for the Conference. Also, we especially thank Valerie Roger, Jean-Louis Duclos, Eliane Fleury, and Jocelyne Richerd, responsible for most of the secretarial duties related to the conference, as well as the members of the Organizing Committee for doing triple duty. In addition to handling organizational details including local arrangements, publicity, and funding, they served as referees and as Program Committee members.

March 1994   M.Cosnard, A.Ferreira, J.Peters

Préface

Avec l'arrivée des années 90, une série réussie de petits séminaires de recherche ont été organisés à Vancouver (Canada) pendant les étés 1990 et 1992 ainsi qu'à Marseille (France) pendant les étés 1991 et 1993. Le sujet principal de ces séminaires était l'étude des communications dans les réseaux d'interconnexion des ordinateurs parallèles à mémoire distribuée.

La première conférence Canada-France sur le Calcul Parallèle est issue de ces séminaires, couplée avec la collaboration bien établie et très productive entre chercheurs Canadiens et Français, dans l'objectif de développer la collaboration entre les théoriciens qui étudient la conception et l'analyse d'algorithmes parallèles et distribués et les réseaux, et les praticiens qui appliquent, adaptent et étendent la recherche théorique pour résoudre des problèmes du monde réel.

Cette conférence, qui est ouverte à tout chercheur de la communauté internationale travaillant dans le domaine du calcul parallèle, est donc une suite logique aux séminaires annuels. Son programme scientifique est constitué de quatre sessions comportant chacune un conférencier invité suivi de 5 intervenants. Les thèmes de ces sessions sont respectivement les algorithmes parallèles et la complexité, les réseaux d'interconnexion et l'informatique distribuée, les algorithmes pour des problèmes non-structurés, et les communications structurées. Les articles qui sont publiés dans ces actes ont été sélectionnés par un Comité de Programme sur la base des rapports des relecteurs. Chaque article a été jugé par au moins deux relecteurs sur la base de son originalité, de sa qualité et de sa pertinence avec les thèmes de la conférence. La contribution invitée de Nick Pippenger a aussi été examinée par le comité de lecture.

Nous souhaitons remercier tous les auteurs qui ont répondu à l'appel à contribution, les conférenciers invités, tous les relecteurs et les membres du comité de programme qui ont jugé de la qualité des articles.

Nous remercions également notre partenaire, le Centre Jacques Cartier, le CRIM à Montréal et le PRC C3 du CNRS Français pour le support financier, le Département d'Informatique et de Recherche Opérationnelle de l'Université de Montréal, le Laboratoire de l'Informatique du Parallélisme, Ecole Normale Supérieure de Lyon pour le support de secrétariat, ainsi que l'Université de Concordia qui a fourni les facilités pour la conférence.

Nous voulons remercier particulièrement Valérie Roger, Jean-Louis Duclos, Eliane Fleury, et Jocelyne Richerd, qui ont assumé la plupart des tâches administratives et de secrétariat liées à la conférence, ainsi que les membres du comité d'organisation pour leur triple responsabilité. En plus de la prise en charge de l'organisation incluant les préparatifs sur place, la publicité et la recherche de financement, ils ont également accepté d'être relecteurs et membres du comité de programme.

Mars 1994   M.Cosnard, A.Ferreira, J.Peters

Committees

Steering Committee

Afonso Ferreira (CNRS – LIP – ENS Lyon, Lyon, France)
Joseph Peters (Simon Fraser University, Burnaby, BC, Canada)

Organising Committee

Afonso Ferreira (CNRS – LIP – ENS Lyon, Lyon, France)
Geña Hahn (Université de Montréal, Montréal, P.Q., Canada)
Jaroslav Opatrný (Concordia University, Montréal, P.Q., Canada)
Joseph Peters (Simon Fraser University, Burnaby, BC, Canada)
Vincent Van Dongen (CRIM, Montréal, P.Q., Canada)

Program Committee

Selim Akl (Queen's University, Kingston, Ont, Canada)
Michel Cosnard – Chair (LIP – ENS Lyon, Lyon, France)
Afonso Ferreira (CNRS – LIP – ENS Lyon, Lyon, France)
Pierre Fraigniaud (CNRS – LIP – ENS Lyon, Lyon, France)
Geña Hahn (Université de Montréal, Montréal, P.Q., Canada)
Marie-Claude Heydemann (Université de Paris-Sud, Paris, France)
Arthur Liestman (Simon Fraser University, Burnaby, BC, Canada)
Jaroslav Opatrný (Concordia University, Montréal, P.Q., Canada)
Prakash Panangaden (McGill University, Montréal, P.Q., Canada)
Joseph Peters (Simon Fraser University, Burnaby, BC, Canada)
Ajit Singh (University of Waterloo, Waterloo, Ont, Canada)
Ivan Stojmenovic (University of Ottawa, Ottawa, Ont, Canada)
Denis Trystram (LMC – IMAG, Grenoble, France)
Vincent Van Dongen (CRIM, Montréal, P.Q., Canada)
Alan Wagner (University of British Columbia, Vancouver, BC, Canada)

Contents

Juggling Networks

Nicholas Pippenger

University of British Columbia, Vancouver, BC V6T 1Z4, Canada

Abstract. Switching networks of various kinds have come to occupy a prominent position in computer science as well as communication engineering. The classical switching network technology has been space-division-multiplex switching, in which each switching function is performed by a spatially separate switching component (such as a crossbar switch). A recent trend in switching network technology has been the advent of time-division-multiplex switching, wherein a single switching component performs the function of many switches at successive moments of time according to a periodic schedule. This technology has the advantage that nearly all of the cost of the network is in inertial memory (such as delay lines), with the cost of switching elements growing much more slowly as a function of the capacity of the network. In order for a classical space-division-multiplex network to be adaptable to time-division-multiplex technology, its interconnection pattern must satisfy stringent requirements. For example, networks based on randomized interconnections (an important tool in determining the asymptotic complexity of optimal networks) are not suitable for time-division-multiplex implementation. Indeed, time-division-multiplex implementations have been presented for only a few of the simplest classical space-division-multiplex constructions, such as rearrangeable connection networks. This paper shows how interconnection patterns based on explicit constructions for expanding graphs can be implemented in time-division-multiplex networks. This provides time-division-multiplex implementations for switching networks that are within constant factors of optimal in memory cost, and that have asymptotically more slowly growing switching costs. These constructions are based on a metaphor involving teams of jugglers whose throwing, catching and passing patterns result in intricate permutations of the balls. This metaphor affords a convenient visualization of time-division-multiplex activities that should be of value in devising networks for a variety of switching tasks.

1. Introduction

In this paper we will present a metaphor for describing the construction and operation of time-division-multiplex networks, and use it to present a new time-division-multiplex implementation of an explicit construction for expanding graphs, which are an essential component in many constructions for switching networks. Both the new metaphor and the main techniques for construction of time-division-multiplex networks will be illustrated in Sect. 2 by a well known construction for rearrangeable connection networks. This construction was described in the context of space-division-multiplex networks by Beneš [6] in 1964. The time-division-multiplex implementation was first described by Marcus [13] in 1970, and has recently been rediscovered by Ramanan, Jordan and Sauer [23]. The resulting implementation is "time-slot interchanger" in the sense of Inose [9]. In Sect. 3 we indicate how these methods can be adapted to other types of switching networks. The main obstacle for such applications is the requirement for "expanding graphs" (and related objects) presented by many constructions for switching networks. In Sect. 4 we present a time-division-multiplex implementation of a well known construction for expanding graphs (and, more generally, for graphs with a prescribed "eigenvalue separation ratio"). This construction was first proposed by Margulis [14] in 1974. Quantitative estimates essential for its application were provided by Gabber and Galil [8] in 1981, and improvements to these estimates have been given by Jimbo and Maruoka [10], whose version of the space-division-multiplex construction we follow. In Sect. 5 we present some open problems prompted by this work.

2. Connectors

Imagine a *juggler* who can with complete reliability throw balls to a fixed height, so that they always return a fixed amount of time after they are thrown. All amounts of time considered in this paper will be multiples of some fixed unit of time that will be called the *pulse*. Suppose that our juggler can take a ball at each pulse from an external agent, the juggler's *source*, and can give a ball at each pulse to another external agent, the juggler's *sink*. Suppose further that at each pulse the juggler can execute either of two *moves*, which will be called the *straight* and *crossed* moves. In the straight move, the juggler rethrows the ball that returns from the air (if any such ball returns), and gives the ball taken from the source to the sink (if any such ball is taken). In the crossed move, the juggler throws the ball taken from the source (if any is taken), and gives the ball that returns from the air to the sink (if any returns).

Now imagine a *chain* of jugglers; that is, a finite sequence of jugglers $J_1, \ldots, J_\mu$ in which J_λ is the source of $J_{\lambda+1}$, and $J_{\lambda+1}$ is the sink of J_λ, for $1 \leq \lambda < \mu$. (The source of J_1 and the sink of J_μ are external to the chain. They will be called the *source* and *sink* of the chain.) We assume that the jugglers may have different "spans" (where the *span* of a juggler is the amount of time between the throw of a ball and its return), but that all of these are multiples of a common pulse. Depending on the spans of the various jugglers, and on the sequence of straight and crossed moves executed by each juggler, the sequence of balls passed by the source of the chain (and empty pulses during which no ball is passed) will be rearranged in some way before being passed to the sink of the chain.

In what follows, we shall regard the span of each juggler as a fixed and unchanging attribute of the juggler, while we regard the sequence of moves as being variable. How does each juggler decide what sequence of moves to execute?

Our assumption will be that each juggler has a partner, called the juggler's *cox*, who calls out the name, "straight" or "crossed", of the move to execute at each pulse. How does the cox decide what sequence of moves to call? Our assumption will be that each cox is also a juggler who juggles a fixed sequence of balls. A cox has no source or sink, and always executes straight moves, rethrowing each ball as it returns from the air. We shall assume that a ball returns at each pulse (there are no empty pulses), so that the number of balls being juggled by the cox is equal to the cox's span (which may be different from the cox's partner's span). Finally, we shall assume that each ball juggled by the cox has one of two *colors*, say *red* for "straight" and *blue* for "crossed", and that the cox calls out the move corresponding to the color of each ball as it is rethrown. Thus each cox calls for a periodic sequence of moves, corresponding to the cyclic sequence of colors of balls in the cox's pattern, with a period that is equal to the cox's span.

We can now give a simple example showing how a coxed chain of jugglers can serve as a model for a time-division-multiplex rearrangeable connection network. Let $n = 2^\nu$ be an integral power of 2. Consider a chain of $2\nu - 1$ jugglers $J_1, \ldots, J_{2\nu-1}$. Suppose that jugglers $J_1, \ldots, J_\nu$ have spans $2^0 = 1, \ldots, 2^{\nu-1} = n/2$, respectively, and that jugglers $J_{\nu+1}, \ldots, J_{2\nu-1}$ have spans $2^{\nu-2} = n/4, \ldots, 2^0 = 1$, respectively. Suppose further that all $2\nu - 1$ coxes have span $2^\nu = n$.

Suppose that the source of the chain just described passes it a sequence of balls at successive pulses. Let us divide the pulses into a sequence of *frames*, with each frame comprising n successive pulses. The sequence of balls passed by the source to the chain may be broken into frames, with each frame of balls comprising the balls passed to the chain during a frame of pulses. The sequence of balls passed by the chain to its sink may be broken into frames in a similar way. Furthermore, we may establish a correspondence between source frames and sink frames in the following way. Imagine that each juggler in the chain executes only crossed moves, so that the stream of balls from the source is passed on to the sink after a fixed delay, equal to the sum of the spans of the jugglers in the chain (which is in this case $3n/2 - 2$). Thus each source frame corresponds to a sink frame that is the series of n pulses during which the balls of the source frame emerge from the chain in this situation. The positions of the n balls within their frame will be called *slots*. We shall index the slots of each frame from $1, \ldots, n$ (slot 1 is the earliest, and slot n the latest, slot of its frame).

Theorem 0. *For every permutation $\pi : \{1, \ldots, n\} \rightarrow \{1, \ldots, n\}$, there exist patterns for each cox that cause each ball that is passed by the source to the chain in slot i of a frame to be passed by the chain to its sink in slot $\pi(i)$ of the corresponding frame.*

The proof of this theorem, which is implicit in the work of Marcus [13] in 1970, is based on the construction of Beneš for rearrangeable connection networks. This space-division-multiplex construction employs $(2\nu - 1)2^{\nu-1}$ switching elements (2×2 crossbar switches), arranged in $2\nu - 1$ *stages*, with each stage comprising

$2^{\nu-1}$ crossbars. In the time-division-multiplex implementation of this construction, each of the $2\nu-1$ jugglers in the chain will simulate the $2^{\nu-1}$ crossbars of the corresponding stage.

The space-division-multiplex construction is usually described recursively. In the drawing resulting from this description, crossbars are depicted as "boxes" and the wires interconnecting them are depicted as "lines" that follow "perfect shuffle" interconnection patterns. It is possible to redraw the this picture, however, so that the wires that carry the signals from the inputs to the outputs remain parallel to each other, with the crossbars of each stage conditionally exchanging the signals on wires separated by a fixed distance (depending upon the stage). This can in fact be done so that the distance in each stage is just the span that we have assigned to the corresponding juggler.

When this redrawing has been done, we see that the task of a juggler for each pair of slots (separated by the span of the juggler) is either to leave them unaffected, or to exchange the balls in these two slots. In the latter case, we need to "delay" the contents of the earlier slot by a number of pulses equal to the span, and to "advance" the contents of the later slot by the same amount. Since we cannot implement negative delays, we add a constant delay, equal to the span, to all slots of the frame. With this adjustment, each juggler's task is either to delay both slots by the span (which can be accomplished by three crossed moves at appropriate pulses), or to delay the earlier slot by twice the span and the later slot not at all (which can be accomplished by a crossed move, followed by a straight move, followed by another crossed move). Thus in any case the juggler can be instructed to perform the appropriate sequence of moves by a suitable pattern for the cox.

We may summarize the import of Theorem 0 by saying that a time-division-multiplex rearrangeable connection network with $n = 2^{\nu}$ slots can be implemented by a *juggling network* with $2\nu - 1 = O(\log n)$ jugglers, overall delay $3n/2 - 2 = O(n)$, and total memory $(3n/2 - 2) + (2\nu - 1)2^{\nu} = O(n \log n)$. (In the expression for the total memory, the term $(3n/2 - 2)$ represents the memory for the principal jugglers in the chain, while the term $(2\nu - 1)2^{\nu}$ represents the memory of the coxes.) This yields an extremely attractive time-division-multiplex implementation, since the only aspect of the cost that grows as fast as the size of the corresponding space-division-multiplex network (as $O(n \log n)$) is the total memory, which can be furnished by relatively inexpensive technology (inertial delay lines), whereas the number of high-speed switching elements (represented by the jugglers) grows much more slowly (as $O(\log n)$).

In our description of juggling networks, we have assumed that jugglers execute their moves instantaneously, so that a ball received by a juggler executing a straight move is passed on at the same pulse. In practice there would be a fixed overhead time for a juggler, which might be a large fixed multiple of the pulse. In the chain of jugglers we have described, and more generally in any juggling network in which all balls are processed by the same number of jugglers, this overhead delay can be ignored in the analysis of the network, and it results merely in the addition of a constant delay per juggler being added to the overall delay. Even in more complicated juggling networks, with different numbers of jugglers on various paths between the source and the sink (as is necessary, for example, for the efficient construction of superconcentrators), this overhead delay can be taken into account by setting up "time zones" for the various jugglers, and introducing extra delays to compensate for differences in time zones. Thus we shall maintain the convenient fiction that jugglers act instantaneously, as it will have no effect on our conclusions and will simplify our analysis.

3. Applications

The great economy and elegance of the construction given in Sect. 2 leads us to seek other applications for these ideas. The natural starting point is the class of switching networks with interconnection patterns similar to that of the Beneš network. Some prominent members of this class are (1) the spider-web interconnection networks (see Pippenger [19,20]), (2) the Cantor non-blocking network [7], and (3) the Batcher bitonic sorting network [5]. The first two of these are externally controlled interconnection networks analogous to the Beneš network, and require no further comment. The Batcher bitonic sorting network, however, is based on comparators, and we should say something about how these devices can be realized by jugglers.

As described by Batcher [5], a comparator is a finite automaton that sorts two records received at its inputs, producing the same two records in sorted order at its outputs. To do this, it receives the records one bit at a time, with the bits of the keys by which the records are to be sorted preceding any other data in the records, and with the bits of the keys being received in order of decreasing significance. As long as the bits of the two input records remain identical, these identical streams of bits are reproduced at the outputs. Once the bits of the input keys differ, the correct sorted order is established, and the remainders of the records are reproduced at the outputs in this order. Viewed as a finite automaton, a comparator requires two bits of state information to keep track of whether or not the sorted order has been established and, if so, what that order is.

A time-division-multiplex implementation of a comparator entails three jugglers: a principal juggler who juggles balls representing the successive bits of the records, an assistant who juggles balls representing the state of the comparator (these balls will be of three distinct colors, representing the three possible states of the automaton), and a cox who instructs the other two jugglers as to which of the larger and smaller records should appear in the earlier and later output slots. In this way one can easily construct a time-division-multiplex implementation of Batcher's bitonic sorting network [5] with $O((\log n)^2)$ jugglers, overall delay $O(n)$ and total memory $O(n(\log n)^2)$.

To go beyond these simple applications, however, it is necessary to employ one of the essential tools of the theory of switching networks: expanding graphs (or, more generally, graphs with favorable eigenvalue separation ratios). Armed with an efficient time-division-multiplex implementation of this tool, we can explore the possible time-division-multiplex analogs of the following kinds of networks: (1) concentrators and superconcentrators, as introduced by Pinsker [16] and Valiant [24] (see also Pippenger [21]), (2) non-blocking connection networks, following Bassalygo and Pinsker [4] (see also Pippenger [17]), (3) sorting networks, following Ajtai, Komlós and Szemerédi [1,2] (see also Pippenger [18]), and (4) self-routing networks, as introduced by Arora, Leighton and Maggs [3] and Pippenger [22]. We shall not delve further into any of these applications here, but will describe in Sect. 4 a time-division-multiplex implementation for expanding graphs that should be of use in attacking all of them.

4. Expanders

This section is devoted to the time-division-multiplex implementation of expanding graphs. Our implementation will be based upon a particular explicit construction for expanding graphs, originated by Margulis [14], with improvements due to Gabber and Galil [8] and Jimbo and Maruoka [10].

We shall construct a basic expanding graph, which is a regular bipartite multigraph $G = (A, B, E)$, in which every vertex (in $A \cup B$) has degree 8 (meets 8 edges in E), and in which A and B each contain n vertices, where $n = m^2$ is a perfect square, and $m = 2^\mu$ is a perfect power of 2 (so that $n = 4^\mu$ is a perfect power of 4).

We shall do this by describing 8 perfect matchings $E_1, \ldots, E_8 \subseteq A \times B$, the union $E_1 \cup \cdots \cup E_8$ of which is E. To describe these matchings, we let $\mathbf{Z}_m$ denote the ring of integers modulo m, and identify both A and B with the direct product $\mathbf{Z}_m \times \mathbf{Z}_m$, which we shall regard as having for its elements the 2-element columns of elements from $\mathbf{Z}_m$. Each of the matchings E_i will then have the form

$$E_i = \{(z, \pi_i(z)) : z \in \mathbf{Z}_m \times \mathbf{Z}_m\},$$

where π_i is a permutation of $\mathbf{Z}_m \times \mathbf{Z}_m$ defined by an affine mapping of the form

$$\begin{pmatrix} x \\ y \end{pmatrix} \mapsto \begin{pmatrix} a & b \\ c & d \end{pmatrix} \begin{pmatrix} x \\ y \end{pmatrix} + \begin{pmatrix} u \\ v \end{pmatrix}.$$

Thus it will suffice to specify, for each $i \in \{1, \ldots, 8\}$, the matrix $\begin{pmatrix} a & b \\ c & d \end{pmatrix}$ and the column $\begin{pmatrix} u \\ v \end{pmatrix}$.

For one particular construction given by Jimbo and Maruoka [10], the matrix $\begin{pmatrix} a & b \\ c & d \end{pmatrix}$ is one of the matrices $\begin{pmatrix} 1 & 2 \\ 0 & 1 \end{pmatrix}, \begin{pmatrix} 1 & 0 \\ 2 & 1 \end{pmatrix}$ or their inverses $\begin{pmatrix} 1 & -2 \\ 0 & 1 \end{pmatrix}, \begin{pmatrix} 1 & 0 \\ -2 & 1 \end{pmatrix}$, and the column $\begin{pmatrix} u \\ v \end{pmatrix}$ is one of the columns $\begin{pmatrix} 1 \\ 0 \end{pmatrix}, \begin{pmatrix} 0 \\ 1 \end{pmatrix}$ or their negatives $\begin{pmatrix} -1 \\ 0 \end{pmatrix}, \begin{pmatrix} 0 \\ -1 \end{pmatrix}$. Thus it will suffice to show how the permutations corresponding to each of these matrices and columns can be implemented by juggling networks, since then the permutations corresponding to the affine transformations can be implemented by connecting two such juggling networks in series, while the basic expanding graph can be implemented by connecting 8 such series combinations in parallel.

One approach to the problem of implementing these permutations would be to observe that, like all permutations, they can be carried out by the network described in Sect. 2, provided the coxes juggle appropriate patterns. The total number of balls juggled by coxes in Sect. 2 is $O(n \log n)$, but it might be possible to reduce this to $O(n)$ by careful analysis of the structure of the permutations. This sort of analysis has been done by Lenfant [11] for the space-division-multiplex implementation of Beneš's rearrangeable connection network. We shall not undertake such an analysis here, but rather will directly implement the required permutations with juggling networks.

First let us consider the map

$$\varrho : \begin{pmatrix} x \\ y \end{pmatrix} \mapsto \begin{pmatrix} x \\ y \end{pmatrix} + \begin{pmatrix} 0 \\ 1 \end{pmatrix} = \begin{pmatrix} x \\ y+1 \end{pmatrix}.$$

We may arrange the elements of $\mathbf{Z}_m \times \mathbf{Z}_m$ in an $m \times m$ array, with $\begin{pmatrix} x \\ y \end{pmatrix}$ in the x-th row (numbered from the top) and the y-th column (numbered from the left). The map ϱ then corresponds to the operation of cyclically rotating each row one position to the right. Let the m^2 entries in this array correspond to the m^2 slots in a frame in "row-major order"; that is, let the first m slots in the frame correspond to the entries in the top row of the array (from left to right), and so forth. The successive rows of the array correspond to successive intervals of m slots (which we shall call "lines"), and to implement the map ϱ, we need to cyclically rotate each line of the frame, so that the last slot of the line is moved to the first slot of that line, and each other slot of the line is moved to the immediately following slot. Since the same operation is to be performed on each line, we may ignore the overarching organization of lines into frames, and consider simply the operation of cyclically rotating a line by one position.

We seek to delay slots 0 through $m-2$ of each line by 1 pulse and to "delay" slot $m-1$ by $-(m-1)$ pulses (that is, to advance it by $m-1$ pulses). We eliminate the negative delay by adding a delay of $m-1$ pulses to every slot of the line: thus we seek to delay slots 0 through $m-2$ by m pulses, and to delay slot $m-1$ by 0 pulses. This pattern of delays can be achieved by a single juggler who passes balls either immediately or after a single toss with a delay of m pulses. The corresponding pattern of straight and crosses moves has a period of m pulses, and thus can be coxed by a juggler with m balls. To summarize, the permutation ϱ can be implemented by a juggling network with $O(1)$ jugglers, $O(m)$ memory, and overall delay $O(m)$.

We can easily generalize the foregoing argument to the map

$$\varrho^k : \begin{pmatrix} x \\ y \end{pmatrix} \mapsto \begin{pmatrix} x \\ y \end{pmatrix} + \begin{pmatrix} 0 \\ k \end{pmatrix} = \begin{pmatrix} x \\ y+k \end{pmatrix},$$

where $1 \leq k \leq m-1$. In this case, we seek to delay the first $m-k$ slots of each line by k pulses, and to "delay" the last k slots by $-(m-k)$ pulses. Adding a constant delay to eliminate negative delays, we find that the resulting pattern of delays can be achieved by the same juggler and cox as before; only the cox's pattern and the overall delay are changed, and the overall delay is *reduced* from its maximum of $m-1$. To summarize, the permutation ϱ^k can be implemented by a juggling network with $O(1)$ jugglers, $O(m)$ memory, and overall delay $O(m)$, where all constants are *independent* of k.

Next let us consider the map

$$\sigma : \begin{pmatrix} x \\ y \end{pmatrix} \mapsto \begin{pmatrix} 1 & 0 \\ 1 & 1 \end{pmatrix} \begin{pmatrix} x \\ y \end{pmatrix} = \begin{pmatrix} x \\ x+y \end{pmatrix} .$$

Using the same organization of frames into lines as was used above, to implement the map σ we need to cyclically rotate the 0-th line not at all, rotate the 1-st line 1 position to the right, and in general rotate the x-th line x positions to the right.

To obtain an efficient implementation of this permutation, we shall assume that $m = 2^\mu$, for some natural number μ, so that each element of $\mathbf{Z}_m$ can be regarded as a μ-bit word (with the usual binary interpretation). Then, instead of subjecting each line to one of m different cyclic rotations, we will subject each line to a different subset of μ different rotations, with amounts of $2^0 = 1$ through $2^{\mu-1} = m/2$. In general, we will subject the x-th line to the rotation 2^λ positions to the right (for $0 \leq \lambda \leq \mu-1$) if the $(\lambda+1)$-st bit in the binary representation of x is 1 (where the 1-st bit is the least, and the μ-th bit is the most, significant).

The permutation σ can thus be implemented by a chain of μ jugglers, each of whom passes each ball to the next juggler in the chain, either directly or after a single toss with a span of m pulses. Since each juggler contributes at most $O(m)$ to the overall delay, the chain contributes at most $O(\mu m) = O(m^2)$ to the overall delay. Each of these jugglers has a cox whose pattern has a period that depends on the position of the juggler in the chain. The cox for the juggler with rotation amount 1 has a period of 2 lines, the cox for the juggler with rotation amount 2 has a period of 4 lines, and in general the cox for the juggler with rotation amount 2^λ has a period of $2^{\lambda+1}$ lines. Summing these periods, we see that the total memory required by the coxes is $O(m)$ lines, or $O(m^2)$ pulses. To summarize, the permutation σ can be implemented by a juggling network with $O(\log m)$ jugglers, $O(m^2)$ memory, and overall delay $O(m^2)$.

We can easily generalize the foregoing argument to the map

$$\sigma^k : \begin{pmatrix} x \\ y \end{pmatrix} \mapsto \begin{pmatrix} 1 & 0 \\ k & 1 \end{pmatrix} \begin{pmatrix} x \\ y \end{pmatrix} = \begin{pmatrix} x \\ kx+y \end{pmatrix} ,$$

where $1 \leq k \leq m-1$. We need only alter the behavior of each juggler to replace a cyclic rotation of 2^λ pulses by one of $k2^\lambda$ pulses (modulo m), for $1 \leq \lambda \leq \mu-1$. This affects the patterns of the coxes, but not their periods or the spans of the jugglers. To summarize, the permutation σ^k can be implemented by a juggling network with $O(\mu)$ jugglers, $O(m^2)$ memory, and overall delay $O(m^2)$, where all constants are *independent* of k.

At this point we have seen how to implement 4 of the 8 permutations of our expanding graph, each with a juggling network of $O(\mu)$ jugglers, total memory $O(m^2)$ and overall delay $O(m^2)$. If we were to use the same strategy for the remaining 4 permutations, we would encounter the following problem: in order to cyclically rotate a column (rather than a row) we need a juggler with a span of $O(m^2)$ (rather than $O(m)$), and thus a chain of μ such jugglers would require a total memory of $O(\mu m^2)$, which exceeds our goal of $O(m^2)$. We shall therefore use a different strategy for these 4 remaining permutations. We shall consider the map

$$\tau : \begin{pmatrix} x \\ y \end{pmatrix} \mapsto \begin{pmatrix} 0 & 1 \\ 1 & 0 \end{pmatrix} \begin{pmatrix} x \\ y \end{pmatrix} = \begin{pmatrix} y \\ x \end{pmatrix}.$$

We shall implement the corresponding permutation using a chain of $O(\mu)$ jugglers, with $O(m^2)$ memory and overall delay $O(m^2)$. We can then implement the permutation corresponding to the map

$$\varrho'^k : \begin{pmatrix} x \\ y \end{pmatrix} \mapsto \begin{pmatrix} x \\ y \end{pmatrix} + \begin{pmatrix} k \\ 0 \end{pmatrix} = \begin{pmatrix} x + k \\ y \end{pmatrix}$$

using the identity $\varrho'^k = \tau \circ \varrho^k \circ \tau$, and the permutation corresponding to the map

$$\sigma'^k : \begin{pmatrix} x \\ y \end{pmatrix} \mapsto \begin{pmatrix} 1 & k \\ 0 & 1 \end{pmatrix} \begin{pmatrix} x \\ y \end{pmatrix} = \begin{pmatrix} x + ky \\ y \end{pmatrix}$$

using the identity $\sigma'^k = \tau \circ \sigma^k \circ \tau$.

The map τ corresponds to the permutation that transposes the array of elements of $\mathbf{Z}_m \times \mathbf{Z}_m$. Our implementation of this permutation will be based on the following identity, in which A, B, C and D denote $(m/2) \times (m/2)$ subarrays of an $m \times m$ array, and a superscript T denotes "transpose":

$$\begin{pmatrix} A & B \\ C & D \end{pmatrix}^T = \begin{pmatrix} A^T & C^T \\ B^T & D^T \end{pmatrix}.$$

This identity suggests a strategy that begins by exchanging the subarrays B and C (without transposing them), then proceeds recursively to transpose all four subarrays.

The operation of exchanging B and C is straightforward, since it reduces to exchanging a sequence of pairs of slots at a fixed distance in each frame. Specifically, we want to delay the last $m/2$ slots of the first $m/2$ lines (the elements of B) by $m(m-1)/2$ pulses ($m/2$ lines minus $m/2$ pulses), delay the first $m/2$ slots of the last $m/2$ lines (the elements of C) by $-m(m-1)/2$ pulses, and delay all other slots by 0 pulses. Adding an overall delay of $m(m-1)/2$ pulses to eliminate the negative delays, we see that the required exchange can be accomplished by a juggler with a span of $m(m-1)/2$ pulses, who passes each ball after 0, 1 or 2 tosses, for a delay of 0, $m(m-1)/2$ or $m(m-1)$ pulses. The juggler is coxed by a partner with a pattern of period 1 frame, or m^2 pulses.

After the exchanges performed by the juggler just described, it remains to transpose each of the subarrays A, B, C and D. To do this we proceed recursively, partitioning each of these subarrays into four $(m/4) \times (m/4)$ subsubarrays, exchanging the two off-diagonal subsubarrays of each subarrays, and so forth. Each level of the recursion will contribute one juggler to a chain of μ jugglers, of which the first (described above) is responsible for exchanging two subarrays, the second is responsible for exchanging four pairs of subsubarrays (one pair in each subarray), and so forth. The λ-th juggler will have a span of $m(m-1)/2^\lambda$ pulses (and will pass each ball after 0, 1 or 2 tosses), and will be coxed by a partner with a period of $m^2/2^{\lambda-1}$ pulses. Adding the contributions of the jugglers in this chain, we see that the permutation corresponding to the map τ is implemented by a juggling network with $O(\mu)$ jugglers, total memory $O(m^2)$ and overall delay $O(m^2)$, as claimed above.

This completes the implementation of our basic expanding graph, since the 8 permutations required for this graph can be fabricated by composing a bounded number of permutations, each of the form ϱ^k, σ^k, or τ. Furthermore, a graph with any desired fixed ratio of eigenvalue separation can be obtained by raising our basic expanding graph to a fixed power (see for example Pippenger [22]). Thus each of the bounded number of permutations required for this desired graph can be fabricated by composing a bounded number of permutations from the basic expanding graph, and we obtain the following theorem.

Theorem 1. *For any desired eigenvalue separation ratio (that is, ratio between largest two absolute values of eigenvalues) R, there exists a natural number $d = 2^\delta$ such that, for every natural number $n = 4^\mu$, there exist d permutations $\pi_1, \ldots, \pi_d$ of n objects such that (1) the sum of the matrices of the permutations $\pi_1, \ldots, \pi_d$ has eigenvalue separation ratio at least R, and (2) each of the permutations $\pi_1, \ldots, \pi_d$ can be implemented by a juggling network with $O(\log n)$ jugglers, total memory $O(n)$, and overall delay $O(n)$.*

5. Conclusion

We have shown in this paper how to construct time-division-multiplex analogues of expanding graphs, which are an essential component in many asymptotically optimal constructions for switching networks. We have described this construction in terms of a juggling metaphor that is useful in its own right as an aid to visualizing the operation of switching networks. Aside from more or less routine applications of this construction to various problems concerning switching networks, some more conceptual problems remain to be addressed.

At this time there are no lower bounds for time-division-multiplex networks except for those inherited in an obvious way from the theory of space-division-multiplex networks. Consider for example the construction of connectors given in Sect. 2. The memory requirement $O(n \log n)$ is clearly best possible, since that much memory, $\Omega\big(\log(n!)\big) = O(n \log n)$, is needed to remember the identity of

1 out of $n!$ possible permutations. Similarly, the overall delay of $O(n)$ is best possible, since routing the last slot of an input frame to the first slot of an output frame clearly requires that the frame be delayed by $\Omega(n)$ pulses. Finally, the bound of $O(\log n)$ switches is best possible, provided we assume an overall delay of $O(n)$, since the number of switches in a time-division-multiplex network, times the overall delay of that network, must be at least as large as the number of switches in a space-division-multiplex network performing the same task. We do not know, however, how to prove a lower bound to the number of switches when the constraint on the overall delay is relaxed (say to $O(n \log n)$), or how to prove a lower bound to the number of switches required to implement specific permutations such as those treated in Sect. 4.

While the construction for expanding graphs used in Sect. 4 suffices to provide any desired eigenvalue separation ratio (given that the degree is no object), there are other constructions that are both more economical from a practical point of view and essential for certain theoretical purposes. The most prominent of these are the Ramanujan graphs introduced by Lubotzky, Phillips and Sarnak [12] and by Margulis [15]. Whether there are efficient time-division-multiplex implementations of these graphs remains an open question.

References

1. M. Ajtai, J. Komlós and E. Szemerédi: Sorting in $c \log n$ parallel steps. Combinatorica **3** (1983) 1–19
2. M. Ajtai, J. Komlós and E. Szemerédi: An $O(n \log n)$ sorting network. Proc. ACM Symp. on Theory of Computing **15** (1983) 1–9
3. S. Arora, T. Leighton and B. Maggs: On-line algorithms for path selection in a nonblocking network. Proc. ACM Symp. on Theory of Computing **22** (1990) 149–158
4. L. A. Bassalygo and M. S. Pinsker: Complexity of an optimal nonblocking switching network without reconnections. Problems of Inform. Transm. **9** (1974) 64–66
5. K. E. Batcher: Sorting networks and their applications. Proc. AFIPS Spring Joint Computer Conf. **32** (1968) 307–314
6. V. E. Beneš: Optimal rearrangeable multistage connecting networks. Bell Sys. Tech. J. **43** (1964) 1641–1656
7. D. G. Cantor: On non-blocking switching networks. Networks **1** (1971) 367–377
8. O. Gabber and Z. Galil: Explicit constructions of linear-sized superconcentrators. J. Comp. and System Science **22** (1981) 407–420
9. H. Inose: Blocking probability in 3-stage time division switching network. J. IECEJ **44** (1961) 935–941
10. S. Jimbo and A. Maruoka: Expanders obtained from affine transformations. Combinatorica **7** (1987) 343–355
11. J. Lenfant: Parallel Permutations of Data: A Beneš network control algorithm for frequently used permutations. IEEE Trans. on Computers **27** (1978) 637–647
12. A. Lubotzky, R. Phillips and P. Sarnak: Ramanujan graphs. Combinatorica **8** (1988) 261–277
13. M. J. Marcus: Designs for time slot interchangers. Proc. National Electronics Conf. **26** (1970) 812–817

14. G. A. Margulis: Explicit construction of concentrators. Problems of Inform. Transm. **9** (1974) 71–80

15. G. A. Margulis: Explicit group-theoretical constructions of combinatorial schemes and their application to the design of expanders and concentrators. Problems of Inform. Transm. **24** (1988) 39–46

16. M. S. Pinsker: On the complexity of a concentrator. Proc. Internat. Teletraffic Congr. **7** (1973) 318/1–4

17. N. Pippenger: Telephone switching networks. Proc. AMS Symp. Appl. Math. **26** (1982) 101–133

18. N. Pippenger: Communication networks. In J. van Leeuwen (ed.), Handbook of Theoretical Computer Science – Volume A: Algorithms and Complexity, Elsevier, Amsterdam, 1990.

19. N. Pippenger: The blocking probability of spider-web networks. Random Structures and Algorithms **2** (1991) 121–149

20. N. Pippenger: The asymptotic optimality of spider-web networks. Discr. Appl. Math. **37/38** (1992) 437–450

21. N. Pippenger: Rearrangeable circuit-switching networks. Proc. Internat. Conf. on Graph Theory, Combinatorics, Algorithms and Applications **7** (1992) (to appear)

22. N. Pippenger: Self-routing superconcentrators. Proc. ACM Symp. on Theory of Computing **25** (1993) 355–361

23. S. V. Ramanan, H. F. Jordan and J. R. Sauer: A new time-domain, multistage permutation algorithm. IEEE Trans. Info. Theory **36** (1990) 171–173

24. L. G. Valiant: Graph-theoretic properties in computational complexity. J. Computer and Sys. Science **13** (1976) 278–285

Optimal Parallel Verification of Minimum Spanning Trees in Logarithmic Time

Brandon Dixon[*,†]

Robert E. Tarjan[*,‡,§]

1 Introduction

The problem of verifying a minimum spanning tree, that is, determining if a given spanning tree of a weighted graph is of minimum cost, is intriguing because it is a problem for which faster algorithms exist to verify a solution than exist to find a solution. In particular, Dixon, Rauch and Tarjan [4] recently gave a sequential algorithm to verify minimum spanning trees in linear time. This is in contrast to the fastest known deterministic algorithm to find minimum spanning trees [5], which runs in $O(m \log \beta(m,n))$ time, where $\beta(m,n) = \min\{i \mid \log^{(i)} n \leq m/n\}$, and $\log^{(i)} n$ is defined recursively by $\log^{(0)} n = n$, $\log^{(i+1)} n = \log \log^{(i)} n$, where m is the number of edges and n is the number of veritices in the graph. In a new result, Klein and Tarjan[7] give a randomized linear time algorithm to find minimum spanning trees. An intriguing feature of their algorithm is that it requires the linear time verification algorithm as a subroutine.

This paper shows how to parallelize the verification algorithm of Dixon, Rauch, and Tarjan to run on a CREW PRAM in $O(\log n)$ time using $\Theta((n + m)/\log n)$ processors. This algorithm uses linear work and is therefore optimal. It is interesting to note that the gap in efficiency between verification and finding minimum spanning trees is greater in the parallel models of computation than in the sequential case. The fastest known parallel algorithm for finding minimum spanning trees without using concurrent writes takes $O((\log n)^{3/2})$ time using

[*] Department of Computer Science, Washington and Lee University, Lexington, VA 24450.

[†] Research partially supported by a National Science Foundation Graduate Fellowship and by DIMACS (Center for Discrete Mathematics and Theoretical Computer Science), a National Science Foundation Science and Technology Center, Grant No. NSF-STC88-09648.

[‡] NEC Research Institute, Princeton, New Jersey 08540

[§] Research at Princeton University partially supported by the National Science Foundation, Grant No. CCR-8920505, the Office of Naval Research, Contract No. N00014-91-J-1463, and by DIMACS (Center for Discrete Mathematics and Theoretical Computer Science), a National Science Foundation Science and Technology Center, Grant No. NSF-STC88-09648.

$\Theta(m + n)$ processors [8]. Thus verification takes a factor of $\Theta((\log n)^{3/2})$ less work. Using concurrent writes, the fastest known algorithm requires $O(\log n)$ time and $\Theta(n + m)$ processors [2], but this is still a factor of $O(\log n)$ more work than verification. Additionally, it may be possible that a parallel variant of the new randomized minimum spanning tree finding algorithm [7] can be constructed using the parallel verification algorithm.

A previous, unpublished result of Alon and Schieber [1] gave an algorithm for verifying minimum spanning trees in $O(\log n)$ time using $\Theta((n+m)\alpha(m, n)/\log n)$ processors, where $\alpha(m, n)$ is the functional inverse of Ackerman's function [14].

2 The sequential algorithm

Let $G = (V, E)$ be a connected, undirected graph with $|V| = n$, $|E| = m$, and each edge having a real-valued cost. Given a spanning tree T of G, the algorithm of Dixon, Rauch and Tarjan [4] verifies whether T is a minimum spanning tree of G using $O(n + m)$ time and space on a $\Theta(\log n)$ word-size RAM.

Their algorithm uses precomputation combined with table look-up. It begins by precomputing a decision tree for every graph and associated spanning tree with fewer than $c(\log n)^{1/3}$ vertices for some constant c. When given the edge costs, the decision trees determine whether the associated tree is a minimum spanning tree for their graph.

The algorithm proceeds by breaking the graph G into connected pieces with fewer than $c(\log n)^{1/3}$ vertices. Each piece is called a *microtree*. A reduced tree T' is constructed from T by shrinking each microtree to a single vertex. Each microtree is verified using the precomputed decision trees, and the reduced tree T' is verified using the almost-linear-time method of Tarjan [12].

We describe a slightly different decomposition of the problem that can be done in parallel, and we also show that a reduced tree T' with $O(n/\log n)$ vertices, computed by applying the decomposition twice, can be verified in $O(\log n)$ time (as we now have a processor for each vertex in T'). This yields an algorithm that runs in $O(\log n)$ time using $\Theta((n + m)/\log n)$ processors.

3 Verification in parallel

For the parallel version of the verification algorithm, we assume that the graph is stored in the shared memory of the machine, and that edges are stored in adjacency lists at the nodes of the graph. Since we use the sequential linear-time result of [4] to verify the microtrees, no precomputation is needed for the parallel portion of the algorithm.

3.1 Decomposition

We decompose T into microtrees somewhat differently than in the sequential case. We first build a compressed tree C (as in [6]) in the following way. We compute $size(v)$ for each vertex in T, where $size(v)$ is the number of nodes in the subtree of v. Define an edge $(v, p(v))$ of T to be *heavy* if $size(v) > \frac{1}{2} size(p(v))$. An edge is considered *light* otherwise. This naturally partitions T into a collection of heavy and light paths. We store each heavy path in C by creating a single node a that represents the highest node on the path. We make all other nodes on the path children of a and we call a the *apex* of the heavy path. The cost of the edge to each child v of a is the maximum edge cost on the heavy path between a and v. This can be computed using prefix–sum [9]. We additionally store each heavy path as a linked list for further processing. Note that $size(p(p(v))) > 2size(v)$ after this construction. Therefore the depth of C is $O(\log n)$. This transformation can be done using $O(\log n)$ time and $\Theta(n/\log n)$ processors using parallel tree contraction[9].

Next we break the compressed tree into the reduced tree T' and the microtrees. In the following decomposition, we cut the microtrees only from the bottom of C. We cut the edge $(v, p(v))$ in C if $size(v) \le (\log n)^{1/2} < size(p(v))$. Each tree cut off from the bottom is a microtree; the reduced tree T' is the remainder, plus the roots of the microtrees. In C, since the size doubles every other level up the tree, there are at most $2n/(\log n)^{1/2}$ nodes with size between $(\log n)^{1/2}$ and $2(\log n)^{1/2}$, at most $n/(\log n)^{1/2}$ nodes with size between $2(\log n)^{1/2}$ and $4(\log n)^{1/2}$, and so on. The number of nodes in each size category is bounded by a geometric series and therefore T' has size $O(n/(\log n)^{1/2})$. We use the root of a microtree, say vertex v, to identify the tree: we label each vertex in the microtree with v. Note that when we cut the microtrees from C we may cut some of the nodes of a heavy path from the apex node of the heavy path, but this poses no problems for the later processing of the heavy paths.

Cutting the microtrees only from the bottom of C helps to simplify the processing of each non-tree edge. We replace each non-tree edge by four edges. To do this, we first compute the nearest common ancestor $nca(x, y)$ for each non-tree edge (x, y) and replace the edge (x, y) by the edges $(x, nca(x, y))$ and $(y, nca(x, y))$. In order to avoid write conflicts, we store the new edges only at the endpoints x and y and not at $nca(x, y)$. We replace each of the new edges in turn; this time we replace an edge $(x, nca(x, y))$ by (x, r) and $(r, nca(x, y))$ where r is the microtree root of x. Note that we can store the new edge $(r, nca(x, y))$ at r by computing a post-fix ordering of the non-tree edges in the microtree of r and writing each edge in an array position indexed by the ordering. After this decomposition there can be many edges from a node v to the root of its microtree; we compute and keep only the minimum–cost edge. We repeat the cutting of the microtrees and replacing of the non-tree edges one more time, cutting at a node v if $size(v) \leq \log n < size(p(v))$. This reduces the size of the remaining problem, yielding a reduced graph T' with $O(n/\log n)$ nodes and $O(m)$ edges.

We have broken the verification of the spanning tree into three pieces: the verification of the microtrees, the verification of the reduced tree T', and the verification of the heavy paths. In the remainder of this section, we show how to solve each of these subproblems.

3.2 Verification of the microtrees

Given this decomposition of the graph, we can verify the minimality of the microtrees in parallel. Each microtree can be verified in time proportional to its size using the linear time sequential algorithm of [4]. In the following, we show how to assign each processor a set of microtrees of size $O(\log n)$ to verify. When we cut the microtrees from C, each processor cuts up to $O(\log n)$ microtrees from C. Thus each processor writes the size of the microtrees that it cuts in an array of length $O(\log n)$. We collect all of the non-zero entries from all of the processors into one large array. Using prefix-sum [9], we can compute a starting and ending array index for each processor so that each range contains microtrees having total size at most $O(\log n)$. Each processor can then verify the microtrees in the array positions between its starting and ending index.

We must still verify the heavy paths and the reduced tree T'. We first describe the verification of the reduced tree T' and we later show how to verify the heavy paths using similar techniques.

3.3 Verification of T'

We now binarize T' in a balanced way as in [6]. This is a recursive procedure that replaces every node v with more than two children by a new binary node. The children of v are split into two sets that become the children of the new binary node. The new tree edges are given cost $-\infty$ so that they do not affect the minimality of T. This procedure is repeated until there are no nodes with more than two children. If we split the children of v into roughly equal-size sets, then the depth of the binarized tree is still $O(\log n)$ [6]. Since we have a processor for each node in the tree, we can assign d processors to every node with d children. Given an initial ordering of the children at every node, each split can be performed in constant time. Since the depth of the tree is increased by at most $O(\log n)$, this binarization can be done in $O(\log n)$ time.

We now have a binary tree of depth $O(\log n)$ with $O(n/\log n)$ nodes to verify. Additionally, each nontree edge is from a node to one of its ancestors. We have a processor for each node in the tree as well as one for every $O(\log n)$ nontree edges. We verify T' using pipelining as follows. We assign each node a level based on their depth. The nodes at the lowest depth are assigned level 0, the nodes one step higher in the tree are assigned level 1, and so on until the root of the tree. At each node in the tree, we bucket sort the nontree edges according to the level of their ancestor endpoint. Note that this only requires $O(\log n)$ time with $\Theta((n+m)/\log n)$ processors[10] and $O(n+m)$ total space. All edges in the same bucket start and end at the same nodes and thus cover the same tree path. Therefore only the minimum-cost edge in the bucket needs to be considered. We can compute the minimum edge in each bucket and discard the rest of the edges in $O(\log n)$ time since we can assign $\Theta(l/\log n)$ processors to each bucket of size l.

We can now pipeline the comparisons of the nontree edges to the tree edges, noting that T is a minimum spanning tree only if every tree edge has cost at least as large as every nontree edge with which it is compared. For ease of description, we assign the edge $(v, p(v))$ to each node v in the tree. The pipelining process starts at the bottom of the tree and works its way toward the root. The nodes that are already active in the process make a comparison and pass an edge up to their parent, and thus at step i nodes at level $i-1$ become active as well. The leaves can have both a tree edge and a nontree edge to their parent, so the process begins by each node at level 0 comparing the costs of those two edges (if the nontree edge exists). In the next step, the nodes at level 0 compare the cost of their tree edge with the nontree edge to their grandparent (if it exists). They then pass this nontree edge up to their parent. At step i, the level 0 nodes compare the cost of their tree edge to their nontree edge to level i (if the edge exists), the nodes at level 1 compare their tree edge with the nontree edge to level $i-2$, and so on. Each node receives two edges that end at the same level, say k, and computes the minimum value of those two edges and its nontree edge to level k (if it exists). This minimum cost edge is used as the edge from thin the next step. Each nontree edge is discarded as soon as it is compared to a smaller nontree edge or it reaches its ancestor endpoint. This continues for $O(\log n)$ steps and finishes when all edges have been discarded. Thus if in every comparison the outside edge is no smaller than the tree edge then T' has been verified.

3.4 Verification of the heavy paths

The verification of the heavy paths is quite similar to the verification of the compressed tree C. We transform a heavy path of length k into a balanced binary tree with k leaves; we call the new binary tree H. For ease of description, we place the edge costs of the heavy path at the leaves of H. The cost of the deepest edge on the path is placed at the leftmost leaf of H, the next edge up the path is assigned to the next leaf of H to the right, and so on until the highest edge of the path is assigned to the rightmost leaf of H. At each internal node of H, we compute and store the maximum value in its subtree. This is an easy form of expression tree evaluation and can be handled in $O(\log n)$ time [9].

The verification of a nontree edge in H involves comparing the nontree edge against some of the max values stored in the internal nodes as follows. We again replace each nontree edge (u, v) in H by two edges: $(u, nca(u, v))$ and $(v, nca(u, v))$. Let u be to the left of v; we define the edge $(u, nca(u, v))$ to be a *right* edge since we compare this edge cost against leaves to the right of u, and similarly, we define $(v, nca(u, v))$ to be a *left* edge. We define a node of H to be *covered* by a nontree edge if the node lies on the cycle created by the nontree edge. To verify a right edge $(u, nca(u, v))$, at every node x that is covered by $(u, nca(u, v))$ and whose left son is also covered by $(u, nca(u, v))$, we compare the max cost stored at the right child of x with the cost of $(u, nca(u, v))$. In the other case, when the left son is not covered, no comparisons are needed. We perform the symmetric comparisons to verify a left edge. When we are finished with the above comparisons, the maximum tree edge on the path from u to v has been compared to the nontree edge (u, v).

These comparisons can be performed using similar methods to those used to verify the compressed tree C. The process is nearly identical: the only difference is that comparisons may not be needed at some levels and that comparisons are done with respect to the value stored in a child of a node instead of an edge at the node. Note that no new heavy paths are created in this process. What remains to be shown is how to store the nontree edges at the appropriate nodes of the heavy paths. Because we wish to perform comparisons as in the compressed tree case, we must store each nontree edge at both of its endpoints on the heavy path. If we cut the children of an apex node of a heavy path to form microtrees, then storing the nontree edges at the nodes of the heavy path is a simple operation. Each node on the heavy path is a microtree root, and we store the nontree edges there as described earlier.

For the other heavy paths, we must first move the nontree edges whose nca is on a heavy path up the tree from their endpoints in T' to their first ancestor on the heavy path. Since we have $O(n/\log n)$ nodes in the remaining graph, we can use the previously described pipelining technique but replacing the min operation with list concatenation. This moves a list of nontree edges up to each node on a heavy path. The heavy-path node is one of the endpoints of each edge stored in the list there, but we do not necessarily know the other endpoint of each edge. Remember that we store $(v, nca(u, v))$ and $(u, nca(u, v))$. To help determine the remaining endpoint of each edge, we store three additional pieces of information with each nontree edge at the heavy path: the endpoints of the edge in T', the nca in T, and the nca in T'. There are two kinds of edges in the list at the heavy-path node v: those whose nca in T is v and those whose nca in T is higher than v.

If the nca in T is higher in the path, no other processing needs to be done;

the other endpoint is simply the *nca*. If the *nca* is v itself, we do not know the other endpoint on the path for this edge. We compute it in the following way. We write at each apex of a heavy path its depth in T'. At each leaf l of T', we store an array of size $O(\log n)$. If the path from l to the root of T' contains an apex a at depth i, then entry i in the array contains the child of a which is on the path from l to the root. In other words we store the node where the path from l to the root enters a heavy path. If a nontree edge where we need to find an endpoint in the heavy path has an endpoint x in T' and the apex of the heavy path is at depth k, then we find the missing endpoint by looking at position k in the array stored at x. The entries in these arrays are easily filled by traversing the path to the root from each leaf, taking $O(\log n)$ time and making use of the fact that we have a processor for each node in the tree. Once all edges are stored at both endpoints, the path is verified by using the bucket sorting and pipelining exactly the same way as we verified T'.

4 Sensitivity analysis

The problem of sensitivity analysis of a minimum spanning tree T in a graph G is to determine how much the cost of each edge in the graph G can change while maintaining the minimality of T. This extension of the verification problem is studied in [13, 4, 3].

We modify the previous algorithm and data structures to solve the sensitivity analysis problem. In [4] it is shown that the sensitivity analysis problem can be solved in linear time using a randomized sequential algorithm, and an optimal deterministic sequential algorithm is also given. The actual running time of the optimal deterministic algorithm is not known. Our results are the parallel equivalents of these results; a randomized $O(\log n)$ time and $\Theta(n + m)$ work algorithm and an optimal deterministic algorithm. The best known upper bound on the running time of the deterministic algorithm is $O(\alpha(m, n) \log n)$, but the exact bound is not known. As in the verification algorithm, we use $\Theta((n + m)/\log n)$ processors.

To perform the sensitivity analysis, we must find the maximum cost tree edge on the tree path determined by each nontree edge and also the minimum cost nontree edge that spans each tree edge. We consider each of these cases separately. We first show how to find the minimum cost nontree edge for each tree edge.

4.1 Sensitivity analysis of the tree edges

For this case, the algorithm closely resembles the parallel verification algorithm of section 3. The decomposition of the tree is exactly as for the verification algorithm. We perform sensitivity analysis of the microtrees using the sequential algorithm from [4]. We modify the parallel verification algorithm so that during the verification of the reduced tree T' we store at every tree edge the minimum nontree edge cost that is compared to the tree edge. This is exactly the information needed for the sensitivity analysis of the tree edges of T'. Note that the minimum cost nontree edge covering a tree edge will be compared to the tree edge cost during the verification algorithm. The only remaining tree edges that need sensitivity analysis information are those stored in the heavy paths.

We again modify the verification algorithm to perform this task. During the verification of the heavy paths, nontree edge costs are compared against maximum costs from subtrees of the balanced binary tree H. We additionally store the minimum nontree edge that is used in a comparison at each internal node of H. When all comparisons have been performed, we propagate the sensitivity analysis information down the tree. Any nontree edge whose cost is stored at a node v in the binary tree covers all leaves of the subtree rooted at v. Therefore each node sends its minimum nontree edge cost to its children. If the children are internal nodes then they compare the received value with the value already stored at their node and send the minimum of these values to their children. We can propagate the sensitivity analysis information down the tree in $O(\log n)$ time, and this completes the sensitivity analysis for the tree edges. The next section describes the sensitivity analysis for the nontree edges.

4.2 Sensitivity analysis of the nontree edges

In this section, we describe how to determine the maximum tree edge on the tree path determined by each outside edge (u, v). The algorithm of [4] can find this information for the microtrees, so we only need to worry about the heavy paths and the reduced tree T'. In both cases we rely heavily on the unpublished result of Alon and Schieber [1]. Using their algorithm, given $O(\log n)$ preprocessing time and $O((n + m)\alpha(m, n)/\log n)$ processors, queries of the form "What is the maximum value on the tree path between u and v?" can be answered in $O(\alpha(m, n))$ time using a single processor.

In this reduced tree, we simply use the Alon and Schieber result to perform sensitivity analysis on T' since $\alpha(m, n/\log n)$ is $O(1)$. The preprocessing therefore takes $O(\log n)$ time and each outside edge can be processed in $O(1)$ time. Since we have $\Theta(m/\log n)$ processors, we can process all of the outside edges in $O(\log n)$ time.

To perform sensitivity analysis on the heavy paths, we use the same binary tree transformation as in the verification case. We make the additional observation that we can apply the sensitivity analysis algorithm on this binary tree recursively. Note that no new heavy paths are created by the recursive solution. Therefore the solution only needs to perform sensitivity analysis on some new microtrees and a new reduced graph T'.

5 Conclusion

We have presented algorithms for verification and sensitivity analysis of minimum spanning trees. Both algorithms are optimal; however, the exact bound on the number of processors for deterministic sensitivity analysis is not known. This leaves the obvious open question from the Dixon, Rauch and Tarjan paper: "What is the best bound on the number of comparisons needed to perform sensitivity analysis?" The algorithms given here are for the CREW PRAM model. Can these algorithms be transformed into an EREW algorithm?

A very interesting question now is whether the new result of Klein and Tarjan [7] can be paralleized using the parallel verification algorithms? The existance of both a deterministic linear time verification algorithm and a randomized linear time algorithm to find minimum spanning trees certainly asks how fast can we find minimum spanning trees deterministicly?

6 Acknowledgements

We would like to thank Bernard Chazelle for reading preliminary versions of this paper. We would also like to thank Jon Bright for pointing out the Alon and Schieber result, and Monika Rauch, Uzi Vishkin, and Neal Young for stimulating discussions.

References

[1] N. Alon and B. Schieber Optimal preprocessing for answering on-line product queries, *Unpublished manuscript*

[2] B. Awerbuch and Y. Shiloach, New connectivity and MSF algorithms for shuffle-exchange network and PRAM, *IEEE Trans. on Computers*, Vol. C-36, No. 10, Oct. 1987, pp. 1258–1263

[3] H. Booth and J. Westbrook, Linear Algorithms for Analysis of Minimum Spanning and Shortest Path Trees in Planar Graphs, Yale University, Department of Computer Science, TR-768, Feb. 1990.

[4] B. D. Dixon, M. Rauch, and R. E. Tarjan, Verification and Sensitivity Analysis of Minimum Spanning Trees in Linear Time, *SIAM J. Comput.* **21**(6) (1992) pp. 1184–1192.

[5] H.N. Gabow, Z. Galil, T. Spencer, and R.E. Tarjan, Efficient Algorithms for Finding Minimum Spanning Trees in Undirected and Directed Graphs, *Combinatorica* **6**(2) (1986) pp. 109–122.

[6] D. Harel and R.E. Tarjan, Fast Algorithms for Finding Nearest Common Ancestors, *SIAM J. Comput.* **13**(2) (1984) pp. 338–355.

[7] P. Klein and R.E.Tarjan, A Randomized Linear Time Algorithm for Finding Minimum Spanning Trees, Personal communication.

[8] D. B. Johnson and P. Metaxas, A parallel algorithm for computing minimum spanning trees, 4th ACM Symposium on Parallel Algorithms and Architectures, 1992, pp. 363–372.

[9] R. Karp and V. Ramachandran, Parallel Algorithms for Shared Memory Machines, *Handbook of Theoretical Computer Science* (1990) Chapter 17.

[10] J. H. Reif An Optimal parallel algorithm for integer sorting, *Proc. STOC 26* (1985) 496-504.

[11] B. Schieber and U. Vishkin, On Finding Lowest Common Ancestors: Simplification and Parallelization, *SIAM J. Comput.* **17**(6) (1988) pp. 1253–1262.

[12] R.E. Tarjan, Applications of Path Compressions on Balanced Trees, *J. Assoc. Comput. Mach.* **26**(4) (1979) pp. 690–715.

[13] R.E. Tarjan, Sensitivity Analysis of Minimum Spanning Trees and Shortest Path Trees, *Information Processing Letters* **14**(1) (1982) pp. 30–33. Corrigendum, Ibid **23** (1986), p.219.

[14] R.E. Tarjan, *Data Structures and Network Algorithms,* Society for Industrial and Applied Mathematics, Philadelphia, PA, 1983.

The Parallel Complexity of Algorithms for Pattern Formation Models (Extended Abstract)

Raymond Greenlaw[1]* and Jonathan Machta[2]**

[1] Department of Computer Science, University of New Hampshire, Durham, New Hampshire 03824, e-mail address: greenlaw@cs.unh.edu
[2] Department of Physics and Astronomy, University of Massachusetts, Amherst, Massachusetts 01003, e-mail address: machta@phast.umass.edu

Abstract. This paper investigates the computational complexity of several important models that are currently used by physicists to study pattern formation. The results represent one of the first applications of the tools of parallel computational complexity theory to complex physical systems. The specific models we study are *diffusion limited aggregation, fluid invasion, invasion percolation,* and *invasion percolation with trapping.* It is known that all of these processes yield complex, fractal patterns. The paper shows that decision problems based on fluid invasion and diffusion limited aggregation are $\mathcal{P}$-complete. In contrast, we give $\mathcal{N}C^2$ algorithms for the two variants of invasion percolation. The results may have a practical significance for numerical simulations and could lead to fundamental insights into the nature of complex pattern forming systems. The $\mathcal{P}$-completeness theorems are interesting because they involve simulations of physical systems and are more intricate than other $\mathcal{P}$-completeness proofs. The $\mathcal{N}C$ algorithms are interesting because they are given for two procedures that seem inherently sequential.

1 Introduction

This paper is concerned with the computational complexity of several pattern formation models that are widely used in statistical physics. Specifically, the models we discuss are *diffusion limited aggregation, fluid invasion, invasion percolation,* and *invasion percolation with trapping*; they are defined below in section 2. Complex patterns formed by nonequilibrium processes have been the subject of intensive investigations by physicists [18]. Previous efforts to characterize these patterns have focused mainly on structural features such as their fractal dimensions. In this research we use the tools of parallel complexity theory to examine these pattern formation processes. To our knowledge this is one of the first applications of these tools to studying complex physical systems.

---
* This research was partially funded by the National Science Foundation Grant CCR-9209184.
** This research was partially funded by the National Science Foundation Grant DMR-9014366 and 9311580.

The motivations for studying the parallel complexity of pattern formation are both practical and theoretical. Since simulations using such models are computationally intensive, it is important to determine whether or not one can take advantage of the power of a parallel machine to conduct the simulations efficiently. Parallel algorithm development and simulations for related physical models are underway. For example, simulations of Ising model clusters on the Connection Machine^{TM} have been explored [1, 5]. The models discussed here generate patterns via a step by step process. Because of their highly sequential nature, it is far from obvious that good parallel algorithms exists for these models.

On a more fundamental level we believe that computational complexity theory will be essential in formalizing the notion of 'physical complexity.' In recent years there has been enormous interest in complex physical systems but little agreement on how to define the concept of complexity in a physical setting. By viewing models of physical processes as solving computational problems, we can invoke the tools of computational complexity theory to characterize these physical systems. The appropriateness of computational complexity as a measure of physical complexity is discussed in [15, 4].

All of the models discussed here are stochastic growth models in which the input is a sequence of random numbers and the output is a cluster of occupied nodes on a grid or more generally, a graph. In general this cluster forms a complex fractal pattern. In the present study we take the input set of numbers as given and investigate the difficulty of generating the output pattern in parallel.

The diffusion limited aggregation model [20] is the simplest and most well known member of a class of growth models used to describe a wide variety of physical processes including fluid flows, crystal growth, and electrical breakdown. In this model a cluster grows by accretion starting from an initial seed. Successive particles start at a source and random walk until they reach the growing cluster, where they stick. A new particle begins its random walk as soon as the previous particle is incorporated into the cluster. If the cluster is grown on a two dimensional grid, the result is a tenuous branched structure with a fractal dimension of 1.72. The *fluid invasion* model, though defined in a quite different way, generates the same distribution of clusters as diffusion limited aggregation and is better suited to theoretical analysis.

The physical setting for both fluid invasion and invasion percolation is a porous material filled with a fluid. The porous medium is viewed as a connected collection of pore spaces with random sizes. A second *invading* fluid is injected into the system at a source. The invading fluid displaces the *defending* fluid sequentially by filling the pores of the porous material. The invading and defending fluids do not mix and the defending fluid leaves the system at a sink. The process comes to completion when the invading fluid forms a connected path from source to sink. Fluid invasion and invasion percolation represent different physical mechanisms for the dynamics of the fluid flow.

The fluid invasion model [6] applies to a situation where the invading fluid (e.g. water) is much less viscous than the defending fluid (e.g. oil) and where flow velocities are high enough that viscosity controls the flow. The next pore space to be filled during the growth of the cluster is determined by the sizes of the pores and the pressure field in the defending fluid which is governed by Laplace's equation. We show that a decision problem based on this model is $\mathcal{P}$-complete. The proof is unlike other $\mathcal{P}$-completeness results (see [11]) and is closely connected to the physics of the fluid flow problem. The result suggests that simulation of the model on large parallel machines is unlikely to be very efficient. Although this model involves a fluid flow, it does not seem related to any standard *flow problems*, many of which are $\mathcal{P}$-complete [11].

Invasion percolation [7, 19] applies to the situation where surface tension effects dominate viscous forces. As the region of invading fluid grows, the next pore to be filled is the smallest pore on the perimeter of the region occupied by the invading fluid. A trapping rule may be added to the model to forbid the invasion of a region if there is no path for the defending fluid to leave the pore and reach the sink. The trapping rule represents the physical requirement that the defending fluid is incompressible.

From an algorithmic point of view, the porous medium can be represented by a graph where the pore sizes determine an ordering of the vertices in the graph. Invasion percolation defines a greedy search, *invasion search*, whose search order is determined by the ordering of the vertices in the graph. Invasion search combines features of both depth-first and breadth-first search. In this paper we show that there are fast parallel ($\mathcal{NC}$) algorithms for simulating invasion percolation with and without trapping. These results are rather surprising considering the highly sequential and greedy way in which the models are defined. The algorithms we develop seem practical and we hope they can be implemented on parallel machines. If so, they will allow fast simulations of the models and thus permit larger problem sizes to be studied.

As a further motivation, invasion percolation seems related to small space bounded algorithms [3, 2] for *s-t* connectivity. In the algorithms from [3] a set of "landmark" vertices is found and a search is carried out from these to help in determining the connectivity of the graph. The actual search strategy used in [3] is a bounded breadth-first search but the overall strategy has the flavor of invasion percolation. In [2] a breadth-first search is also used to partition the input graph. There it is suggested that a strategy other than breadth-first search might work even better for partitioning the graph. It may be worth investigating strategies like invasion search in this context.

The remainder of this paper is outlined below. In section 2 formal definitions of the models are presented. Our main results are described in section 3. Conclusions are presented in section 4. *Additional proof details may be found in [12, 15].*

2 Statement of Problems

Fluid Invasion

The first problem we formulate is based on the fluid invasion model mentioned in the introduction. We begin with some preliminaries. The input to the problem is a graph $G = (V, E)$ with nonnegative real edge "conductances" K_{ij} and nonnegative real "capacities" ϕ_i assigned to each vertex in V. There is a distinguished source node s and a sink s'. The vertices in V are numbered from 1 to $|V|$. We call the defending fluid O (corresponding to oil) and the invading fluid W (corresponding to water). As a function of time t, the volume of fluid O at vertex i is denoted $\sigma_i(t)$

and the pressure at vertex i is denoted $p_i(t)$. At every time step the set of vertices in V is partitioned into three sets:

$$C(t) = \{i \in V \mid \sigma_i(t) = 0\}$$
$$\partial C(t) = \{i \in V \mid i \notin C(t) \ and \ \exists j \in C(t) \ such \ that \ \{i,j\} \in E\} \qquad (1)$$
$$F(t) = V - C(t) - \partial C(t)$$

$C(t)$ represents the *cluster* of pores already completely filled with W at time t. We use C^* to denote the completed cluster. $\partial C(t)$ is the *perimeter*, consisting of sites that contain O but are adjacent to nodes filled with W. $F(t)$ consists of nodes still completely filled with O *and* not adjacent to any member of $C(t)$. The pressure in the invading fluid is one and is zero at the sink. The pressure in O satisfies the following discrete, inhomogeneous Laplace equation:

$$i \in C(t) \cup \partial C(t) \Rightarrow p_i(t) = 1$$
$$p_{s'}(t) = 0 \qquad (2)$$
$$i \in F(t) \Rightarrow p_i(t) = \frac{\sum_j p_j(t) K_{ij}}{\sum_j K_{ij}}$$

The evolution of the fluid configuration is governed by Darcy's law:

$$\dot{\sigma}_i(t) = -\sum_j K_{ij} [p_i(t) - p_j(t)] \qquad (3)$$

with the initial condition that the cluster is the source vertex and every other vertex is filled with O,

$$C(0) = \{s\} \qquad and \ for \ all \ i \in E - \{s\}, \ \sigma_i(0) = \phi_i. \qquad (4)$$

The cluster grows until there is a path from the source to the sink, i.e. s' joins the perimeter. The growth of the cluster is simulated in discrete time steps using the following iterative procedure:

Fluid Invasion Algorithm

begin
    initialize using Equation 4;
    $t \leftarrow 0$;
    $v \leftarrow s$;
    $cluster \leftarrow v$;
    while $s' \notin \partial C(t)$ do
        for each vertex in $F(t)$ compute its pressure by solving the set of
            linear equations in Equation 2;
        for each vertex in $\partial C(t)$ compute its emptying rate $\dot{\sigma}(t)$ using
            Equation 3;
        $v \leftarrow$ the value of $i \in \partial C(t)$ such that $\sigma_i(t)/|\dot{\sigma}_i(t)|$ is a minimum;
        $t \leftarrow t + \min_{i \in \partial C(t)} \{\sigma_i(t)/|\dot{\sigma}_i(t)|\}$;
        $cluster \leftarrow cluster \cup \{v\}$;
        add neighbors of v to $\partial C(t)$;
        for each vertex in $F(t) \cup \partial C(t)$ update its capacity using Equation 3;
end.

We can now state the formal decision problem based on this model.

Definition 1. Fluid Invasion Problem
Given: A graph $G = (V, E)$ with nonnegative real edge "conductances" K_{ij}, nonnegative real "capacities" ϕ_i assigned to each vertex in V, and a numbering on the vertices in V. Three distinguished nodes source s, sink s', and "output" node o.
Problem: Is node o an element of the cluster, C^*, formed by the invading fluid according to the fluid invasion algorithm?

Since we wish to show that fluid invasion is hard to simulate in parallel, it is desirable to obtain results for more restrictive versions of the problem. In the following we describe a restriction of the Fluid Invasion Problem to a two-dimensional grid with a single value of the conductance. This version of the problem generates the same distribution of patterns as diffusion limited aggregation in two dimensions if the capacities are taken to be independent, identically distributed random variables chosen from an exponential distribution [14]. Two dimensional diffusion limited aggregation has been the subject of numerous large scale simulations [18].

Definition 2. Two Dimensional Fluid Invasion Problem
Given: A graph $G = (V, E)$ that represents an $L \times L$ two-dimensional square grid. Edge "conductances" K_{ij} of value one, nonnegative real "capacities" ϕ_i assigned to each vertex in V, and a numbering on the vertices in V. Three distinguished nodes source s, sink s', and "output" node o.
Problem: Is node o an element of the cluster, C^*, formed by the invading fluid according to the fluid invasion algorithm?

Invasion Percolation

The next model we present is *invasion percolation*. Invasion percolation is defined on a graph $G = (V, E)$ with nodes numbered in an arbitrary way from 1 to $|V|$. The invasion process uses as its source node vertex number 1. During the invasion process, a cluster is formed. At each succeeding step in the process, the lowest numbered vertex that is adjacent to the cluster (directly connected to a node that is already in the cluster) is added to the cluster. The vertex is labeled at that point. The label of vertex i is denoted $inv(i)$ and indicates the order in which the vertex was visited. The following algorithm captures the procedure formally.

Invasion Percolation Algorithm

Input: An undirected graph $G = (V, E)$ in which the nodes in V are numbered from 1 to $|V|$.
Output: Nodes are labeled consecutively as they are added to the cluster.

```
begin
    CLUSTER ← {1};
    inv(1) ← 1;
    for t ← 2 to |V| do
        v ← lowest numbered non-cluster vertex that is adjacent to a vertex of
            the CLUSTER;
        inv(v) ← t;
        CLUSTER ← CLUSTER ∪ {v};
end.
```

Note that the invasion percolation order may yield a greedy depth-first search or a greedy breadth-first search based on the particular vertex numbering of the underlying graph. The invasion percolation process gives rise to a search strategy that we call *invasion search*. It is apparent that this search strategy falls somewhere between a depth-first and a breadth-first search. Another similar search strategy called breadth-depth search that combines features of both breadth-first and depth-first search was proposed in [13]. Its parallel complexity was analyzed in [10, 9].

Invasion percolation as we have defined it is more general than the conventional formulation of the problem. In the physics literature invasion percolation is defined on a d-dimensional lattice with random weights assigned to the bonds of the lattices. During the invasion process the bond on the perimeter of the cluster with the smallest weight is invaded next. This (bond) version of the problem can be reduced to our (site) version by choosing a graph in which each node on the lattice is represented by a vertex with a low number. Each bond is represented by a vertex and two edges that form a path of length two between the nodes connected by the bond. The numbers assigned to these auxiliary vertices are all greater than those assigned to the original nodes and are ordered according to the ranking of the random weights assigned to the bonds. Thus the parallel algorithm that we present for the general case may be applied to the conventional definition of the problem.

Invasion Percolation with Trapping

Invasion percolation with trapping is a variant of invasion percolation which incorporates the requirement that the defending fluid is incompressible. This process is also defined on a graph $G = (V, E)$ in which the nodes are numbered consecutively from 1 to $|V|$. There are two distinguished vertices — a source s and a sink s'. We take s equal to 1 and s' as any other node. During invasion percolation with trapping, nodes are added to the cluster as in invasion percolation, except now a node v to be added must also satisfy the requirement that there is a path from v to s' consisting entirely of vertices that are not yet in the cluster. Each node i for which there is such a path is given a label, $invt(i)$, indicating the order it was visited among such nodes. Nodes for which a non-cluster path to the sink does not exist are given the label $|V| + 1$ and are referred to as *trapped*. The following algorithm specifies the procedure formally.

Invasion Percolation with Trapping Algorithm

Input: An undirected graph $G = (V, E)$ in which the nodes in V are
numbered from 1 to $|V|$ and a sink node s'.
Output: Nodes are labeled consecutively as visited by invasion percolation
with trapping, i.e. the order in which they are added to the cluster.
Trapped nodes are labeled $|V| + 1$.

```
begin
    CLUSTER ← {1};
    invt(1) ← 1;
    for i ← 2 to |V| do invt(i) ← |V| + 1;
    NEXT ← 2;
    S ← {2, ..., |V|};
```

```
for i ← 2 to |V| do
    v ← lowest numbered vertex in S that is adjacent to the CLUSTER;
    if there is a path P from v to s' such that all vertices on P are not in
        the CLUSTER then
            invt(v) ← NEXT;
            NEXT ← NEXT + 1;
            CLUSTER ← CLUSTER ∪ {v};
    if v = s' then HALT;
    S ← S − {v};
end.
```

It is not immediately obvious that the if condition "there is a path P from v to s' such that all vertices on P are not in the CLUSTER" is in $\mathcal{P}$, since there are exponentially many paths to be checked. By deleting all cluster vertices from the graph and then checking the connectivity of v and s', one can solve this problem in polynomial time. It follows that the overall algorithm runs in polynomial time. We define natural decision problems based on the previous two percolation models.

Definition 3.
Given: An undirected graph $G = (V, E)$ in which the nodes in V are numbered from 1 to $|V|$, a source node s equal to 1, a sink node s', a designated vertex u, and a bound B.
Invasion Percolation Problem Is $inv(u) < B$?, i.e. is vertex u labeled less than B by the invasion percolation algorithm?

Invasion Percolation with Trapping Problem Is $invt(u) < B$?, i.e. is vertex u labeled less than B by the invasion percolation with trapping algorithm?

The models defined in this section are all used to study physical systems and have practical significance. We have presented decision problems based on the models and in the next section apply the tools of computational complexity theory to determine their complexity.

3 Statement of Results

Our first result shows that the Fluid Invasion Problem is $\mathcal{P}$-complete. It is clear that the Fluid Invasion Problem is in $\mathcal{P}$. The completeness result follows via a reduction from a variant of the NOR Circuit Value Problem with gates numbered in topological order [11]. Each NOR gate is represented by a gadget consisting of a chain of nodes with appropriate capacities. Small conductance edges between gadgets permit the flow in one gadget to control the flow in another gadget. In effect, the reduction consists of designing a fluidic computer. The correctness of the reduction is based on a lemma proved in [15]. Combining all of these ideas we obtain the following theorem.

Theorem 4. The Fluid Invasion Problem is $\mathcal{P}$-complete under log-space reducibility.

The Two Dimensional Fluid Invasion Problem is a very restricted version of the Fluid Invasion Problem and is thus in $\mathcal{P}$. We have not found a way of utilizing the reduction in Theorem 4 to prove $\mathcal{P}$-completeness of the two dimensional version of the problem and the reduction involved here is quite different. It is from a version of the planar OR and NOT Circuit Value Problem that is shown to be $\mathcal{P}$-complete by modifying a reduction given in [8]. The fundamental building block in the reduction is a *wire*. A wire is a connected sequence of sites having small capacities in a background of sites having very large capacities. Wires are able to direct the growth of the cluster. Gadgets representing logic gates are constructed from wires connecting nodes of intermediate capacities. We state the result below.

Theorem 5. *The Two Dimensional Fluid Invasion Problem is $\mathcal{P}$-complete under log-space reducibility.*

The invasion percolation process sometimes mimics a depth-first search or a breadth-first search. In light of results showing ordered depth-first search is $\mathcal{P}$-complete [17], a variant of breadth-first search is $\mathcal{P}$-complete [10], and breadth-depth search is $\mathcal{P}$-complete [10, 9], it seems plausible that the Invasion Percolation Problem is $\mathcal{P}$-complete. If the Invasion Percolation Problem were $\mathcal{P}$-complete, certainly Invasion Percolation with Trapping would also be $\mathcal{P}$-complete. One could directly connect all of the nodes to an added sink node numbered $|V| + 2$. If, on the other hand, the Invasion Percolation Problem were in $\mathcal{NC}$, one still might conjecture that Invasion Percolation with Trapping is $\mathcal{P}$-complete. In fact, both of these processes seem to be highly sequential and characterized by many of the same properties as other problems that are $\mathcal{P}$-complete [11]. Based on these observations, it is interesting that we are able to prove both problems can be resolved in $\mathcal{NC}^2$.

Below we specify an $\mathcal{NC}^2$ algorithm for the Invasion Percolation Problem. The key idea in the algorithm is to convert G to a directed graph and assign edge weights such that minimum weight paths from the source node using these weights reflect the invasion percolation order.

Parallel Invasion Percolation Algorithm

Input: An undirected graph $G = (V, E)$ in which the nodes in V are
           numbered from 1 to $|V|$.
Output: Nodes are labeled consecutively in the order in which they are
           visited by the sequential invasion percolation algorithm.
begin
    for each edge $\{j, k\} \in E$ in parallel do
        replace $\{j, k\}$ by (j, k) and (k, j) to form a directed graph $G' = (V, E')$;
    for each edge $(j, k) \in E'$ in parallel do assign (j, k) a weight of 2^k;
    for each $i \in \{1, \ldots, |V|\}$ in parallel do
        $W(i) \leftarrow$ the weight of the minimum weight path from 1 to i;
    sort the $W(i)$'s in increasing order and label the nodes according to their
        rank in this order;
end.

It is easy to see how the algorithm can be used to solve the Invasion Percolation Problem. We present two lemmas pertaining to the correctness of the algorithm. The first lemma states that all vertices on a minimum weight path from the source to a given node v are added to the cluster before v. Let $W(i)$ denote the weight of the minimum weight path from vertex 1 to vertex i, and $W(i, j, P)$ denote the weight of the subpath of P from vertex i to vertex j.

Lemma 6. *Let $G = (V, E)$ be an undirected graph with the vertices numbered from 1 to n, where $n = |V|$. Let i be such that $1 \leq i \leq n$. Let $G' = (V, E')$ be the directed, weighted graph computed in the first step of the parallel invasion percolation algorithm. Let P denote the path in G corresponding to the minimum weight path in G' from vertex 1 to vertex i. When i is added to the invasion percolation cluster originating from vertex 1, all other vertices appearing on P have already been added to the cluster.*

Proof. Suppose for contradiction that the statement in the lemma is false. When i was added to the cluster, there must have been a path $P' \neq P$ contained entirely within the cluster and connecting 1 to i. Let $j \neq i$ be the first vertex on P that does not appear on P' and is not contained in the cluster when i is added. Let k denote the last vertex before j on P such that P and P' are the same up to and including k. Note that k could equal 1.

We claim that j is greater than the maximum vertex number occurring on P' after k. If the claim is true, then because of the manner in which weights are assigned to edges in E' and the fact that P was a minimum weight path, it follows that $W(k, j, P)$ is greater than $W(k, i, P')$. This implies that in G' the weight of P' is less than that of P. This contradicts the fact that P is the minimum weight path from 1 to i in G'. Thus, subject to the proof of the claim the lemma is proved.

The proof of the claim follows. Let l denote the maximum vertex number on P' appearing after k. By assumption all vertices on P' were added to the cluster before i was added. Using this together with the definition of j, it follows that all vertices on P up to j are added to the cluster before i is added. Therefore, l must be numbered greater than all vertices appearing between k and j on P. If it were not, these vertices would not get put in the cluster before i as we just observed they do. It follows that j is greater than l and the claim is proved. □

The next lemma proves that we can use minimum weight paths to determine the relative order vertices are visited.

Lemma 7. *Let G and G' be as specified in Lemma 6. Let i, j, u, and v be such that $1 \leq i, j, u, v \leq n$. Let P_i (P_j) denote the minimum weight path in G' from vertex 1 to vertex i (j). If $i \neq j$ then $W(i) \neq W(j)$. If $W(i) > W(j)$ then $\mathrm{inv}(i) > \mathrm{inv}(j)$, and vice versa when $W(i) < W(j)$.*

Proof. We first prove that if $i \neq j$ then $W(i) \neq W(j)$. Note that i (j) is on P_i (P_j). This combined with the observation that for any l, $1 \leq l \leq n$, an edge weighted 2^l can occur at most once on any minimum weight path yields the fact that the i-th $(j$-th$)$ bit in the binary representation of $W(i)$ $(W(j))$ is a 1. Since P_i and P_j are simple paths terminated by i and j, respectively, and since $i \neq j$ by assumption, it follows that either i is not on P_j or j is not on P_i. Thus, $W(i) \neq W(j)$.

We now prove the second part of the lemma. Without loss of generality, suppose $W(i) > W(j)$. The case when $W(i) < W(j)$ is symmetrical. We need to show that j is visited before i in the invasion search beginning from vertex 1. Since P_i and P_j are the minimum weight paths to i and j, respectively, it is not hard to see that there exists a vertex m, $1 \le m \le n$, such that P_m is a subpath of both P_i and P_j; however, after passing through m, P_i and P_j share no common vertices.

Since $W(i) > W(j)$, it is clear that $i \ne m$. Using the fact that $W(i) - W(m) > W(j) - W(m)$ and the observation that the weighting scheme is such that the k-th bit in the binary representation of a path's weight specifies whether or not the k-th vertex is on the path, there exists a vertex k (possibly equal to i) on the path from m to i along P_i that has a greater number than any vertex on the subpath of P_j from m to j. When m equals j, P_j is a subpath of P_i and the lemma follows directly from Lemma 6. In the case m is not equal to j, applying Lemma 6, it is easy to see that m will be added to the cluster before k. The remainder of P_j from m onward will be added to the cluster before k is added since all vertices on this path have numbers less than k. Using Lemma 6, we conclude that $inv(j) < inv(i)$. □

Using the two lemmas presented above and several additional observations, we obtain the following:

Theorem 8. *The parallel invasion percolation algorithm can be used to solve the Invasion Percolation Problem in $\mathcal{N}C^2$.*

In contrast to the result proved above, we saw that the Fluid Invasion Problem is $\mathcal{P}$-complete. The Invasion Percolation with Trapping Problem is intermediate between these two problems. Our original intuition suggested this problem would be $\mathcal{P}$-complete. Below we specify an $\mathcal{N}C^2$ algorithm solving it.

Parallel Invasion Percolation with Trapping Algorithm

Input: An undirected graph $G = (V, E)$ in which the nodes in V are
        numbered from 1 to $|V|$.
Output: Nodes are labeled as they are visited by the invasion percolation
        with trapping algorithm. Trapped nodes are labeled $|V| + 1$.
begin
    $L \leftarrow \{1, \ldots, |V|\}$;
    run the parallel invasion percolation algorithm on G to compute the
        invasion percolation order, $inv(i)$, for each node i;
    for each node i in parallel do
        $w(i) \leftarrow 2^{-inv(i)}$;
        $W(i) \leftarrow$ the minimum weight path from i to s', where the path is
                computed using the node weights and the weight of i
                is not counted;
        if $W(i) > 2^{-inv(i)}$ then $invt(i) \leftarrow |V| + 1$ and delete i from L;
    sort the remaining nodes in L based on their invasion percolation labels,
        $inv(i)$, and let $invt(i)$ be the rank in this list;
end.

The following lemma is needed for proving the correctness of the algorithm presented above.

Lemma 9. *Let $G = (V, E)$ be an undirected graph in which the nodes in V are numbered from 1 to $|V|$. Let s equal to 1 be the source and s' be the sink. Let u and v be two vertices in V that are labeled with values less than or equal to $|V|$ by the parallel invasion percolation with trapping algorithm on input G. Then $\mathrm{inv}(u) < \mathrm{inv}(v)$ if and only if $\mathrm{invt}(u) < \mathrm{invt}(v)$.*

Proof. The difference between the sequential invasion percolation algorithm and the sequential invasion percolation with trapping algorithm is that in invasion percolation (invasion percolation with trapping) trapped vertices are (are not) added to the cluster. Because of the definition of trapped, it is clear that from trapped vertices only other trapped vertices can be "reached," i.e. if a trapped vertex is considered as vertex v in the first step of the main for loop in the invasion percolation with trapping algorithm, it is impossible for there to be a path from vertices in the current set S to a vertex that will eventually be classified as not trapped. Thus, in invasion percolation with trapping algorithm adding vertices to the cluster as they are trapped or as they would be added if adjacent to other trapped vertices (already added to the cluster) can not affect the relative order that untrapped vertices are added to the cluster. It will in general affect which vertices are numbered less than $|V| + 1$ and the order that trapped vertices are visited. Since invasion percolation uses the same greedy rule as invasion percolation with trapping in looking for the next vertex adjacent to the cluster, it follows that $inv(u) < inv(v)$ if and only if $invt(u) < invt(v)$. $\square$

Combining the above lemma with several additional observations, we obtain the following theorem:

Theorem 10. *The parallel invasion percolation with trapping algorithm can be used to solve the Invasion Percolation with Trapping Problem in $\mathcal{N}C^2$.*

4   Conclusions

This paper has applied the tools of computational complexity theory to several different pattern formation models that are widely used in statistical physics. The results obtained, which are to our knowledge some of the first relating parallel complexity theory and statistical physics, shed some light on why it is difficult to parallelize the simulation of the fluid invasion model and also the diffusion limited aggregation model. The counter intuitive results obtained for invasion percolation both with and without trapping indicate that these processes can potentially be simulated efficiently in parallel. From a technical point of view, the $\mathcal{P}$-completeness results are of special interest because they occur in a new problem domain; the $\mathcal{N}C$ algorithms are interesting because they apply to problems that appear highly sequential. We hope the parallel algorithms developed can be implemented on a parallel machine and that they will provide better performance than is currently available for simulations.

The interested reader is referred to [16] for some recent results on the parallel complexity of several other growth models.

References

1. A. Apostolakis, P. Coddington, and E. Marinari. A multi-grid cluster labeling scheme. *Europhysics Letters*, 17(3):189–194, 1992.
2. G. Barnes, J.F. Buss, W.L. Ruzzo, and B. Schieber. A sublinear space, polynomial time algorithm for directed $s - t$ connectivity. In *Proceedings of the 7th Annual Structure in Complexity Theory Conference*, pages 27–33, 1992.

3. G. Barnes and W.L. Ruzzo. Deterministic algorithms for undirected $s - t$ connectivity using polynomial time and sublinear space. In *Proceedings of the 23rd ACM Symposium on the Theory of Computing*, pages 43–53, 1991.

4. C. H. Bennett. How to define complexity in physics, and why. In W. H. Zurek, editor, *Complexity, Entropy and the Physics of Information*, pages 137–148. Addison-Wesley, 1990.

5. R.C. Brower, P. Tamayo, and B. York. A parallel multi-grid algorithm for percolation clusters. *Journal of Statistical Physics*, 63(1/2):73–88, 1992.

6. D.Y.C. Chan, B.D. Hughes, L. Paterson, and C. Sirakoff. Simulating flow in porous media. *Physical Review A*, 38(8):4106–4120, 1988.

7. R. Chandler, J. Koplick, K. Lerman, and J. F. Willemsen. Capillary displacement and percolation in porous media. *Journal of Fluid Mechanics*, 119:249–267, 1982.

8. L. M. Goldschlager. The monotone and planar circuit value problems are log space complete for P. *SIGACT News*, 9(2):25–29, 1977.

9. R. Greenlaw. Breadth-Depth search is $\mathcal{P}$-complete. *Parallel Processing Letters*, 3(3). To appear.

10. R. Greenlaw. A model classifying algorithms as inherently sequential with applications to graph searching. *Information and Computation*, 97(2):133–149, 1992.

11. R. Greenlaw, H.J. Hoover, and W.L. Ruzzo. *Topics in Parallel Computation: A Guide to P-completeness Theory*. Computing Science Series, editor Z. Galil. Oxford University Press. To appear.

12. R. Greenlaw and J. Machta. On the parallel complexity of invasion percolation. Technical Report 93-04, University of New Hampshire, 1993.

13. E. Horowitz and S. Sahni. *Fundamentals of Computer Algorithms*. Computer Science Press, Rockville, Md., 1984.

14. Z. Koza. The equivalence of the DLA and a hydrodynamic model. *Journal of Physics A: Mathematical and General*, 24(20):4895–4905, 1991.

15. J. Machta. The computational complexity of pattern formation. *Journal of Statistical Physics*, 70(3/4):949–966, 1993.

16. J. Machta and R. Greenlaw. The parallel complexity of growth models. Technical Report 94-05, University of New Hampshire, 1994.

17. J. Reif. Depth-first search is inherently sequential. *Information Processing Letters*, 20(5):229–234, 1985.

18. T. Vicsek. *Fractal Growth Phenomena*. World Scientific, Singapore, 1989.

19. D. Wilkinson and J. F. Willemsen. Invasion percolation: A new form of percolation theory. *Journal of Physics A: Mathematical and General*, 16(14):3365–3376, 1983.

20. T. A. Witten and L. M. Sander. Diffusion-limited aggregation, a kinetic critical phenomenon. *Physical Review Letters*, 47(19):1400–1403, 1981.

On the parallel complexity of iterated multiplication in rings of algebraic integers

Stephan Waack*

Institut für Numerische
und Angewandte Mathematik
Georg–August–Universität Göttingen
Lotzestr. 16–18
37083 Göttingen
Germany

Abstract. The parallel complexity of iterated multiplication in an arbitrary but fixed ring of algebraic integers is studied. Boolean circuits of fan–in 2 are used. It is shown that polynomial time uniform circuits of logdepth can be constructed to solve this problem algorithmically.

<u>Keywords</u>: uniform boolean circuits of logarithmic depth, iterated multiplication

1 Introduction

Much effort has been done in the last years to study the complexity of arithmetic operations. It is a well–known fact that iterated addition, subtraction and multiplication of integers on modern computers are significantly faster operations than iterated multiplication and division. Until recently there seemed to be some theoretical reasons: Whereas it has been known for a long time that iterated addition and multiplication can be computed by logdepth Boolean circuits, this was open for iterated multiplication. But in [1] logdepth circuits for this problem are constructed using Chinese remaindering. These circuits are only polynomial time uniform instead of logspace uniform. Consequently, it does not follow from [1] that iterated multiplication belongs to DSPACE(log n)*. (If K denotes a complexity class of formal languages defined by Turing machine–based resource bounds, then K^* denotes the corresponding class of functions.)

In [9] logspace uniform circuits for iterated multiplication of polynomial size and depth $\log n \cdot \log \log n$ are constructed.

In this paper we extend the result of Beame, Cook, and Hoover [1] to rings of algebraic integers. The basic idea of the proof is the same as in [1]: modulear arithmetic is used. The situation is as follows. Assume that we are given an algebraic number field $K_F = \mathbf{Q}[X]/(F(X))$, where $\mathbf{Q}$ is the field of rational numbers. Without loss of generality we may assume that the irreducible polynomial $F(X) = X^\delta + \sum_{i=0}^{\delta-1} f_i X^i$ has integer–coefficients. The ring of algebraic integers of K_F, i.e. the integral closure of the ring of integers $\mathbf{Z}$ in K_F, which we denote by I_F, is a free $\mathbf{Z}$–module of rank δ.

---

* Research was supported in part by the Heisenberg Programm der Deutschen Forschungsgemeinschaft, Bonn.

This means, that there are integers $d_1, \ldots, d_\delta$, and polynomials $H_1, \ldots, H_\delta \in \mathbb{Z}[X]$ of degree less than δ, such that each element of I_F has a unique representation as $a_1 \frac{H_1}{d_1} + \cdots + a_\delta \frac{H_\delta}{d_\delta}$, where the a_i are integers. If d is the least common multiple of the numbers $d_1, \ldots, d_\delta$, then each element of I_F has a representation $\frac{R(X)}{d}$, where $R(X) \in \mathbb{Z}[X]$, and degree$(R) \leq \delta - 1$. This motivates to consider the ring $\mathbb{Z}[X]/(F(X))$ instead of I_F. (In general $\mathbb{Z}[X]/(F(X))$ is only a subring of I_F, although the quotient field of $\mathbb{Z}[X]/(F(X))$ equals K_F.)

Each element R of $\mathbb{Z}[X]/(F(X))$ has a unique representation as $(r_0, \ldots, r_{\delta-1})$, where $R(X) = \sum_{i=0}^{\delta-1} r_i X^i$. Iterated multiplication in $\mathbb{Z}[X]/(F(X))$ is the following problem, which we denote by MULT_n^F.

Input: $R_1(X), \ldots, R_N(X)$,
  where the coefficients $r_{k,i}$ ($i = 0, \ldots, \delta - 1$, $k = 1, \ldots, N$) of the R_i are bounded above in absolute value by $2^N - 1$, and are given in binary.
Output: $R(X) = \sum_{i=0}^{\delta-1} r_i X^i$,
  such that $R(X) \equiv R_1(X) \cdot \ldots \cdot R_N(X) \pmod{F(X)}$, where the coefficients r_i are represented in binary.

Observe that the total input length n, which is the index of the problem instance, equals δN^2.

In this paper the following result is proved.

RESULT *There is a logdepth Boolean circuit family, which is P-uniform, computing* $\boldsymbol{MULT} = MULT_n^F$.

The interested reader can find a good overview on computations with matrices and polynomials in [11].

2 The parallel computation model and some previous results

We assume familiarity with what might be termed "standard" complexity theory such as can be found in [7] and [14].

We adopt the usual definition of an fan-in two Boolean circuit family in which the nth circuit has n inputs and $h(n)$ outputs where $h(n)$ is a nondeacreasing polynomially bounded function. Observe, that with this definition depth $O(\log n)$ implies polynomial size.

We define logspace uniformity of a circuit family $< \alpha_n >$ according to [12].

Definition 1 logspace uniformity. A family $< \alpha_n >$ of circuits is logspace uniform provided that some deterministic logarithmic space bounded Turing machine can compute the transformation $1^n \rightarrow \overline{\alpha_n}$, where $\overline{\alpha_n}$ is the standard encoding [12] of α_n.

The following complexity classes are of outstanding interest in the theory of parallel computing.

Definition 2. 1. U–SIZE,DEPTH$(n^{O(1)}, d(n))$ is the set of all functions $f : \{0,1\}^* \rightarrow \{0,1\}^*$ computable by a logspace uniform Boolean circuit family $< \alpha_n >$ in which α_n has size polynomial in n and depth $O(d(n))$.

2. $NC^k = $ U-SIZE,DEPTH$(n^{O(1)}, (\log n)^k)$, and $NC = \bigcup_k NC^k$.

Intuitively, NC is the set of all functions computable superfast on a parallel computer of feasible size. The correspondence between uniform circuit size and Turing machine classes is among others given by the inclusions

$$NC^1 \subseteq DSPACE(\log n)^* \subseteq NC^2 \subseteq \ldots NC \subseteq P^* = DTIME(n^{O(1)})^*.$$

For details see [3], [4], [5], and [6].

Following [5] in the context of a particular problem instance of binary length n, we say that an integer m is *tiny* if $|m| \leq n^c$ where c is a constant. As to inputs, we assume that integers are specified in binary notations, except when an integer is tiny, in which case unary notation is used. As to outputs, integers are represented in binary.

Definition 3. The function $f : \{0,1\}^* \rightarrow \{0,1\}^*$ is NC^1 reducible to the function $g : \{0,1\}^* \rightarrow \{0,1\}^*$ if and only if there exists a logspace uniform circuit family $< \alpha_n >$ which computes f with depth$(\alpha_n) = O(\log n)$ where, in addition to the usual gates, oracle gates for g are allowed. An oracle gate is a node which has some sequence $y_1, \ldots, y_u$ of input edges and some sequence $z_1, \ldots, z_v$ of output edges with associated function $(z_1, \ldots, z_v) = g(y_1, \ldots, y_u)$.

For the purpose of defining depth, the oracle node counts as depth $\lfloor \log(u + v) \rfloor$.

Now we mention some results concerning the parallel complexity of arithmetic operations, some of them we need later.

Theorem 4 Iterated Addition [2]. *Finding the sum of N integers of N bits each belongs to NC^1.*

Theorem 5 Multiplication, Division [2], [9], [13]. *Let a and b be two n-bit integers.*
NC^1 contains the problem of computing $a \cdot b$, whereas $\lfloor a/b \rfloor$ is contained in

$$U\text{-}SIZE, DEPTH(n^{O(1)}, \log n \cdot \log\log n).$$

Proposition 1 ([1]) *Let a be any integer and let m be a tiny integer. Computing $a \bmod m$ is contained in NC^1*

Theorem 6 [1]. *Let $a_1, \ldots, a_N$ be any integers, and let m be a tiny integer. NC^1 contains the problem of computing $a_1 \cdot \ldots \cdot a_N \bmod m$.*

Theorem 7 [9]. *Finding the product of N integers of N bits each can be done by logspace uniform circuits of polynomial size and depth $O(\log n \cdot \log\log n)$, where $n = N^2$*

The best known algorithms for integer arithmetic (improving the size) can be found in [10].

Although logspace uniformity is desirable for theoretical reasons, there is a weaker kind of uniformity which provides a natural condition on circuit families. The builder of computer hardware may simple want to have fast circuits which are easy to construct. Once a circuit has been constructed it will be used over and over again.

Definition 8 *P*–uniformity. A family $< \alpha_n >$ of circuits is *P*–uniform provided some deterministic polynomial time bounded Turing machine can compute the transformation $1^n \to \overline{\alpha_n}$, where $\overline{\alpha_n}$ is the standard encoding [12] of α_n.

Theorem 9 [1]. *Finding the product of N integers of N bits each can be done by P–uniform circuits of logarithmically bounded depth.*

The major tool in proving the latter theorem is the so–called INTEGER CHINESE REMAINDERING problem (ICR_n). We need the following definition in order to formulate it.

For any integer m we define the set

$$\{-\lfloor (m-1)/2 \rfloor, \ldots, \lfloor m/2 \rfloor\}$$

to be *the set of canonical representatives* mod m.

Now let us turn to ICR_n.

Input: $c_1, \ldots, c_N, x \bmod c_1, \ldots x \bmod c_N$,
where $c_1, \ldots, c_N$ are relatively prime and the $x \bmod c_i$ are given as canonical representatives.

Output: $x \bmod \prod_{i=0}^{n} c_i$
as canonical representative.

A trivial modification of lemma 5.1 in [1] yields the following proposition.

Proposition 2 (Integer Chinese Remaindering) ICR_n *for relatively prime* $c_1, \ldots, c_N$, *where* $1 < c_1 <, \ldots, c_N \leq N^{O(1)}$, *is* NC^1–*reducible to the problem of computing* $c = \prod_{i=1}^{N} c_i$.

3 The algebraic background

The idea of this section is to provide the algebraic equipment necessary to prove the result. Proofs which are not carried out can be found in [8].

A ring is always a commutative one. Rings are denoted by capital italic letters (A or B) from the beginning of the alphabet, elements of them by small italic letters. Ideals are denoted by calligraphical letters. Polynomials are denoted by capital italic letters from the midddle of the alphabet (F, G or R), whereas variables are capital italic letters from the end (X, Y, Z). A field is denoted by K. $\mathbb{Z}$, $\mathbb{Q}$, $\mathbf{F}_{p^r}$ denote respectively the ring of rational integers, the field of rational numbers, and the finite field of exactly p^r elements.

3.1 The Euclidean algorithm

It is well–known that if K is a field, then $K[X]$ ia an Euclidean ring. More precisely, if $F(X), G(X) \in K[X]$, then there are unique polynomials $Q(X), R(X) \in K[X]$ such that $G = F \cdot Q + R$. Hereby either the degree of R is less than the degree of F or $R = 0$.

If A is an entire ring contained in K, if $F(X), G(X) \in A[X]$, and if $F(X)$ is *monic*, i.e. the leading coefficient of F is equal to 1, then the coefficients of $Q(X)$ as well as of $R(X)$ belong to A. The polynomial $R(X)$ is called the canonical representative of the residue class $G(X)$ mod $F(X)$. Assume that $F(X) = X^\delta + \sum_{i=0}^{\delta-1} f_i X^i$. The following lemma is useful to estimate the values of the coefficients of the polynomial R as well as to see that certain computations can be done in logspace.

Lemma 10. Let $G(X) = \sum_{i=0}^{\Delta} g_i X^i$, $\Delta \geq \delta - 1$, $F(X) = X^\delta + \sum_{i=0}^{\delta-1} f_i X^i$, and $R(X) = \sum_{i=0}^{\delta-1} r_i X^i$ as before, i.e. $R(X) \equiv G(X) \pmod{F(X)}$. Then

$$
\begin{pmatrix} r_0 \\ r_1 \\ \vdots \\ r_{\delta-1} \end{pmatrix} = \begin{pmatrix} g_0 \\ g_1 \\ \vdots \\ g_{\delta-1} \end{pmatrix} + \sum_{i=1}^{\Delta-\delta+1} \begin{pmatrix} 0 & 0 & \cdots & 0 & -f_0 \\ 1 & 0 & \cdots & 0 & -f_1 \\ \vdots & \vdots & & \vdots & \vdots \\ 0 & 0 & \cdots & 1 & -f_{\delta-1} \end{pmatrix}^i \cdot \begin{pmatrix} 0 \\ 0 \\ \vdots \\ g_{\delta+i-1} \end{pmatrix}
$$

Proof. By the Euclidean Algorithm $A(X)/(F(X))$ is a free A–module of rank δ. The sequence $1, X, X^2, \ldots, X^{\delta-1}$ forms a basis. Multiplying by the element X supplies an A–endomorphism ϕ of the A–module $A[X]/(F(X))$. The matrix

$$
\begin{pmatrix} 0 & 0 & \cdots & 0 & -f_0 \\ 1 & 0 & \cdots & 0 & -f_1 \\ \vdots & \vdots & & \vdots & \vdots \\ 0 & 0 & \cdots & 1 & -f_{\delta-1} \end{pmatrix}
$$

represents this endomorphism with respect to the above basis. It remains to remark that

$$
\begin{aligned}
g_{\delta+i-1} X^{\delta+i-1} &= X^i \cdot g_{\delta+i-1} X^{\delta-1} \\
&\equiv \phi^i \left(g_{\delta+i-1} X^{\delta-1} \right) \pmod{F(X)}
\end{aligned}
$$

Now assume that we are given the input–situation of MULT_n^F, i.e. assume that we are given polynomials $R_k(X) = \sum_{i=0}^{\delta-1} r_{k,i} X^i$, where $|r_{k,i}| \leq 2^{N-1}$. Using lemma 10 there is a straightforward way to prove

Lemma 11. If $\sum_{i=0}^{\delta-1} r_i X^i \equiv \prod_{k=1}^{N} R_k(X) \pmod{F(X)}$, then $|r_i| = 2^{O(n)} = 2^{O(N^2)}$.

3.2 Extension of ideals

Let $\psi : A \to B$ be a ring homomorphism. If $\mathcal{A}$ is an ideal in A, the extension $\mathcal{A}^e$ of $\mathcal{A}$ is defined to be $B \cdot \psi(\mathcal{A})$. Obviously, ψ induces a ring homomorphism $\overline{\psi} : A/\mathcal{A} \to B/\mathcal{A}^e$. If ψ is onto, $\mathcal{A}^e = \psi(\mathcal{A})$.

Proposition 3 If $\mathcal{A}_1, \ldots, \mathcal{A}_N$ are pairwise coprime ideals then $A \big/ \bigcap_{i=1}^{N} \mathcal{A}_i \cong A/\mathcal{A}_1 \times \ldots \times A/\mathcal{A}_N$.

The homomorphism ψ can be extended to a homomorphism $A[X] \to B[X]$ by $G(X) = \sum_{i=0}^{\Delta} g_i X^i \mapsto G^\psi(X) = \sum_{i=0}^{\Delta} \psi(g_i) X^i$. Let $F(X) = X^\delta + \sum_{i=0}^{\delta-1} f_i X^i \in \mathbb{Z}[X]$ be a monic irreducible polynomial, and let, for p a prime number,

$$\sigma_p : \mathbb{Z} \longrightarrow \mathbf{F}_p$$

and

$$\pi_F : \mathbb{Z}[X] \longrightarrow \mathbb{Z}[X]/(F(X))$$

be the canonical projections. From σ_p we get the projection

$$\pi_p : \mathbb{Z}[X]/(F(X)) \longrightarrow \mathbf{F}_p[X]/(F^{\sigma_p}(X)).$$

We shall study the ideals

$$m\mathbb{Z}[X]^e = \pi_F(m\mathbb{Z}[X]),$$

for $m \in \mathbb{Z}$, of the ring $\mathbb{Z}[X]/(F(X))$. First we describe the elements of $m\mathbb{Z}[X]^e$ in terms of a property of the corresponding canonical representative in $\mathbb{Z}[X]$.

Lemma 12. Let $R(X) = \sum_{i=0}^{\delta-1} r_i X^i \in \mathbb{Z}[X]$, and let $G(X) = Q(X) \cdot F(X) + R(X)$. Then

$$G(X) \bmod F(X) \in m\mathbb{Z}[X]^e \iff R(X) \in m\mathbb{Z}[X].$$

Proof. The direction ($\Longleftarrow$) is trivial. Let us turn to ($\Longrightarrow$) Since π_F is onto, $m\mathbb{Z}[X]^e = \{G(X) \bmod F(X) | G(X) \in m\mathbb{Z}[X]\}$. Consequently, there is a $G(X) \in m\mathbb{Z}[X]$ and a $Q(X) \in \mathbb{Z}[X]$ such that $G(X) = Q(X)F(X) + R(X)$. We have to show that $R(X) \in m\mathbb{Z}[X]$. This is done by induction on degree(G). If degree(G) $\leq \delta - 1$, everything is clear. Assume that for all polynomials $G(X)$ of degree less than or equal to k the implication is proved. Let $G(X)$ be a polynomial of degree $k + 1$ now. We get

$$\begin{aligned} G_1(X) &= G(X) - g_{k+1} X^{k+1-\delta} F(X) \\ &\equiv R(X) \pmod{F(X)}, \end{aligned}$$

where $G(X) \in m\mathbb{Z}[X]$, especially $m | g_{k+1}$. Consequenly, $G_1(X) \in m\mathbb{Z}[X]$. Clearly, degree($G_1$) $\leq k$ and the induction hypothesis can be applied.

Second we apply lemma 12 to easily get the following two lemmas.

Lemma 13. The kernel of π_p equals $p\mathbb{Z}[X]^e$.

Lemma 14. If m_1 and m_2 are coprime integers, then

1. $m_1 \mathbb{Z}[X]^e + m_2 \mathbb{Z}[X]^e = \mathbb{Z}[X]/(F(X))$, i.e. the ideals $m_1 \mathbb{Z}[X]$ and $m_2 \mathbb{Z}[X]$ are coprime, too.
2. we have the equations

$$\begin{aligned} (m_1 m_2)\mathbb{Z}[X]^e &= m_1 \mathbb{Z}[X]^e \cdot m_2 \mathbb{Z}[X]^e \\ &= m_1 \mathbb{Z}[X]^e \cap m_2 \mathbb{Z}[X]^e \end{aligned}$$

Now we can prove

Proposition 4 (Chinese Remaindering) *Let $p_1 < \ldots < p_N \leq N^{O(1)}$ be an ordered sequence of prime numbers. Then*

1. **Algebraic Part:** *The factor-ring of $\mathbb{Z}[X]/(F(X))$ by $\left(\left(\prod_{i=1}^N p_i\right)\mathbb{Z}[X]\right)^e$ is canonically isomorphic to*

$$\prod_{i=1}^N \mathbf{F}_{p_i}[X]/\left(F^{\sigma_p - i}(X)\right).$$

2. **Algorithmic Part:** *If CR_n^F, where $n = \delta \sum_{i=1}^N \left(\lceil \log_2 p_i \rceil + 2\right)$, is defined to be the following problem*
   Input: *Sequences $(r_{i,0}, \ldots, r_{i,\delta-1})$,*
       for $i = 1, \ldots, N$, of canonical representatives $\mathrm{mod}\, p_i$
   Output: *$(r_0, \ldots, r_{\delta-1})$*
   of canonical representatives $\mathrm{mod}\, \prod_{i=1}^N p_i$ such that, for all $i = 1, \ldots, N$,

$$\sum_{j=0}^{\delta-1} r_j X^j \equiv \sum_{j=0}^{\delta-1} (r_{i,j} \bmod p_i) X^i \quad (\bmod\, p_i \mathbb{Z}[X]).$$

then CR_n^F is NC^1–reducible to the problem of computing $\prod_{i=1}^N p_i$.

Proof. Claim 1 follows from proposition 3, lemma 13, and lemma 14. Let us turn to claim 2. From claim 1 together with lemma 12 it follows that

$$\forall j \in \{1, \ldots, \delta - 1\} : r_j = \mathrm{ICR}(p_1, \ldots, p_N, r_{1,j}, \ldots, r_{N,j})$$

Now claim 2 follows from proposition 2.

3.3 The discriminant, finite fields

Let A and B be entire rings, and let $\psi : A \to B$ be a ring homomorphism. Let $F(X) = X^\delta + \sum_{i=0}^{\delta-1} f_i X^i$ be an element of $A[X]$. The discriminant is defined to be $D(F) = \prod_{i<j}(\alpha_i - \alpha_j)^2$, where $\alpha_1, \ldots, \alpha_\delta$ are the not necessarily distinct roots of the polynomial F in an algebraic closure of the quotient field of A. Obviously, F has no multiple roots iff $D(F) \neq 0$. Since $D(F)$ is symmetric in the $\alpha_1, \ldots, \alpha_\delta$, $D(F)$ can be represented as a polynomial in the $f_0, \ldots, f_{\delta-1}$ with integer coefficients. Consequently, $D(F^\psi) = \psi(D(F))$.

In particular we get that if ψ is equal to the canonical projection $\sigma_p : \mathbb{Z} \to \mathbf{F}_p$, then $F^{\sigma_p}(X)$ has multiple roots iff $p|D(F)$.

We know that in a fixed algebraic closure $\overline{\mathbf{F}_p}$ of $\mathbf{F}_p$ there is exactly one finite field $\mathbf{F}_{p^r}$, for each natural number r, which has exactly p^r elements. It is the splitting field of the polynomial $Z^{p^r} - Z$. Hence $\mathbf{F}_{p^r}$ is a normal extension of $\mathbf{F}_p$. Moreover, this extension is separable. The multiplicative group $(\mathbf{F}_{p^r})^*$ is cyclic.

If $\mathbf{F}_{p^s}$ is another finite field of characteristic p, then we have the tower of finite extensions $\mathbf{F}_p \subseteq \mathbf{F}_{p^r} \subseteq \mathbf{F}_{p^s}$ if and only if r divides s.

The finite field $\mathbf{F}_{p^r}$ can be represented as follows. Since $\mathbf{F}_{p^r}$ is the unique extension of $\mathbf{F}_p$ of degree r, for any irreducible polynomial $\Phi_{p,r}(Z) \in \mathbf{F}_p[Z]$ of degree r we have $\mathbf{F}_{p^r} = \mathbf{F}_p[Z]/(\Phi_{p,r}(Z))$. All polynomials $\Phi_{p,r}(Z)$ are separable ones, and each irreducible polynomial $\Psi(Z) \in \mathbf{F}_p[Z]$ the degree of which divides r splits over $\mathbf{F}_{p^r}$ in to linear factors. Moreover, by the Euclidean algorithm each element of $\mathbf{F}_{p^r}$ can be uniquely represented by a polynomial of degree less than r with coefficients in $\mathbf{F}_p$.

Let δ be a positive natural number. Define $\delta^* = \text{l.c.m.}(2, \ldots, \delta)$. It follows from the above considerations that $\mathbf{F}_{p^{\delta^*}}$ is the splitting field of all polynomial of degree less than or equal to δ with coefficients in $\mathbf{F}_p$.

4 The proof

We use the notations in the formulation of MULT_n^F in section 1.

It is well–known, that there is a constant λ' such that $\prod_{p \leq k} p \geq 2^{\lambda' k}$, where the product ranges over prime numbers only. This remains valid if the finite set of primes $\{p \mid p \text{ divides } D(F)\}$ is excluded.

Let $PRIME_{N,\lambda}$ be the set of all prime numbers not dividing $D(F)$ which are less then or equal to λN^2. By lemma 11 and by the foregoing remark it follows that there is a constant λ such that

$$\prod p > 2 \max\{|r_i|; \ i = 0, \ldots, \delta - 1\},$$

where the product ranges over all $p \in PRIME_{N,\lambda}$. These prime numbers $p \in PRIME_{N,\lambda}$ and their product are computed by a polynomial time Turing machine and are hardwired in the circuit.

Because of the fact that δ and consequently δ^* are constants, we can do the following. If $\mathbf{F}_{p^{\delta^*}} = \mathbf{F}_p[Z]/(\Phi_{p,\delta^*}(Z))$, then Φ_{p,δ^*} can be computed by brute force with $O(\log n)$ space. Moreover, $\Phi_{p,\delta^*}(Z)$ can be stored within $O(\log n)$ space on a Turing tape. Lemma 10 ensures that the multiplication of elements of $\mathbf{F}_{p^{\delta^*}}$ can be done on a $O(\log n)$ space bounded Turing tape. Consequently, the roots $\alpha_{p,1}, \ldots, \alpha_{p,\delta}$ of $F^{\sigma_p}(X)$ and a generator ω_p of $(\mathbf{F}_{p^{\delta^*}})^*$, for all prime numbers $p \in PRIME_{N,\lambda}$ can be precomputed by a log n–Turing machine. Thus, for each prime $p \in PRIME_{N,\lambda}$, each $G(X) \in \mathbf{F}_p[X]$, $\text{degree}(G) \leq \delta - 1$, and each $i = 1, \ldots, \delta$, $G(\alpha_{p,i})$ can be computed in $O(\log n)$ space and hardwired into the circuit. The same can be done with the discrete logarithm function l_p respect to ω_p. These tables can be used in either direction. (In the case of evaluating all polynomial $G(X) \in \mathbf{F}_p[X]$, $\text{degree}(G) \leq \delta-1$, where $p \in PRIME_{N,\lambda}$ the invertibility, i.e. the interpolation, follows from the fact that the polynomials $F^{\sigma_p}(X)$ have no multiple roots and thus Vandermond's Determinant in these cases is not equal to zero.)

The algorithm then proceeds as follows, where p ranges over all elements from $PRIME_{N,\lambda}$, and k ranges over $\{1, \ldots, N\}$.

Step 1: Compute for all p and k in parallel

$$(r_{k,0} \bmod p, \ldots, r_{k,\delta-1} \bmod p)$$

by proposition 1.

Step 2: Compute for all p and all k

$$((R_k \bmod p)(\alpha_{p,1}), \ldots, (R_k \bmod p)(\alpha_{p,\delta}))$$

by table look–up.

Step 3: Compute for all p, k, and $i = 1, \ldots, \delta$ in parallel

$$l_p\left((R_k \bmod p)(\alpha_{p,i})\right)$$

by table look–up.

Step 4: Compute for all p, and for all $i = 1, \ldots, \delta$ in parallel

$$\sum_{k=1}^{N} l_p\left((R_k \bmod p)(\alpha_{p,i})\right) \bmod \left(p^{\delta^*} - 1\right)$$

by theorem 4 and proposition 1.

Step 5: Compute for all p and for all $i = 1, \ldots, \delta$ in parallel

$$l_p\left((R \bmod p)(\alpha_{p,i})\right) \text{ and } (R \bmod p)(\alpha_{p,i})$$

by table look–up.

Step 6: Compute for all p in parallel

$$(r_0 \bmod p, \ldots, r_{\delta-1} \bmod p)$$

from

$$((R \bmod p)(\alpha_{p,1}), \ldots, (R \bmod p)(\alpha_{p,\delta}))$$

by table look–up.

Step 7: Compute

$$(r_0, \ldots, r_{\delta-1})$$

by proposition 4.

Because of the fact that $\prod p > 2 \max\{|r_i|; i = 0, \ldots, \delta - 1\}$, where the product ranges over all $p \in \text{PRIME}_{N,\lambda}$, the result of Chinese Remaindering gives the exact values.

References

1. P. W. Beame, S. A. Cook, H. J. Hoover, *Logdepth circuits for division and related problems*, SIAM J. Computing 1986, **15**(4), pp. 993–1003.
2. A. Borodin, S. A. Cook, N. Pippenger, *Parallel computation for well–endowed rings and space bounded probabilistic machines*, TR 162/83, University of Toronto.
3. S. A. Cook, *The classification of problems which have fast parallel algorithms*, in: Lecture Notes in Computer Sci. **158**, Springer–Verlag, Berlin, 1983.
4. S. A. Cook, *A taxonomie of problems with fast parallel algorithms*, Information and Control 1985, **64**, pp. 2–22.
5. S. A. Cook, P. McKenzie, *The parallel complexity of abelian permutation group problems*, SIAM J. Computing 1987, **16**(2) pp. 880–909.
6. S. A. Cook, P. McKenzie, *Problems complete for deterministic logspace*, J. of Algorithms 1987, **8** pp. 385–394.
7. J. E. Hopcroft, J. D. Ullman, *Introduction to automata theory, languages, and computation*, Addison–Wesley, Reading, 1979.
8. S. Lang, *Algebra*, Addison–Wesley, Reading 1965.

9. J. H. Reif, *Logarithmic depth circuits for algebraic functions*, SIAM J. Computing 1986, **15** (1), pp. 231–242.
10. J. H. Reif, S. Tate, *Optimal size integer division circuits*, in: Proc. 21st ACM STOC 1989, pp. 264–273.
11. V. Pan, *Complexity of computations with matrices and polynomials*, SIAM review 1992, **34-2**, pp. 225–262.
12. W. L. Ruzzo, *On uniform circuit complexity*, J. Comput. System Sci. 1981, **22**, pp.365–383.
13. J. E. Savage, *The complexity of computing*, John Wiley, New York 1976.
14. I. Wegener, *The complexity of Boolean functions*, Wiley–Teubner Series in Comput. Sci., 1987.

$\mathcal{N}C^2$ Algorithms Regarding Hamiltonian Paths and Circuits in Interval Graphs (Extended Abstract)

Y. Daniel Liang[1]* and Raymond Greenlaw[2]** and Glenn Manacher[3]

[1] Department of Computer Science, Indiana Purdue University at Fort Wayne, Fort Wayne, IN 46805, e-mail address: liangy@panda.ipfw.indiana.edu
[2] Department of Computer Science, University of New Hampshire, Durham, NH 03824, e-mail address: greenlaw@cs.unh.edu
[3] Department of Mathematics, Statistics, and Computer Science, University of Illinois at Chicago, Chicago, IL 60680

Abstract. This paper describes an $O(\log^2 n)$ time, n processor EREW PRAM algorithm to determine if there is a Hamiltonian path in an interval graph. If there is a Hamiltonian path, we can find it within the same resource bounds. Many graph theoretic problems including finding all maximal cliques, optimal coloring, minimum clique cover, minimum weight dominating set, and maximum independent set were previously known to be in $\mathcal{N}C$ when restricted to interval graphs. However, the Hamiltonian path problem was open and resisted classification until now. We also show that testing whether an interval graph has a Hamiltonian circuit can be done in $\mathcal{N}C^2$. If the intervals are presorted, our parallel approach leads to an $O(n\alpha(n))$ sequential algorithm, where $\alpha(n)$ is the inverse of Ackermann's function. This improves on the previous bound of $O(n \log \log n)$ for the sequential case with presorted intervals.

1 Introduction

The *Hamiltonian path* (HP) problem asks whether there is a path in a graph that passes through each node exactly once. If such a path exists, it is called a HP of the graph. The importance of this problem is well-known and widely documented in the literature. Below we mention a few of the results regarding HP. Note that throughout this paper we will be concerned only with undirected graphs.

The problem of deciding whether a graph G has a HP is $\mathcal{N}P$-complete [10]. The problem remains $\mathcal{N}P$-complete if G is bipartite or planar, cubic, 3-connected, and has no face with fewer than five edges [6, page 199]. If either the starting point, ending point, or both are specified the problem is still $\mathcal{N}P$-complete [6, page 200]. The problem can be solved in polynomial time if G has no vertex with degree greater than two or if G is a line graph [6, page 199]. In the former case the problem is easily seen to be in $\mathcal{N}C^2$. The problem can also be solved in time $O(n + m)$ if G is an

* This research supported by a grant from the Purdue University Research Foundation.
** This research was partially funded by the National Science Foundation grant CCR-9209184.

interval graph, where n is the number of intervals and m is the number of edges in G [12]. If the intervals are presorted, the problem can be solved in $O(n \log \log n)$ time [16].

It is known that if one restricts the inputs to be interval graphs then for many graph properties the resulting problems fall into the class $\mathcal{NC}$. For example, $\mathcal{NC}$ algorithms have been developed on chordal graphs, a superclass of interval graphs, for finding the following: all maximal cliques, an intersection graph representation, an optimal coloring, a perfect elimination scheme, a maximum independent set, a minimum clique cover, and the chromatic polynomial [17]. Several of these results were improved in [9]. In [1] weighted versions of several of these and related problems are shown to be in $\mathcal{NC}$ for circular-arc graphs, another generalization of interval graphs. A *proper interval graph* is a special type of interval graph in which no interval is completely contained in another. For proper interval graphs it is known that finding a HP is in $\mathcal{NC}$ [1]. The problem of finding a HP in an interval graph though has remained open until now.

A *Hamiltonian circuit* (HC) is a HP where the first and last nodes on the path are connected. In this paper we prove that the question of whether an interval graph has a HP or HC is in $\mathcal{NC}^2$. The result is proved by exhibiting a reduction to a restricted form of the maximum matching problem. Finding a maximum matching in an arbitrary graph (or a bipartite graph) is known to be in $\mathcal{RNC}$ but not $\mathcal{NC}$ [8]. Therefore, to place the HP problem in $\mathcal{NC}$ it was necessary for us to reduce it to a restricted form of maximum matching. Our reduction is to the maximum matching problem for *convex bipartite graphs*; it is known that this restricted maximum matching problem is in $\mathcal{NC}^2$ [4].

Our algorithm computes the same HP as the sequential algorithm given in [16]. This algorithm, which we describe in section 2, seems to be highly sequential. The algorithm always makes a greedy choice of which interval to add next to the path being constructed. It always chooses the interval that overlaps the last interval added to the path and that has the rightmost left endpoint. Thus, in a sense the algorithm finds a canonical HP. This is important in the development of our algorithm. Frequently, algorithms that employ such greedy strategies are difficult to parallelize and many greedy algorithms have decision problems based on them that are $\mathcal{P}$-complete [8]. Thus, in this case it is somewhat surprising that the greedy algorithm does parallelize well. There are several hurdles that we overcome in showing the HP problem is in $\mathcal{NC}^2$ for interval graphs. Details of proofs that are omitted may be found in [13].

The remainder of this paper is outlined as follows: in section 2 we describe some background material and several results necessary for obtaining our algorithm, in section 3 we give the reduction from the HP problem in interval graphs to the maximum matching problem in convex bipartite graphs, the description and analysis of the algorithm are presented in section 4, related results are presented in sections 5 and 6, and open problems are given in section 7.

2 Preliminaries

We assume the reader is familiar with basic complexity theory, the complexity class $\mathcal{NC}$, and the EREW PRAM model of computation. For background material in these areas, see for example one of the following: [5, 8, 11]. As is customary we call an $\mathcal{N}C$ *algorithm*, an algorithm that runs on a PRAM in $\log^k n$ time for some constant k while using only a polynomial number of processors. A common approach in complexity theory for studying problems that are $\mathcal{N}P$-complete on arbitrary graphs is to restrict the problems to a large subclass of graphs. Intervals graphs are an important class of graphs that we focus our attention on. They arise in resource allocation problems and in a number of scheduling problems. Interval graphs are formally defined below.

Definition 1. Let I be a finite set of closed intervals on the real line. A graph $G = (V, E)$ is called an **interval graph** for I if there is a one-to-one correspondence between I and V such that two intervals have nonempty intersection (**overlap**) in I if and only if the corresponding vertices in V are adjacent. I is called an **interval model** of G.

Let $I = \{I_1, \ldots, I_n\}$ be a set of intervals with each interval I_i represented by $[l_i, r_i]$, where l_i and r_i denote the *left endpoint* and the *right endpoint* of I_i, respectively. Figure 1 depicts an interval model with n equal to 9. Without loss of generality in what follows, we may assume that all the endpoints are different. For any pair of intervals $I_i, I_j \in I$, we sometimes denote $l_i < l_j$ by $I_i < I_j$. If $C = \{C_1, \ldots, C_p\}$ with $C_1 < \cdots < C_p$ and $D = \{D_1, \ldots, D_q\}$ with $D_1 < \cdots < D_q$ are two sets of intervals, then $C < D$ $(D < C)$ means $C_1 < D_1$ $(D_1 < C_1)$.

[htb]

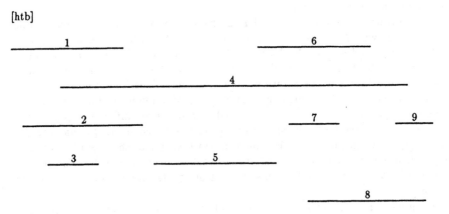

Fig. 1. Example of an interval model I with nine intervals.

The HP that our algorithm computes is the same one that is computed by the sequential algorithm given in [16]. Their algorithm is described by the following pseudo-code:

Sequential Algorithm for Finding a HP in an Interval Graph

Input: A set of intervals $I = \{I_1, \ldots, I_n\}$.
Output: A path P.

begin
    let i be such that $I_j < I_i$ for all $j \neq i$, $1 \leq j \leq n$;
    $P \leftarrow I_i$;
    $S \leftarrow I - \{I_i\}$;
    COMMENT: $last(P)$ returns the interval most recently added to P.
    while there exists an $I_i \in S$ such that I_i overlaps $last(P)$ do
        let I_t be the interval with the largest left endpoint of those in S
            overlapping $last(P)$;
        $P \leftarrow P \cup I_t$;
        $S \leftarrow S - \{I_t\}$
    endwhile
end.

Initially, the path P contains the interval with the largest endpoint in I. The remaining interval I' intersecting $last(P)$ with the largest left endpoint is always chosen next and added to P. This greedy process is repeated until no more intervals can be found to connect with $last(P)$. The sequential algorithm is such that a HP is found if P includes all the intervals in I. In addition, [16] implies no Hamiltonian path exists in I if and only if P is a *partial path*, i.e. $|P| < n$. Hence, I has a HP if and only if P as constructed above is a HP.

    For example, in Fig. 1 the path P constructed is $9-8-7-6-5-4-3-2-1$. P is a HP because it includes all intervals in I. Note, if we replace the first statement of the while loop by the two statements

1. let I_t be the interval with the largest left endpoint of those in S; and
2. if I_t does not overlap $last(P)$ then return P;

then a different path is computed, call the new path P'. P' is a form of greedy path and can easily be found in $\mathcal{N}C^2$ using pointer doubling techniques [11]. If all the intervals in a model I have the same length, then I has a HP if and only if P' is a HP. Thus, for models in which all intervals are the same length, it is easy to compute a HP in $\mathcal{N}C^2$. In fact, such a result will hold for any proper interval graph (see section 1 for the definition).

    The reduction in our main theorem is to the maximum matching problem in convex bipartite graphs. Below we define this restricted type of graph.

Definition 2. A convex bipartite graph G is a triple (A, B, E), where $A = \{a_1, \ldots, a_p\}$ and $B = \{b_1, \ldots, b_q\}$ are disjoint sets of vertices, and E satisfies the following two properties:

1. If $\{u, v\}$ is an edge of E, then either $u \in A$ and $v \in B$ or vice versa; i.e. no edge joins two vertices in A or two in B.

2. If $\{a_i, b_j\} \in E$ and $\{a_i, b_{j+k}\} \in E$, then $\{a_i, b_{j+1}\}, \{a_i, b_{j+2}\}, \ldots, \{a_i, b_{j+k-1}\}$ are also edges in E.

The first condition in the definition is the usual bipartite property; the second condition is the convexity property. We make use of the following theorem proved in [4].

Theorem 3. *The maximum matching problem restricted to convex bipartite graphs can be solved on an EREW PRAM in time $O(\log^2 n)$ using n processors.*

3 Reduction from Hamiltonian Path to Maximum Matching

In this section we reduce the problem of finding a HP in an interval graph to that of finding a maximum matching in a convex bipartite graph. Since our reduction can be performed quickly in parallel (as shown in section 4) and since a maximum matching in a convex bipartite graph can be found quickly in parallel (Theorem 3), we are able to develop a fast parallel algorithm for computing a HP in an interval graph. In attempting to parallelize the sequential algorithm shown in section 2, a difficulty arises in how to quickly choose the next interval to add to the path. The following notion turns out to be critical in selecting the next interval.

Definition 4. A **maximal clique** of I is a set S of mutually overlapping intervals such that no more intervals can be added to S without violating the mutual overlap property.

The overlapping in a maximal clique is pairwise. Each interval must overlap every other interval. It is known that all such maximal cliques of I can be linearly ordered as stated in the following corollary of a result due to Gilmore and Hoffman [7].

Corollary 5. *Let $I = \{I_1, \ldots, I_n\}$ be a set of intervals. The maximal cliques of I can be linearly ordered using $<$ such that for every interval $I_i \in I$, the maximal cliques containing I_i occur consecutively.*

Let $(C, <) = (C_1, \ldots, C_k)$ be the linearly ordered list of all maximal cliques as given from Corollary 5. For the set of intervals in Fig. 1, we have $C_1 = \{1, 2, 3, 4\}$, $C_2 = \{4, 5, 6\}$, $C_3 = \{4, 6, 7, 8\}$, and $C_4 = \{4, 8, 9\}$. Here k equals 4 and $C_1 < C_2 < C_3 < C_4$. For example, $C_1 < C_2$ since the left endpoint of 1 is further to the left than the left endpoint of 4, i.e. $1 < 4$. Note, how the maximal cliques containing 4, 6, and 8 occur consecutively. The following theorem shows that the HP problem in a set of intervals I can be reduced to finding a maximum matching in a convex bipartite graph.

Theorem 6. *Let $I = \{I_1, \ldots, I_n\}$ be an interval model. I can be transformed into a convex bipartite graph $G = (A, B, E)$ such that I has a HP if and only if G has a maximum matching of size $|B|$.*

Proof. We construct a convex bipartite graph $G = (A, B, E)$ for I based on the maximal cliques $(C, <) = \{C_1, \ldots, C_k\}$. Let A be the set of intervals that appear in more than one maximal clique in $(C, <)$ and $B = \{C_2, \ldots, C_k\}$. For each $a \in A$, if a appears in C_{i-1} and C_i, then add edge $\{a, C_i\}$ to E. Note, if an interval belongs to both C_i and C_j for $i < j$, then it belongs to every maximal clique C_l for $i \leq l \leq j$. This is because maximal cliques containing the same interval occur consecutively according to Corollary 5. It is clear that G is bipartite and this proves that G is also convex. Figure 2 shows the convex bipartite graph constructed from the interval model in Fig. 1.

[htb]

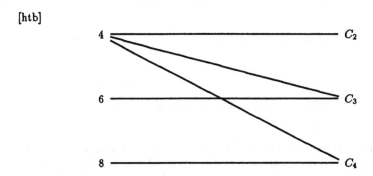

Fig. 2. The convex bipartite graph G for the interval model I of Figure 1.

We need to prove I has a HP if and only if G has a maximum matching of size $|B|$.

(only if) Assume I has a HP. Then $P = p_1 \ldots p_n$ as constructed by the sequential algorithm on page 4 is a HP. We show G has a matching of size $|B|$ corresponding to P by constructing a matching that includes endpoints $C_k, \ldots, C_2$.

Let $I[j]$ denote $I - C_1 - \cdots - C_j$ and $I[0] = I$. We begin by matching C_k. Let $C'_k = I[k-1]$. It is easy to see that $C'_k \neq \emptyset$. All intervals in C'_k mutually overlap, and the left endpoints of the intervals in C'_k are larger than the left endpoints of other intervals in $I - C'_k$. Suppose the intervals in C'_k are labeled in decreasing order of their left endpoints. Let p_i be the first interval in P that belongs to both C_k and C_{k-1}. There must be such a p_i since P is a HP. We see that $p_1, \ldots, p_{i-1}$ appear only in C_k. Thus, C'_k must equal $p_1 \ldots p_{i-1}$. We match C_k with p_i. Next, we find a match for C_{k-1}. Let $C'_{k-1} = I[k-2] - C'_k - \{p_i\}$. Suppose the intervals in C'_{k-1} are labeled in decreasing order of their left endpoints. As was done for C'_k, let p_j be the first interval in $p_{i+1} \ldots p_n$ that belongs to both C_{k-1} and C_{k-2}. There must be such a p_j since P is a HP. Then $C'_{k-1} = p_{i+1} \ldots p_{j-1}$. We match C_{k-1} with p_j. By continuing the same process, $C_{k-2}, \ldots, C_2$ can all be matched in a similar manner. Clearly, the matching constructed has size $|B|$ and is maximum.

(if part) Assume that all vertices in B are matched in G. Let a_i denote the vertex in A that matches C_i in B and let $I'[j] = I[j] - \{a_2, \ldots, a_k\}$. We define C'_i as follows:

$$C'_i = I'[i-1] - C'_{i+1} - \cdots - C'_k, \qquad \text{for} \quad i = 1, \ldots, k-1$$
$$C'_k = I'[k-1]$$

Suppose all intervals in C'_i are labeled in decreasing order of their left endpoints. It follows that $C'_k a_k C'_{k-1} \ldots a_3 C'_2 a_2 C'_1$ is a HP in I using the definitions of G and the C''s. □

Remark. In the proof of Theorem 6, C_i' could be empty for any $i < k$. For example, in the interval model shown in Fig. 3 we have $C_1 = \{1, 2\}$, $C_2 = \{2, 3\}$, and $C_3 = \{3, 4\}$. In this case $P = 4 - 3 - 2 - 1$, and $C_3' = \{4\}$ and $C_2' = \emptyset$. $\{3, C_3\}$ and $\{2, C_2\}$ are matched by the construction.

[htb]

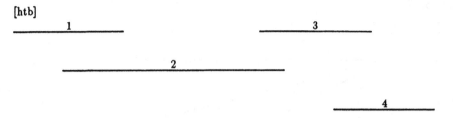

Fig. 3. Example of an interval model I where $C_2' = \emptyset$.

We present two more examples concerning Theorem 6. Consider the graph G shown in Fig. 2. Since G has a matching of size three, I has a HP. The HP is $C_4' a_4 C_3' a_3 C_2' a_2 C_1'$, where $a_4 = 8$, $a_3 = 6$, $a_2 = 4$, $C_4' = \{9\}$, $C_3' = \{7\}$, $C_2' = \{5\}$, and $C_1' = \{1, 2, 3\}$. Now consider the interval model shown in Fig. 4. There $C_1 = \{1, 2, 3\}$, $C_2 = \{1, 2, 4\}$, $C_3 = \{1, 4, 5\}$, $C_4 = \{4, 6\}$, $C_5 = \{6, 7\}$, and $C_6 = \{6, 8\}$. G is shown in Fig. 5. Since no matching of cardinality five can be found in G, no HP exists in I.

[htb]

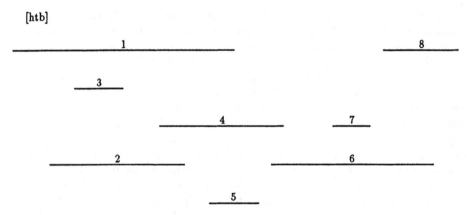

Fig. 4. Another sample interval model I.

In the next section we describe sequential and parallel algorithms based on Theorem 6 for finding a HP in an interval graph.

4 Statement and Analysis of the Main Algorithm

Theorem 6 implies a procedure for finding a HP in an interval model I when a maximum matching of size $|B|$ can be found in the graph $G = (A, B, E)$ constructed in the theorem. Below we describe a sequential algorithm based on this procedure. Following this we present a series of lemmas showing how each step in the algorithm can be implemented quickly in parallel. This analysis leads to a parallel algorithm that runs in $O(\log^2 n)$ time using n processors on an EREW PRAM.

Algorithm for Finding a HP in an Interval Graph Using a Reduction to Maximum Matching

Input: A set of intervals $I = \{I_1, \ldots, I_n\}$.
Output: A HP path P if one exists and otherwise "no."

```
begin
    construct the maximal cliques C₁,...,Cₖ of I;
    construct G = (A, B, E) as in Theorem 6;
    compute M a maximum matching for G;
    if |M| < k − 1 then return "no";
    let a₂,..., aₖ in A be matched to C₂,...,Cₖ in B via M;
    compute C′ₖ,...,C′₁ as in Theorem 6;
    return C′ₖaₖCₖ₋₁...a₃C′₂a₂C′₁
end.
```

[htb]

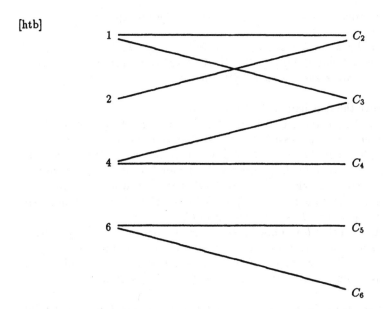

Fig. 5. The graph G for I in Fig. 4. Since no matching of size five exists, the corresponding interval model has no HP.

We now turn our attention to a parallel implementation of the main algorithm. A brute-force approach would be to find $C_1, \ldots, C_k$ and explicitly construct G. Such a strategy would not lead to an $\mathcal{NC}$ algorithm. In what follows we show how to avoid such a construction and obtain an $\mathcal{NC}$ algorithm.

The essential information needed to build G is to know the first and last cliques to which an interval belongs. To compute these intervals quickly in parallel, we introduce the notion of breakpoints. Let b_1 (b_{k+1}) be the smallest left (largest right) endpoint in C_1 (C_k), and for $2 \leq i \leq k$, let b_i be the smallest left endpoint in $C_i - C_{i-1}$; b_i is called a *breakpoint* of C_i. Instead of finding $C_1, \ldots, C_k$ explicitly, we identify $b_1, \ldots, b_{k+1}$. We make use of the following technical fact to construct G.

Lemma 7. *Suppose $b_1, \ldots, b_{k+1}$ are the breakpoints for an interval model I with maximal cliques $C_1, \ldots, C_k$. Let $I' = [l', r']$ be an interval in I. Suppose b_i (b_j) is the first breakpoint equal to or to the left (right) of l' (r'). Then I' belongs to exactly $C_i, \ldots, C_{j-1}$.*

Next we explain how to compute the breakpoints in parallel. Let $d_1, \ldots, d_{2n}$ be the endpoints in I listed in increasing order. An interval I_i is said to *contain* a point d if $l_i < d < r_i$. We use prefix sums to compute the breakpoints quickly in parallel. Assign a value of $+1$ to each left endpoint and a value of -1 to each right endpoint. Let $X[d_j]$ denote the prefix sum of all points less than d_j, $1 \leq j \leq 2n$. For a left endpoint l_j, $X[l_j]$ specifies the number of intervals containing l_j. For example in Fig. 1, $X[l_1] = 0$, $X[l_2] = 1$, $X[l_3] = 2$, $X[l_4] = 3$, $X[l_5] = 1$, $X[l_6] = 2$, $X[l_7] = 2$, $X[l_8] = 3$, and $X[l_9] = 2$.

The following lemma plays a key role in identifying the breakpoints.

Lemma 8. *Let i and j be such that $2 \leq i \leq n$ and $1 < j < k$. l_i is a breakpoint of some clique C_j if and only if $X[l_i] \leq X[l_{i-1}]$. All breakpoints can be computed in $O(\log n)$ time using n processors on an EREW PRAM.*

Proof. We only prove the first half of the lemma here.

(if part) Let S_r denote the set of intervals containing l_r, $1 \leq r \leq n$. Since $X[l_i] \leq X[l_{i-1}]$, there exists at least one interval that contains l_{i-1} but does not contain l_i. This implies that $S = S_{i-1} - S_i \neq \emptyset$. Furthermore, for each interval I_t in S_i, either $I_t = I_{i-1}$ or I_t contains l_{i-1} since $t \leq i - 1$ and I_t contains l_i. Therefore, $S_{i-1} \cup S_i \cup \{I_{i-1}\}$ constitutes a clique C_{j-1}. Recall that there is at least one member in C_{j-1} that does not contain l_i. Thus, I_i is the interval with the smallest left endpoint in $C_j - C_{j-1}$. Hence, l_i is a breakpoint of C_j.

(only if part) Assume that l_i is a breakpoint for some clique C_j, $1 < j \leq k$. Note that d_1 is the breakpoint for C_1. Then l_i is the smallest left endpoint in $C_j - C_{j-1}$, $I_{i-1} \notin C_j$, and $I_{i-1} \in C_{j-1}$. Since no left endpoints are between l_{i-1} and l_i, any intervals containing l_i must also contain l_{i-1}. Therefore, $X[l_i] \leq X[l_{i-1}]$. □

Note, the breakpoints in Fig. 1 are l_1, l_5, l_7, and l_9. For example, l_7 is a breakpoint because $2 = X[l_6] \leq X[l_7] = 2$.

In the second step of the main algorithm the graph $G = (A, B, E)$ is built. G can be represented by a set of triples as was done in [4]. Let $T = \{(a_i, C_{s_i}, C_{h_i})\}$, where $s_i = \min\{j \mid \{a_i, C_j\} \in E\}$ and $h_i = \max\{j \mid \{a_i, C_j\} \in E\}$. In this triple representation, the lowest and highest index vertices in B to which each a_i is connected is recorded. For example, in Fig. 2,

$$T = \{(4, C_2, C_4), (6, C_3, C_3), (8, C_4, C_4)\}.$$

We show below that the representation of G by triples can be computed very fast in parallel.

Lemma 9. *The representation T of G by triples can be constructed in $O(\log n)$ time using n processors on an EREW PRAM.*

Proof. Let $D = \{d_1, \ldots, d_{2n}\}$ ($K = \{b_1, \ldots, b_k\}$) be the set of endpoints (breakpoints except b_{k+1}) in I listed in increasing order. Our goal is to construct the triple representation T for G. We need to identify the intervals that are contained in more than one maximal clique. Note, if I' contains b_i ($i > 1$), then I' is contained in multiple cliques and is in A; therefore, $\{I', C_i\}$ is an edge in G.

We would like to be able to apply Lemma 7 and so need to determine the breakpoints bracketing a given interval. Let δ be one half of the minimum distance between any two endpoints in I. Let $K' = \{b_1 - \delta, \ldots, b_k - \delta\}$. We merge K' and D to form the sorted list D'. Define $rank(S, p)$ to be the rank of point p in a total order S. For each interval I_i in I, let $l = rank(D', l_i) - rank(D, l_i)$ and $r = rank(D', r_i) - rank(D, r_i)$. ¿From Lemma 7, it follows that I_i belongs to $C_l, \ldots, C_r$. If r is greater than l, then (i, C_l, C_r) is an element in T.

We analyze the resources required by the procedure described above. By Lemma 8, K can be computed within the resource bounds stated in lemma. δ can be computed and subtracted from each element in K on an EREW PRAM in $O(\log n)$ step using $n/\log n$ processors. K' and D can easily be merged within the bounds stated in the lemma. The rank function can be computed in $O(\log n)$ using $n/\log n$ processors on an EREW PRAM [11]. Therefore, the triple representation for G can be constructed in $O(\log n)$ time using n processors on an EREW PRAM. □

We can now state our main theorem.

Theorem 10. *Let I be an interval model with $|I| = n$. In $O(\log^2 n)$ time using n processors on an EREW PRAM one can determine if I has a HP. If I does have a Hamiltonian path it can be found in $O(\log^2 n)$ time using n processors on an EREW PRAM.*

Proof. Lemma 9 shows how to compute the triple representation for G in $O(\log n)$ time using n processors on an EREW PRAM. The next step in the main algorithm (see page 8) is to compute a maximum matching M for G. By Theorem 3 a maximum matching for a convex bipartite graph can be computed in $O(\log^2 n)$ time using n processors on an EREW PRAM. Their algorithm makes use of the triple representation that we have produced for G in Lemma 9.

We can easily compute the size of M within the bounds of the theorem. If $|M| < k - 1$ then the algorithm returns "no." Suppose G has a maximum matching of cardinality $k - 1$. Let $a_2, \ldots, a_k$ denote the vertices in A that match $C_2, \ldots, C_k$, respectively. Let $I' = I - \{a_2, \ldots, a_k\}$. Let C_i' consist of all intervals whose left endpoints lie in the interval $[b_i, b_{i+1})$, for $1 \leq i \leq k$. The C_i''s can be found in $O(1)$ time using n processors. According to Theorem 6, a HP in I is given by $C_k' a_k \ldots C_2' a_1 C_1'$. Note that the intervals in each C_i' can occur in any order since $C_i' \cup \{a_k\}$ is a clique. □

5 Testing Interval Graphs for Hamiltonian Circuits

In this section we show that testing whether an interval graph has a HC can be done in $O(\log^2 n)$ time using n^2 processors on an EREW PRAM. The result follows from the lemma stated below and Theorem 6.

Lemma 11. *I has a Hamiltonian circuit if and only if $I - \{I_i\}$ has a HP for any $I_i \in I$.*

Proof. (only if part) If I has a HC, then clearly $I - \{I_i\}$ has a HP for any $I_i \in I$.

(if part) We assume that I has no HC but always has a HP regardless of which single interval is deleted from I. In Theorem (HC) of [16], they show that if I has no HC, then a HP P can be represented as follows:

- $P = A_0 c_0 A_1 c_1 A_2 \dots A_k c_k A_{k+1}$ for some $k \geq 0$, where A_i is a sequence of intervals and c_i is a single interval that connects A_i with A_{i+1}.
- For any $i \neq j$ no interval in A_i overlaps an interval in A_j.

Therefore, $I - \{c_0\}$ consists of $k+2$ disjoint sets $A_0, \dots, A_{k+1}$ and only k intervals $c_1, \dots, c_k$ that can be used to connect them. Thus, $I - \{c_0\}$ has no HP contradicting our initial assumption. □

Theorem 12. *Testing for a Hamiltonian circuit in an interval model can be done in $O(\log^2 n)$ time using n^2 processors on an EREW PRAM model.*

6 Improvement of the Sequential Algorithm for HPs

It is rare when a straightforward simulation of a parallel algorithm leads to an improved sequential algorithm. In this section we show that a direct simulation of our parallel algorithm leads to an improved sequential algorithm for the problem of finding a HP in an interval graph with presorted intervals.[4] The new sequential algorithm requires $O(n\alpha(n))$ time, whereas, previously the best known algorithm required $O(n \log \log n)$ time [16]. Here $\alpha(n)$ denotes the inverse of Ackermann's function.

Theorem 13. *Let I be an interval model with $|I| = n$. If endpoints are presorted in I, a Hamiltonian path can be found sequentially in $O(n\alpha(n))$ time.*

Proof. If the endpoints are presorted, G can be computed in $O(\log n)$ time using $n/\log n$ processors on an EREW PRAM (see Lemma 9). Therefore, G can be constructed in $O(n)$ time using a single processor. Note, the breakpoints merged in Lemma 9 do not need to be sorted explicitly since they can be obtained in sorted order using the construction in Lemma 8. In [15], they showed that a maximum matching in a convex bipartite graph $H = (A, B, E)$ can be found sequentially in

---

[4] Recently, we learned that a linear time sequential algorithm for this problem has been developed [2].

$O(|A|+|B|\alpha(|B|))$ time. This becomes $O(n\alpha(n))$ time for the graph G we construct. Thus, finding a HP in an interval model can be done in sequential time $O(n\alpha(n))$.

$\square$

7 Conclusions and Open Questions

The main result of this paper is an $\mathcal{N}C^2$ algorithm for finding a HP in an interval graph. We also showed that testing for a HC in an interval graph is in $\mathcal{N}C^2$. Whether or not a HC can be exhibited in $\mathcal{N}C$ remains an open problem. It would be interesting to develop $\mathcal{N}C$ algorithms for the HP and HC problems in circular-arc graphs. It would also be interesting to exhibit the smallest possible class of graphs (or for that matter any class of graphs) for which the HP problem is $\mathcal{P}$-complete.

References

1. A. A. Bertossi and S. Moretti. Parallel algorithms on circular-arc graphs. *Information Processing Letters* **33(6)** (1990) 275–281.
2. M. S. Chang, S. L. Pang, and J. L. Liaw. Deferred-query – an efficient approach for some problems on interval graphs. Manuscript, 1993.
3. R. Cole. Parallel merge sort. *SIAM Journal on Computing* **17(4)** (1988) 770–785.
4. E. Dekel and S. Sahni. A parallel algorithm for convex bipartite graphs. *Proceedings International Conference on Parallel Processing* (1982) 178–184.
5. Faith E. Fich. The complexity of computation on the parallel random access machine. In Reif [18], chapter 20, pages 843–899.
6. M. R. Garey and D. S. Johnson. *Computers and Intractability: A Guide to the Theory of NP-Completeness*, W. H. Freeman and Company, San Francisco, 1979.
7. P. C. Gilmore and A. J. Hoffman. A characterization of comparability graphs and of interval graphs. *Canadian Journal of Mathematics* **16** (1964) 539–548.
8. R. Greenlaw, H. J. Hoover, and W. L. Ruzzo. *Topics in Parallel Computation: A Guide to the Theory of P-completeness*. Oxford University Press, New York, to appear.
9. C-W. Ho and R. C. T. Lee. Efficient parallel algorithms for finding maximum cliques, clique trees, and minimum coloring on chordal graphs. *Information Processing Letters* **28(6)** (1988) 301–309.
10. R. M. Karp. Reducibility among combinatorial problems, in R. E. Miller and J. W. Thatcher, eds., *Complexity of Computer Computations*, Plenum Press, New York, (1972) 85–103.
11. R. M. Karp and V. Ramachandran. Parallel algorithms for shared-memory machines. In van Leeuwan [19], chapter 17, pages 869–941.
12. J. M. Keil. Finding Hamiltonian circuit in interval graphs. *Information Processing Letters* **20(4)** (1985) 201–206.
13. Y. D. Liang, R. Greenlaw, and G. K. Manacher. $\mathcal{N}C^2$ Algorithms Regarding Hamiltonian Paths and Circuits in Interval Graphs. Technical report 93-11, University of New Hampshire, 1993.
14. Y. D. Liang, G. K. Manacher, C. Rhee, and T. A. Mankus. An $O(n \log n)$ algorithm for finding Hamiltonian paths and circuits in circular-arc graphs. Manuscript, August, 1992.
15. W. Lipski Jr. and F. P. Preparata. Efficient algorithms for finding maximum matchings in convex bipartite graphs and related problems. *Acta Informatica* **15(4)** (1981) 329–346.

16. G. K. Manacher, T. A. Mankus, and A. J. Smith. An optimum $O(n \log n)$ algorithm for finding a canonical Hamiltonian circuit in a set of intervals. *Information Processing Letters* **35(4)** (1990) 205–211.

17. J. Naor, M. Naor, and A. A. Schäffer. Fast parallel algorithms for chordal graphs. *Proceedings 19th Ann. ACM Symposium on Theory of Computing* (1987) 355–364.

18. John H. Reif, editor. *Synthesis of Parallel Algorithms.* Morgan Kaufman, San Mateo, CA, 1993.

19. Jan van Leeuwan, editor. *Handbook of Theoretical Computer Science*, volume A: Algorithms and Complexity. M.I.T. Press/Elsevier, 1990.

Concurrency in an $O(\log \log N)$ Priority Queue

Brandon Dixon*†

1 Introduction

In this paper we consider a pipelined model of parallel computing, and we present two optimal pipelined algorithms for the bounded universe priority queue problem. Our goal is to increase the throughput of a priority queue data structure by pipelining the operations. In our model each operation is executed by a single processor and the time for the operation to complete is the same as in the sequential version, but our new pipelined data structures allow a new operation to begin every time step. Using multiple processors, this gives a way to increase the throughput of the data structure without any increase in the latency time of a single operation.

A priority queue is a data structure supporting insertions, deletions, and find min operations [1]. Using balanced trees, these operations can be supported in $O(\log n)$ time where n is the number of elements in the queue. If the universe of elements is restricted to the range $[0 \ldots N]$, then these operations can be performed in $O(\log \log N)$ time using the van Emde Boas priority queue [6, 7]. This result requires the addressing power of the RAM model of computing, and the running time no longer depends on the number of elements in the queue. The van Emde Boas algorithm is therefore an improvement whenever $n > \log N$.

We are concerned with versions of the priority queue which allow parallel processing of the priority queue operations. This form of the priority queue is important in various parallel algorithms, e.g. multiprocessor scheduling, graph search, and branch-and-bound. If the size of the universe of the priority queue is unrestricted, it is known how to perform queue operations concurrently [10, 4]. In these algorithms, multiple operations are allowed to access the queue, and each operation requires $O(\log n)$ time. The throughput is increased by the use of multiple processors. In this paper, we show how to perform concurrent priority queue operations in a restricted universe. Like the previous algorithms, our

---

*Department of Computer Science, Washington and Lee University, Lexington, VA 24450.
†Research partially supported by a National Science Foundation Graduate Fellowship and by DIMACS (Center for Discrete Mathematics and Theoretical Computer Science), a National Science Foundation Science and Technology Center, Grant No. NSF-STC88-09648.

goal is to increase the throughput of the data structure given some number of processors.

In our algorithms, we assume a synchronized, shared memory parallel computing environment. We define a time step to be a constant amount of time, whose length we define later. Each of the concurrent operations execute one unit of work every time step, and one new operation is allowed to begin each time step. Two concurrent algorithms are presented in this model. In the first algorithm, each individual insertion, deletion, or find min operation completes in $O(\log \log N)$ amortized time, while in the second algorithm, the operations complete in $O(\log \log N)$ worst–case time. These concurrent algorithms yield parallel algorithms for maintaining a priority queue. Given $\Theta(\log \log N)$ processors, a sequence of m operations can be performed in $O(\log \log N + m)$ time, which is an optimal amount of work due to the lower bound in [2]. We describe the sequential version of each algorithm before giving the concurrent version. The amortized algorithms appear in sections 2 and 3, and the worst–case versions appear in sections 4 and 5.

2   An amortized $O(\log \log N)$ priority queue

This section presents a priority queue where the amortized time of an operation is $O(\log \log N)$.

We use a two level scheme to form the data structure. Let t be the smallest power of 2 larger than $N/\lceil \log N \rceil$. The top level is a fixed complete binary tree, T_{top}, with t leaves. The depth of T_{top} is $\lceil \log N - \log \lceil \log N \rceil \rceil$. We subdivide the universe $\{0, \ldots, N\}$ into contiguous blocks of size $\lceil \log N \rceil$ and the blocks are assigned to the leaves of the tree. We assign the ith block to the ith leaf of T_{top} starting from the left.

The lower level is comprised of up to $N/\lceil \log N \rceil$ balanced trees of size $O(\log N)$. Each leaf of the fixed tree contains a pointer that is either null or points to a lower level tree. We call a leaf of T_{top} active if it has a non–null pointer, and we call the lower level tree to which it points its *lower tree*. The internal nodes of T_{top} have a mark bit that is set for all ancestors of active nodes, as well as pointers to the rightmost and leftmost active leaves in their subtree. Initially the leftmost leaf points to an empty tree, the pointers at the other leaves are null, and only the leftmost path in T_{top} is marked. To quickly find neighboring active leaves, we keep a doubly linked list of the active leaves.

If a leaf l is active, then its lower tree contains the elements of its block that are in the queue, and as is shown below, we allow additional elements from the queue to be stored in the lower trees. Consider the insertion of an element x, and let l be the leaf of T_{top} assigned to the block containing x. If l is active, we insert x into its lower tree. Otherwise, we search in T_{top} for the first active leaf,

l', to the left of l. To find l', we do binary search on the tree path from l to the root of T_{top} to find the lowest marked ancestor of l. At this node, either the pointer to the leftmost active leaf points to the first active leaf to the right of l or the pointer to the rightmost active leaf points to l'. We can determine which case holds in constant time. If we reach the active leaf to the right of l, then we use the doubly linked list to reach l'. Then we insert x into the lower tree of l'. Thus each lower tree can contain elements from the queue that are from blocks to its right. We allow this because we cannot afford to mark the path from a leaf of T_{top} to its root for every insertion, so we instead use the first non–empty tree to the left.

We also cannot afford for the lower trees to become too large, so we split a lower tree t when it contains $3 \log N$ elements into two trees, t_1 and t_2, where all elements of t_1 are smaller than those of t_2. We want that the resulting trees have roughly equal size and do not contain elements from the same block. To achieve this, we split t only along block boundaries. Thus the size of t_1 and also t_2 is at least $logN$ and at most $2 \log N$. The leaf that pointed to t now points to t_1, and the leaf whose block contains the smallest element of t_2 points to t_2. The later leaf becomes active and the path from it to the root is marked and the pointers updated.

To search for an element in the queue, we perform binary search on the path in T_{top} as in the insertion, followed by a standard tree search in the lower tree. To perform a deletion, we simply search for the element and then perform a delete in the lower tree. When we delete the element, we check if the lower tree is now empty. If it is, we remove the empty tree and update the path from its leaf to the root appropriately.

The operations consist of a binary search on a path of length $O(\log N)$ and a search, insertion, or deletion in a balanced tree of size $O(\log N)$, each of which can be performed in $O(\log \log N)$ time. Some insert operations additionally mark a path from a leaf to the root and split a lower tree. This takes time $O(\log N)$, which can be amortized over the last $\Theta(\log N)$ insertions into the tree since it was last split. The cost of unmarking the path due to an empty tree is amortized against the $\Theta(\log N)$ deletions, since its last split. Thus each operation takes $O(\log \log N)$ amortized time.

3 Concurrent operations in the amortized queue

The previous scheme gives a way to maintain a priority queue in amortized $O(\log \log N)$ time. The goal of this section is to achieve the same time bounds for each individual operation, but modify the operations so that several operations can access the priority queue concurrently. The data structures remain basically the same for the concurrent version, but the algorithms for accessing the data structure change to allow concurrent access.

3.1 Concurrent searching

We begin by describing the search algorithm and later describe the insert and delete operations. The T_{top} structure remains exactly as in the sequential version, so to start a search operation, we perform binary search on the path in T_{top} to find the proper lower tree. We force the following additional constraint on the binary search procedure. Since the distance from a leaf to the root of T_{top} is $\lceil \log N - \log \lceil \log N \rceil \rceil$, the binary search takes at most $\log \log N$ time steps to find the lowest marked node on this path if we execute one iteration of the binary search per time step. We force all such searches to take exactly $\log \log N$ time steps by requiring operations that finish early to wait. This synchronization implies that at each time step, at most one new operation begins at the lower trees.

The lower tree reached by the search algorithm may not be the correct tree due to relaxations in the marking of internal nodes and the splitting of lower trees. The details of these relaxations appear below, but for now we simply assert that the correct tree can be reached in constant additional time.

We specifically choose the lower trees to be 2–4 trees [1] since it is know how to perform concurrent insertions and deletions in 2–4 trees [4, 5, 8], and we present algorithms to perform concurrent split and join operations in appendix A.

3.2 Concurrent insertion

As in the search algorithm, we find the correct lower tree for an item x to be inserted. Then we perform a top–down insertion of the item x which is accomplished easily since we have chosen the lower trees to be 2–4 trees[11]. The synchronization of the binary search in T_{top} guarantees that at most one new operation begins at a lower tree at each time step. The top–down insertions are, therefore, never blocked.

When an insert completes, we must split the lower tree if it becomes too large. In the sequential case, we were allowed to spend time proportional to the size of the tree to perform the split. In the concurrent version, however, the split must be performed while other insert and delete operations continue to act on the tree. Thus we create a top–down splitting procedure, that fulfills the same requirements as in the sequential case: the resulting trees must be of roughly equal size and they may not both contain elements from a common block. We choose a element s, called the split value, and partition a lower tree t into t_1 and t_2 so that all elements in t_1 are smaller than s and all elements in t_2 are at least as large as s. If we always choose s to be a multiple of $\lceil \log N \rceil$, then s is a block boundary, and we are guaranteed that t_1 and t_2 will not contain elements from a common block.

In the concurrent version, we must know the split value when we begin a split operation, because after a constant number of time steps, a subsequent search operation must be able to determine whether to continue the search in t_1 or in t_2. If we do not know the split value until the end of the split, then search operations might be blocked. To help find a split value, we keep a pointer in each lower tree to a leaf that will be roughly in the middle of the tree when the tree is split. We now describe how to maintain this pointer for a lower tree t that has never been split and later we describe how to handle a tree that has just been split. If t has never been split, then the pointer stays at the rightmost leaf of t until t has size $\lceil 3/2 \log N \rceil$. Once t has this size, we maintain as an invariant that there are $\lceil 3/2 \log N \rceil$ nodes to the left of the pointer. If the inserted item is smaller than the item at our pointer, then we move our pointer one leaf to the left, otherwise it remains in place. Similarly, if a node to the left of the pointer is deleted then the pointer is moved one step to the right. To be able to move the pointer as necessary, we keep a doubly linked list of the leaves of the lower trees. Thus there are $\lceil 3/2 \log N \rceil$ nodes to the left of our pointer during these operations, and therefore when it is time to split t, the pointer is at a node near the middle of t as described below.

When a split operation begins, there can be other operations acting on the tree. Because the pointer is moved only when an insert or delete operation completes, the pointer location does not reflect these uncompleted operations. Since the tree is of size $3 \log N$ when the split begins and only 1 operation can be at each level in the tree, there can be at most $\log 3 + \log \log N$ operations acting on the tree. Thus the size of the trees resulting from the split can be increased or decreased by $\log \log N + O(1)$. Since we have chosen to split only along block boundaries, we get the actual split value from the pointer by rounding the value stored at the pointer to the nearest multiple of $\lceil \log N \rceil$. This split value guarantees that the resulting trees have size between $\log N - \log \log N - O(1)$ and $2 \log N + \log \log N + O(1)$, and have depth $\log \log N + O(1)$. Thus our split value is roughly in the middle of the tree if we are given an tree that has never been split.

Now we must consider the case of a lower tree t that has just been split. Such a lower tree may be larger than $\lceil 3/2 \log N \rceil$ nodes, and thus the pointer is not at the correct location if we place it at the rightmost leaf of t and run the above algorithm. To correct the pointer location, we keep a counter that is initially $size(t) - \lceil 3/2 \log N \rceil$, or in other words, the counter tells us how many steps to the left the pointer should move in order to maintain our invariant. At each time step we move the pointer one extra step to the left in addition to the regular pointer movement, and we decrement the counter. We continue the extra pointer movement until the counter is zeroed. Since the largest tree that we can be given has $2 \log N + \log \log N + O(1)$ nodes, the largest that the counter can be is $\frac{\log N}{2} + \log \log N + O(1)$. Therefore we know that the counter can be reduced to zero before the next split operation takes place.

We have shown how to find the split value, so now the remaining task is to actually split the 2–4 tree from the top down. The split procedure, described in Appendix A, maintains two trees while performing the split operation: the left tree containing the values smaller than the split value and the right tree containing the rest. At each time step, the split procedure moves one step down the initial tree and therefore finishes after $\log \log N$ time steps.

Since it is possible that either the left or the right tree is empty for some number of time steps, some operations that wish to traverse an empty tree will have to wait. This does not cause a problem, however, because each time step during which an operation waits, the height of the tree that the operation must traverse is reduced by one. This means that the total waiting time together with the traversal time is still $\log \log N + O(1)$. Note also that once the tree is non–empty, the operation never waits; the splitting or joining process stays ahead of the traversal.

Once the split is performed, the leaf of T_{top} that pointed to the original tree now points to the left tree, and the leaf of T_{top} whose block contains the smallest element of the right tree points to the right tree. The later leaf becomes active and the path from it to the root is marked from the bottom–up and the pointers updated. Since other operations are still searching in T_{top} while this path is being marked, it is possible that searches do not find the correct tree. Therefore, when a search completes, we test to see if the proper tree was located, and if not, we use the doubly linked list to move to the correct tree. Note that the path will be marked after $\log N$ time steps and that a new split operation cannot begin at either the left or the right tree in this time. This guarantees that a search is never off by more than one tree to the left or to the right.

3.3 Concurrent deletion

As in the sequential algorithm, we find the correct lower tree and delete the element from the lower tree. In the concurrent version we use a top–down 2–4 tree deletion [11] to remove the element. If a lower tree becomes empty as a result of a deletion, the path from the leaf of the T_{top} tree to the root needs to be updated. Because other operations are accessing T_{top} during the unmarking, it is possible that a binary search can reach a leaf in T_{top} that is no longer active. Therefore, we maintain a pointer from the emptied lower tree to its predecessor, and the processor that is unmarking the path updates this pointer every time step as well. Thus, when an operation reaches an inactive leaf, it finds the correct lower tree using this pointer. Note that checking the predecessors requires concurrent reads, since many inactive leaves may have the same predecessor. After $\log N$ time steps, the updating of the path is complete and the predecessor pointer at the inactive leaf is no longer needed.

This completes the description of the amortized version of this data structure. The next two sections present the worst–case versions of both the sequential and concurrent variants of the data structure.

4 Worst case $\log \log N$ priority queue

This section presents a sequential algorithm for performing inserts, deletes, and find min operations in $O(\log \log N)$ worst–case time where the universe is the range $[0 \ldots N]$. The algorithm and the data structures are quite different than those presented in the previous two sections. For the description of the algorithm, we use the notation presented in Mehlhorn [9], although the data structure was first presented in [6, 7].

To perform insertions, deletions, and membership queries, a bit vector is an efficient data structure since we can perform each of the operations in $O(1)$ time. A priority queue must support find min operations as well, so, one possible solution is to keep a sorted linked list of the elements in the queue. The problem if you only keep a linked list is that insertions are not efficient. However, if we combine the data structures using pointers from the bit vector to the linked list, we can reduce our problem as follows. If we wish to insert element x into the list, and we can find the largest element $y < x$ that is in the queue, then we use the pointer from the bit vector entry for y to find y in the list. We insert x into the list as the successor to y. Thus we have reduced the problem to that of finding predecessors in our bounded universe.

An outline of the recursive searching technique used to find predecessors follows. The idea is roughly to break the current search range, call it $[0 \ldots N]$ into $\sqrt{N}$ blocks of size $\sqrt{N}$. We have two cases when searching for the predecessor of x. If the block for x already contains an element smaller than x, then we recurse on the block of x to find its predecessor. If the block containing x is empty or does not contain an element smaller than x, then we recurse on the set of blocks themselves. In this case, the largest element in the first non–empty block to the left of x will be its predecessor. In each case, we recurse on only one $\sqrt{N}$ size sub–problem, so the recurrence relation describing the running time of the algorithm is $S(N) = S(\sqrt{N}) + O(1)$, which is $O(\log \log N)$[9]. The insertions and deletions use a similar strategy and achieve the same running time. The algorithms are presented in detail below.

We formally define the data structure as a recursive structure T. Since we may assume that N is a power of 2, let $2^k = N$. We define T_k to be a recursive data structure for the range $[0 \ldots 2^k]$. It contains an ordered doubly linked list

L of the items stored in T_k and a bit vector V from $[0 \dots N]$ where entry i in the bit vector is set to 1 if and only if i is in the queue. Each non–zero entry in the bit vector also contains a pointer to its corresponding item in L, and we have pointers to the largest and the smallest elements of L as well. T_k additionally contains an integer *size* that represents the number of elements present in the structure, an array $T_{bot}[0 \dots \lceil k/2 \rceil]$ of pointers to $T_{\lfloor k/2 \rfloor}$ structures, and a $T_{\lceil k/2 \rceil}$ structure called T_{top}. Each sub–structure T_i for some i is a data structure supporting predecessor queries, insertions, and deletions in $O(\log i)$ time and supports find min and find max operations in constant time.

The first check when searching for the predecessor of x is to see that x has a predecessor in T (i.e. it is not smaller than all elements of T), and if not then we return an appropriate message. If x has a predecessor, then let x_{high} be x div $2^{\lceil k/2 \rceil}$ (the high–order $\lceil k/2 \rceil$ bits of x) and x_{low} be $x \bmod 2^{\lceil k/2 \rceil}$ (the low–order bits of x). We examine the top level structure T_k, and if the structure pointed to by $T_{bot}[x_{high}]$ is non–empty and the smallest element in that structure is smaller than x, then we recursively search in that $T_{\lfloor k/2 \rfloor}$ structure for the predecessor of x_{low}. If the structure pointed to by $T_{bot}[x_{high}]$ is empty or its smallest element is larger than x, then we recursively search T_{top} to find the predecessor of x_{high}. Let p be the predecessor of x_{high} in T_{top}. Then p is the largest index such that $p < x_{high}$ and $T_{bot}[p]$ is non–empty. Therefore the largest element in the structure pointed to by $T_{bot}[p]$ is the predecessor of x, which can be retrieved in constant time using the largest element pointer. When the size of a structure is one, no recursion is needed, we simply examine the element and determine if that element is the predecessor.

The insert and delete operations are similar to finding predecessors. If x is the smallest element, it is added to L and V in constant time, otherwise, we first find the predecessor y of x as described above. We insert x into L following the entry for y, and we set the bit $V[x]$ to 1. Next, we examine the T_k structure, incrementing the size by one and updating the smallest and largest element pointers as needed. If the structure at $T_{bot}[x_{high}]$ is non–empty, then we recursively insert x_{low} into the $T_{\lfloor k/2 \rfloor}$ structure stored there. If this structure is empty, we insert x_{low} as a single element there (without recursion) and recursively insert x_{high} into the T_{top} structure since the entry at $T_{bot}[x_{high}]$ is no longer empty. In both cases, only one recursive call is made.

Note that during the recursive insertion calls, no further predecessor searches are required. Whenever a predecessor of x is needed (i.e. x is not the smallest element of the sub–structure), then the predecessor can be computed from y. This follows from the fact that the recursive insertion and predecessor queries follow exactly the same path in T until x is the smallest element of its structure. Thus all required predecessors can be found in the same time as one predecessor search operation.

The insertion of x into a sub–structure t which has size 1 is a special case. Let y be the element already stored at t, and remember that y has not been stored recursively inside t. First we store y_{high} in t_{top} and y_{low} in a new structure at $t_{bot}[y_{high}]$. Note that both of these operations can be done in constant time since no recursion is required. Now, we can insert x_{high} and x_{low} appropriately. The critical point is that only one of the previous insertions requires recursion. If $x_{high} = y_{high}$, then t_{top} is a one element structure and $t_{bot}[x_{high}]$ is a two element structure where recursion is required. If $x_{high} \neq y_{high}$, then t_{top} is the two element structure, but both $t_{bot}[x_{high}]$ and $t_{bot}[y_{high}]$ are one element structures. The time bound follows from the fact that only one recursion is needed as in the case of finding predecessors.

The deletion algorithm is the inverse of the insertion algorithm, and at each recursive step returns the structure to exactly as it would be if the insertion had never taken place.

5   Concurrent operations in the worst case queue

Now we present the version of the worst–case data structure that allows concurrent operations. The data structures and algorithms are similar to the previous sequential version, and our model of concurrency is identical to the one used for the amortized version. We use synchronization to help achieve concurrency, and we again define operations in terms of a time step. The duration of the time step is determined by the longest set of operations that we wish to perform in one time step, and will, as in the amortized version, be some constant amount of time.

The predecessor search operations proceed as in the sequential version. During each time step, the search determines the proper sub–structure to recursively search in, and makes exactly one recursive call. When the search procedure begins to return from the recursive calls, it returns from exactly one call per time step. Note that it is only when the search returns that the predecessor at that sub–structure is known. A predecessor search can take $\log \log N$ time steps to recurse to the bottom of the structure and another $\log \log N$ time steps to return to the top level. We further synchronize the operations by requiring all operations to take $2 \log \log N$ time steps as follows. We say that a sub–queue is at depth d if it can be reached after d recursive steps, i.e. the top level queue is at depth 0 and its T_{top} structure is at depth 1, and so on. Operations that complete their recursion at a depth $d < \log \log N$ wait $2(\log \log N - d)$ time steps to begin returning from their recursive calls, and thus all operations take exactly $2 \log \log N$ time steps.

The insertion and deletion procedures are synchronized in much the same manner. The insertion cannot first find its predecessor and then insert the item, because other insertion operations might cause the predecessor to change before the insertion takes place. The predecessor is, therefore, found while inserting the element from the top–down. This means that at each time step, an insert x operation can update the size of the sub-structure, change the value $V[x]$ in the bit vector, and determine the proper sub-structure for the recursive insertion. If a one element tree is created, then this is also accomplished during the single time step. However, it remains to insert x into the list L. This can only be accomplished once the predecessor of x in the sub-structure is known. It is possible that the predecessor search must recurse into a different sub-structure than the insertion (when x is a new smallest element of its sub-structure). In this case, the predecessor search and the insertion take place simultaneously, and we note that no further predecessors of x are needed in its sub-structure since x is a new smallest element. Because of the imposed synchronization, the insertion and the predecessor query both return at the same time step, and thus the insertion can continue with adding x to L. In all other cases, the predecessor is computed in the same sub-structure as the insertion and thus is known when the insertion returns from its recursion.

To be able to decide whether the predecessor search is in the same sub-structure as the insertion, we must be able to determine the smallest element in a sub-structure. Since there may be elements which are waiting to be inserted into L, we introduce an additional data structure stored at each sub-structure to help answer find min queries. We store the items waiting to be inserted into L using the result of Ajtai, Fredman, and Komlós [3]. They give a solution to the bounded universe priority queue problem where insert, delete, and find min operations take constant time if the queue size is limited to $O((\log N)^c)$ for any constant c. Their technique involves $O(N^\epsilon)$ storage and preprocessing time to build a look–up table. They store an encoding of the set using $O(\log N)$ bits, which we call the AFK–encoding, used to index the look–up table. We keep an AFK–encoding at each sub–structure to store the items that have not yet been inserted into L. We insert the element x into the AFK–encoding at a sub–structure when the insert x operation reaches the sub–structure. We, therefore, determine the smallest element in a sub–structure by checking both its list L and its AFK–encoding. When the insert x operation returns from its recursion, we delete x from the AFK–encoding and insert x into L.

The delete operation is not an exact inverse of the insert operation. A delete x operation makes one recursive call per time step, decrementing the size of the sub–structure, and unmarking the bit $V[x]$ before making the recursive call. Because the sub–structure must support find min and find max operations in constant time, the element x must be removed from L or the AFK–encoding, depending on where it is stored. Removing x from the AFK–encoding is a simple constant time operation. The insert x operation which has not yet completed can detect that x has been deleted by examining the bit $V[x]$. Removing x from L presents the following problem. Suppose that an insert y operation just returns from its recursive call and thinks that x is its predecessor. This is possible because x has not been removed from lower sub-structures. The insert y operation will try to insert y following x in L (i.e. y becomes the successor of x), but x has been removed.

To fix this problem, we remove x from L, but we leave the pointer from x to its successor. Using this pointer, an insert y operation can find its proper location in L in constant time. When the insert y operation changes the successor of x, we update the pointer from x. Note that the successor of x cannot change unless x is found as the predecessor. When the deletion returns from its recursive call, x can be removed entirely. If a delete x operation deletes the largest or the smallest elements of the list L, then the largest or smallest element pointers must be updated appropriately.

Using the above data structures and algorithms we maintain the invariant that at a sub-queue T_i for some i, an insertion or deletion completes in $2 \log i$ time steps. Additionally, find min or find max operations take constant time, reflecting the value of all insertions or deletions which have reached the sub-queue but possibly not have yet completed. This invariant guarantees the correctness of our algorithm and shows that a find min or find max operation for the entire queue is easily answered in $O(1)$ time.

6   Conclusion and open problems

This paper has presented two schemes for performing concurrent operations in a bounded universe priority queue. Each priority queue operation requires $O(\log \log N)$ time and a new operation can begin every $O(1)$ time steps. Our algorithms yield optimal parallel algorithms for maintaining a priority queue, performing a sequence of m operations in $O(\log \log N + m)$ time using $\Theta(\log \log N)$ processors. The amortized version is conceptually simpler and is easier to implement. The worst-case version uses the AFK-encoding [3] to maintain the largest and smallest elements for the sub-queues. The full power of the AFK result is not needed, however, since only $O(\log \log N)$ elements ever need to be stored, so an interesting question is whether the dependence on their data structure can be eliminated?

The synchronization of the presented algorithms is of a global nature, because operations must synchronize based on a clock, but less strict synchronization might be possible. In the worst-case algorithm, an insertion must sometimes wait on the result of a predecessor query in a different sub-structure. Can the imposed synchronization be relaxed while avoiding problems such as deadlock?

7 Acknowledgements

We would like to thank Monika Rauch for helpful discussions, and Bob Tarjan for suggesting the problem.

References

[1] A. Aho, J. Hopcroft, and J. Ullman, The Design and Analysis of Computer Algorithms, Addison–Wesley, Reading, MA 1974.

[2] M. Ajtai, A lower bound for finding predecessors in Yao's cell probe model, *Combinatorica*, 8(3) (1988) pp. 235–247.

[3] M. Ajtai, M. Fredman, and J. Komlós , Hash functions for priority queues, *Information and Control*, **63** (1984) pp. 217–225.

[4] M. Carey and C. Thompson, An efficient implementation of search trees on $\lceil \log N + 1 \rceil$ processors, *IEEE Trans. on Computers*, Vol. c–33, No. 11, Nov. 1984, pp. 1038–1041.

[5] C. Ellis, Concurrent search and insertion into 2-3 trees, *Acta Informatica*, 14 (1980) pp.63–86.

[6] P. van Emde Boas, R. Kaas, and E. Zijlstra, Design and implementation of an efficient priority queue, *Math. Systems Theory*, **10** (1977) pp.99–127.

[7] P. van Emde Boas, Preserving order in a forest in less than logarithmic time and linear space, *Information Processing Letters*, Vol. 6, No. 3 (1977) pp.80–82.

[8] U. Manber and R. Ladner, Concurrency control in a dynamic search structure, *ACM Trans. on Database Sys.*, Vol. 9, No.3 (1984) pp. 439–455.

[9] K. Mehlhorn, Data Structures and Algorithms 1: Sorting and Searching, Springer–Verlag, Berlin, 1984.

[10] V. Rao and V. Kumar, Concurrent access of priority queues, *IEEE Trans. on Computers*, Vol. 37, No. 12, Nov. 1988, pp. 1657–1665.

[11] R. Sedgewick, Algorithms, Addison–Wesley, 1988.

Appendix A: The top–down split procedure

This appendix presents an algorithm for performing top–down splits of a 2–4 tree. Techniques required to perform top–down joins are presented as well.

Given a split value s, we create two trees, the left tree containing values $< s$ and the right tree containing values $\geq s$. We assume that some child of the root of the 2–4 tree belongs in the left tree so that we may view the operation as splitting subtrees from the right tree and joining subtrees to the left tree. Call x the node that we are currently splitting. While doing the split, we maintain the following invariant: the parent of x is not a 2 node. This allows us to restructure the tree after some children of x are split away. Since the parent of x, $p(x)$, has at least 3 children, $p(x)$ must have at least 5 grandchildren after x is split. We can therefore restructure $p(x)$ and its children so that x contains at least 3 nodes after x is split. Note that $p(x)$ may become a 2 node in this restructuring, but this is not important since we are now splitting the left child of x. This maintains the invariant, but we have not shown how to initialize it. Suppose that when the root of the tree is split, 2 children remain. If more than 2 remain, then the invariant is initialized, and if only one remains then the root may be deleted and the child considered as the root. Let x be the child of the root that is being split. Once x is split, we have two cases to consider. Suppose that the root has only 3 or 4 grandchildren once x is split, then the root is deleted, and x merges with its sibling, becoming a 3 or a 4 node. If the root has 5 or more grandchildren after the split, then they can be restructured so that x becomes at least a 3 node. This initializes the invariant.

In the left tree, we join the split subtrees from the top down as described below. This takes place concurrently with the splitting, and because the height of the subtrees decreases as we proceed down the right tree, the splitting and joining can proceed down their respective trees in lockstep. Let y be the rightmost node in the left tree that is at the same height as the node x in the right tree (x is the node we are splitting in the right tree). The pieces split from x are added to the node y. This increase the degree of y by up to 3, and y may now have to be split. We, therefore maintain the invariant that $p(y)$ is not a 4 node. Since the degree of y is at most 7 after the new children are added, we can split y such that the rightmost child of $p(y)$ has degree at most 3, thus maintaining the invariant.

Embedding k-D meshes into Optimum Hypercubes with Dilation 2k-1
(Extended Abstract)

Said Bettayeb[1], Zevi Miller[2], Tony Peng[3] and Hal Sudborough[4]

Abstract.
It is shown that, for every k, and for every k-dimensional mesh M, there is a one-to-one embedding of M into its optimum size hypercube with dilation at most 2k-1.

1. Introduction

Embedding structures into hypercubes has, after a relatively short time, an extensive collection of results [L]. The basic idea is as follows. The hypercube represents an actual network topology. The embedded structure represents an alternative network/data structure one wishes to simulate efficiently (or even a representation of processes in a computation to be executed efficiently). An embedding is an explicit description of how objects of a guest structure are to be assigned to host units in the hypercube network. The *dilation* of such an embedding is the maximum distance in the hypercube between images of adjacent objects of the embedded structure. We consider one-to-one embeddings of multi-dimensional meshes, say with n nodes, into the smallest possible hypercube, *i.e.* the so-called *optimum* hypercube of dimension $\lceil \log_2 n \rceil$

Previous results show that every 2-D mesh can be embedded into its optimum hypercube with dilation at most 2 [C], every 3-D mesh can be embedded into its optimum hypercube with dilation 5 [C2], and, for all k>3, every k-D mesh can be embedded into its optimum hypercube with dilation at most 4k+3 [C]. [BMS] shows dilation k is possible for many k-D meshes and [RSU] shows many 3-D meshes can be embedded with dilation 2.

Previous results, and those contained here, describe upper bounds on dilation for embedding k-D meshes into their optimum hypercube. The only lower bound known shows that dilation 2 is sometimes required. That is, for example, a 2-D mesh M with m

---

[1] Department of Computer Science, Louisiana State University, Baton Rouge, Louisiana
[2] Department of Mathematics and Statistics, Miami University, Oxford, Ohio
[3] Mathematics and Computer Science, Creighton University, Omaha, Nebraska
[4] Computer Science Program, EC 3.1, University of Texas at Dallas, Richardson, Texas 75083-0688

rows and n columns is a subgraph of its optimum hypercube if and only if $\lceil \log m \rceil + \lceil \log n \rceil = \lceil \log mn \rceil$ [L].[5]

Many-to-one embeddings have also been considered. [MS] shows, for every $t>1$, there is a (2^t+1)-to-1 dilation 1 embedding for every 2-D mesh into a hypercube of size 2^{-t} times optimum, there is a 3-to-1 dilation 2 embedding of all 3-D meshes into a hypercube of size 1/2 times optimum, and there is a 5-to-1 dilation 3 embedding of all 3-D meshes into a hypercube of size 1/4 times optimum.

Our result improves the asymptotic upper bound on dilation for 1-to-1 embeddings. We show that, for all $k>2$, at most dilation $2k-1$ is needed for one-to-one embeddings of k-D meshes into their optimum hypercubes. In particular, all 3-D meshes can be embedded with dilation at most 5 (which agrees with the upper bound given recently in [C2]).

2. Basic Technique and Dilation 5 for 3-D Meshes

The technique is iterative (on the number of dimensions) and begins with an optimum embedding of 2-D meshes into their optimum hypercubes. For this we shall take the embedding described in [C]. This embedding can be viewed as a mapping of a 2-D mesh, say with m rows and n columns, into a subgraph of a hypercube with $R = 2^{\lfloor \log m \rfloor}$ rows and $C = 2^{\lceil \log mn \rceil - \lfloor \log m \rfloor}$ columns with $B = 2^{\lceil \log mn \rceil} - mn$ *blanks*. That is, a blank is a position in the host hypercube where no mesh node is mapped. (Observe that $B < 2^{\lceil \log mn \rceil - 1}$; otherwise, the 2-D mesh could be embedded into a smaller hypercube.) Moreover, the embedding described in [C] assigns all blanks to a set of contiguous locations, where each host row receives a number of blanks that differs at most by one from the number of blanks received by any other host row.

We consider moving the B blanks to different positions. Our goal is to move them so that entire columns of the host are blank, with one (or possibly two) column(s) being partially blank. For this purpose we compute the number of entire columns needed for blanks: let $D = \lfloor B/C \rfloor$ and $E = B - RD$. Then, after the movement, D columns of the host will be entirely blank, with E blank positions in one other host column. Moreover, we select these D+1 columns, which are to contain blanks, so that they are evenly distributed across the C columns of the host. (Observe that $D+1 \leq C/2$. For if $D+1 >$

---

[5] Logarithms throughout this paper are taken base 2.

C/2, then $B > 2^{\lceil \log mn \rceil - 1}$ and again one should be embedding M into a smaller hypercube.) For instance, let $F = \lfloor C/(D+1) \rfloor$, then the D+1 columns (that initially contain blanks) can be chosen to be columns F, 2F, 3F, ..., (D+1)F. (By the observation that D+1 < C/2, it follows that $F \geq 2$. That is, the columns selected above are (at worst) every other column.)

Our purpose for moving blanks is now explained. Let M be a 3-D mesh, say with m rows, n columns, and p *pages*. That is, M is a [m×n×p] mesh. We view M as consisting of p pages of a 2-D [m×n] mesh and use the same embedding for each of its p pages except for a varying movement of blanks. (In this p page process, mesh nodes are mapped to hypercube locations, called *initial hypercube addresses* (iha's).) The actual location for a node of M, say one which is the ith mapped to an *iha* x, is obtained by concatenating the ith element of an appropriate length binary reflected Gray code, y, to the bit string x. (The appropriate length of y is the smallest power of two larger than the maximum number of nodes mapped to any *iha*.) We shall call these bit strings y *tag* strings. Thus, we can view the entire procedure as stacking up mesh points at *iha's*, splitting them at the end by attaching *tag* strings and thereby mapping them into unique hypercube locations.

The embedding of page 1 nodes results in each *iha* having either 0 or 1 mesh points. The embedding of page 2 nodes, due to a different arrangement of blanks, results in each *iha* having a total of either one or two mesh points. That is, each *iha* has either one node from page one of the mesh, one node from page 2 of the mesh, or two nodes: one from page one and one from page two of the mesh.

In general, blanks are moved from page-to-page so that, for all i>0, after i pages of M have been mapped, *the total number of mesh points received by any iha differs by at most one from the number received by any other iha*. (Or, equivalently, the number of blanks assigned to one *iha* differs by at most one from the number assigned to any other *iha*.)

The property indicated in the last paragraph is sufficient to ensure that at the end of the process M is embedded into its optimum hypercube. The only conceivable way for M to use a hypercube with dimension larger than required is for a tag string to be too long, *i.e.* longer than appropriate to obtain a minimum length hypercube address. However, the only way for a tag string to be too long is for one of the *iha's* to have a number of points larger than an appropriate power of 2, say 2^b, for some b. (This might result in b+1 bits in every tag string, although b bits are sufficient.) However, if an *iha* receiving the maximum number of mesh nodes has more than 2^b nodes, then, as all other *iha's* must receive at least 2^b nodes, by the above property, length b+1 tag strings are actually necessary.

We claim that each new embedding, obtained from the initial 2-D embedding through the process of moving blanks to desired positions, has dilation at most 4.

We see in the next few paragraphs why the dilation is at most 4. A column of blanks can be moved from their initial position to a desired host column via an iterative shifting of blanks along their original rows to a desired column. As blanks are shifted from one column to the next along this row, mesh points displaced by the blank are shifted in the opposite direction by one column. (That is, the blank and the image of some guest node simply exchange positions.) When all blanks for the desired column have been shifted as described, observe that distances between images of adjacent mesh nodes have been increased at most by one. That is, the image of one node may have been shifted by one column and the image of an adjacent node may not have been shifted. So the distance increases by one. Thus, shifting blanks along a row (as part of the process used in filling all positions in an entire desired column) makes the dilation at most 3. (Recall that the original 2-D mesh embedding had dilation 2.)

This process of shifting blanks to fill up a new column is done iteratively, *i.e.* column by column. However, as the desired columns for blanks are selected according to a uniform distribution, the dilation is not increased beyond 3. That is, columns chosen for blanks are always separated by other columns containing mesh points (and no blanks). So, the distances between images of adjacent mesh points on the same page is at most 3 after this movement of blanks.

However, there is another type of movement of blanks that is also required. That is, we also create one (or perhaps two) partially blank columns. The initial embedding for a 2-D page has specified rows with an extra blank. These extra blanks may not be in the same rows we desire and may need to be shifted within a column from row to row to move to our desired location. This movement may also increase the dilation, but not by more than one. That is, the dilation may become 4.

Recall that the stated dilation for the entire embedding (not just for a page) is 5. What we have just seen is that the maximum distance between *iha's*, say x and y, containing images of adjacent mesh nodes *from the same page* is 4. Note that the indices of these images within x and y may differ by one, but no more than one. That is, let the adjacent mesh nodes be from some page t. By the stated property, the total number of mesh points assigned to x after t-1 pages differs by at most one from the total number of mesh points assigned after t-1 pages to y. Therefore, the tag strings for the images of these adjacent points may differ in at most one bit. So, the distance between their final hypercube addresses is at most five, as the distance between their *iha's* is at most 4 and the tag strings differ in at most one bit.

We note also that the distance between images of corresponding nodes on consecutive pages of the 3-D mesh is at most 5. Recall that the same initial embedding is used for each page. Thus, node (r,s) of page t is assigned by this initial embedding to exactly the same *iha* as node (r,s) of page t+1. The only difference is due to the movement of blanks and the variation of this movement page t to page t+1. However, due to the even distribution of blank columns, the host column assigned for node (r,s) in page t differs by at most one from the host column assigned for node (r,s) in page t+1. And, similarly, the host rows for these points differ by at most two. So, the distance between the *iha* assigned to node (r,s) of page t and the *iha* assigned to node (r,s) of page t+1 is at most 3. The indices of these two nodes within their respective *iha's*, say x and y, may differ by two, as x may have one less node assigned than y after page t. That is, if x has q nodes after page t and y has q+1 nodes, then the node (r,s) of page t, say assigned to *iha* x, has index q, while the node (r,s) of page t+1, say assigned to *iha* y, will receive index q+2. So, the distance between their images' final locations may be at most 5.

3. Extending the Technique to k-D Meshes, for k>3.

The procedure to embed k-D meshes into their optimum hypercubes, for k>3, is an extension of that described in the last section for 3-D meshes. We describe the extension inductively. For any k>3, assume a procedure to embed one-to-one with dilation at most 2k-3 an arbitrary (k-1)-D mesh M into a (k-1)-D mesh representation Q of its optimum hypercube, where every dimension of Q is a power of 2. Each position in Q is referred to as an *initial hypercube address (iha)*.

Now let M be a k-D mesh, say one whose i^{th} dimension is d_i. Consider the (k-1)-D mesh M', whose i^{th} dimension is d_i, for i<k. We view M as d_k copies of M', in which, of course, corresponding points in the ith and $(i+1)^{th}$ copies are connected by an edge. We refer to each copy of M' as a page. As before, we compute the number of blanks per page P. That is, subtract the product of all of the first k-1 dimensions, *i.e.* d_1 through d_{k-1}, from 2^s, where 2^s is the smallest power of 2 larger than this product. This is the number of blanks needed per page, denoted by n(P).

The embedding of each page P is a 1-to-1 embedding with dilation at most 2k-3 into a (k-1)-D mesh Q, in which every dimension is a power of 2, say one in which the

i^{th} dimension is $2^{t(i)}$, for i<k. As described in the last section for 3-D meshes, we use this embedding as an initial, say canonical, map, but the actual embedding of each page, and hence each mesh point's *iha*, varies due to the movement of blanks.

Again, as in the last section, the final location for a node of M, say one which is the ith mapped to an *iha* x, is obtained by concatenating the ith element of an appropriate length Gray code, y, to the bit string x. We call these bit strings y *tag* strings. Thus, we view the entire procedure as before, as one of stacking up mesh points at *iha's*, splitting them at the end by attaching *tag* strings and thereby mapping them into unique hypercube locations.

The embedding of page 1 nodes results in each *iha* having either 0 or 1 mesh points. The embedding of page 2 nodes, due to a different arrangement of blanks, results in each *iha* having a total of either one or two mesh points. That is, each *iha* has either one node from page one of the mesh, one node from page 2 of the mesh, or two nodes: one from page one and one from page two of the mesh.

As before, blanks are moved from page-to-page so that, for all i>0, after i pages of M have been mapped, *the total number of mesh points received by any iha differs by at most one from the number received by any other iha.* (Or, equivalently, the number of blanks assigned to one *iha* differs by at most one from the number assigned to any other *iha*.)

As indicated in the last section for 3-D meshes, this property is sufficient to ensure that at the end of the process M is embedded into its optimum hypercube. As was the case in the last section in the embedding of 3-D meshes, blanks need to be evenly distributed. There is a relatively straightforward technique to ensure an even distribution of blanks. Consider the product of the first k-2 dimensions of M, *i.e.* the product of d_1 through d_{k-2}, denoted by p(P). We shall call the (k-2)-D mesh, in which the ith dimension is d_i, for all i<k-1, a *page of dimension k-2*. (This corresponds to a column in the technique described earlier for 3-D meshes.) Divide the number of blanks, n(P), by the size of a page of dimension k-2, namely p(P). This is the number of pages of dimension k-2 which are completely blank. The remainder from this division is the number of blanks that will need to be assigned to an extra page of dimension k-2. That is, one may need partially blank pages, just as one needed partially blank columns in the 3-D embedding.

The pages of dimension k-2 which are either completely blank or partially blank are spaced uniformly. This ensures that corresponding points in successive pages P (of dimension k-1) are assigned to positions of Q at distance at most 2k-1. (This is more easily satisfied as k gets larger, as an image of a given mesh point changes position from page-to-page only due to the movement of blanks. That is, it is independent of distance between images of adjacent points from the same page.)

In fact, we show that any given configuration of blanks for one (or possibly two) partial pages of blanks, satisfying an even distribution requirement, is possible at the cost of adding one to the dilation. Thus, the distance between images of adjacent mesh nodes from the same page increases at worst from 2k-3, in the initial canonical embedding, to 2k-2, due to the movment of blanks. Then, as indicated in the last section, the distance is increased once more by one, due to the attachment of tag strings. That is, inductively, the dilation grows by 2 with each increase in the number of dimensions in the mesh: the canonical initial embedding of each page, which is a (d-1)-D mesh, has dilation 2k-3, one is added due to the movement of blanks, and one more is added due to the attachment of tag strings.   Otherwise, the basic observation is that the procedure for embedding 3-D meshes, which uses the 2-D mesh embedding described in [C], can be generalized. For instance, our embedding of a 4-D mesh uses the embedding of 3-D mesh pages the same way our 3-D mesh embedding uses the embedding of 2-D mesh pages. The embedding of a k-D mesh uses the embedding of a (k-1)-D mesh page in the same way. Thus, we obtain:

Theorem
For all k>2 and all k-D meshes M, there exists a dilation 2k-1 one-to-one embedding of M into its optimum hypercube.

We believe the upper bound should, in fact, be much smaller than that indicated, but that dilation 2 is not possible for all multi-dimensional meshes.

References

[BMS] S. Bettayeb, Z. Miller, and I. H. Sudborough, "Embedding Grids into Hypercubes", **J. Computer and System Sci.** (1992).

[C] M.-Y. Chan, "Embedding of Grids into Optimal Hypercubes", **SIAM J. Computing**, 20,5 (1991), pp. 834-864.

[C2] M.-Y. Chan, F. Chin, S. Xu, and F. He, "Dilation 5 Embedding of 3-Dimensional Meshes into Hypercubes", **Proc. 1993 IEEE Symp. on Parallel and Distributed Computing.**

[L]   F.T. Leighton, **Introduction to Parallel Algorithms and Architectures: Arrays, Trees, Hypercubes,**

[MS] Z. Miller and I. H. Sudborough, "Compressing Grids into Hypercubes", to appear in **Networks**.

[RSU] M. Roettger, U.P. Schroeder, W. Unger, "Embedding 3-Dimensional Grids into Optimal Hypercubes", these proceedings.

Embedding 3-dimensional Grids into Optimal Hypercubes*

Markus Röttger, Ulf-Peter Schroeder, Walter Unger

Department of Computer Science
University of Paderborn
D-33095 Paderborn, Germany

Abstract. The hypercube is a particularly versatile network for parallel computing. It is well-known that 2-dimensional grid machines can be simulated on a hypercube with a small constant communication overhead. We introduce new easily computable functions which embed many 3-dimensional grids into their optimal hypercubes with dilation 2. Moreover, we show that one can reduce the open problem to recognize whether it is possible to embed every 3-dimensional grid into its optimal hypercube with dilation at most 2 by constructing embeddings for a particular class of grids. We embed some of these grids, and thus for the first time one can guarantee that every 3-dimensional grid with at most $2^9 - 18$ nodes is embeddable into its optimal hypercube with dilation 2.

Key words: embedding, hypercubes, 3-dimensional grids, dilation, *GrayCode*

1 Introduction

Various parallel computer architectures have gained favour and are in use today. The hypercube is emerging as a popular network architecture for parallel machines and algorithms. One of the key features of the hypercube is a rich interconnection structure which permits many important network topologies, such as grids, to be efficiently simulated. Hence grid-based algorithms can be executed efficiently on hypercubes. This was our motivation to study the following open problem:

> Is it possible to embed every 3-dimensional grid into its optimal hypercube (the smallest hypercube with at least as many nodes as the grid) with dilation at most 2?

Dilation 2 would be optimal, because for every d there exist some d-dimensional grids that are not subgraphs of their optimal hypercubes. For more details see [MS]. Chan has described a technique for constructing embeddings of 2-dimensional grids into their optimal hypercubes with dilation at most 2 ([Ch1], [Ch4]). Thus the corresponding problem for 2-dimensional grids is solved.

* This work was partially supported by the German Research Association (DFG) and by the ESPRIT Basic Research Action No. 7141 (ALCOM II).

We will now present the formal definitions used through this paper.

Definition 1. The *hypercube* of dimension $n \in \mathbb{N}$, denoted by $Q(n)$, is the undirected graph (V, E) with: $V := \{0, 1\}^n$ and $E := \{\{e_1, e_2\} : e_1, e_2 \in V,\ Bitdiff(e_1, e_2) = 1\}$, where *Bitdiff* is the Hamming distance, i. e. the number of bits where the corresponding sequences differ.

Definition 2. The *3-dimensional grid* with side lengths x_1, x_2, x_3, ($x_i \in \mathbb{N}$), denoted by $x_1 \times x_2 \times x_3$, is the undirected graph (V, E) with: $V := \{(a_1, a_2, a_3) : a_i \in \mathbb{N}, a_i \le x_i\}$ and $E := \{\{e_1, e_2\} : e_1, e_2 \in V,\ e_1, e_2$ differ exactly in one entry by one$\}$.

Definition 3. Let $G = (V, E)$ and $H = (V', E')$ are finite undirected graphs. An *embedding* f of G into H is an injective mapping $f : V \longrightarrow V'$. G is called the *guest* graph and H is called the *host* graph of the embedding f. The *dilation* of the embedding f is the maximum distance in the host between the images of adjacent guest nodes.

In this paper we consider only embeddings of each 3-dimensional grid G into the smallest hypercube — which is called the *optimal hypercube* for G — that has at least as many nodes as G. Thus the optimal hypercube for the $x \times y \times z$ grid has dimension $\lceil \log(xyz) \rceil$. Through this paper we denote by log the logarithm based on 2.

Various papers were devoted to embeddings of grids of arbitrary dimension into hypercubes ([BMS], [Ch3], [Ch4]), but the dilation that was achieved for 3-dimensional grids is equal to 2 only for a small class of grids. We list the most important results.

Proposition 4. *Scott and Brandenburg showed in [SB]:*
If $\lceil \log x \rceil + \lceil \log y \rceil + \lceil \log z \rceil = \lceil \log(xyz) \rceil$ with $x, y, z \in \mathbb{N}$ holds, then the $x \times y \times z$ grid is a subgraph of its optimal hypercube of dimension $\lceil \log(xyz) \rceil$.

The following is an easy consequence implied by the pipelining of Chan's embedding ([Ch1], [Ch4]) of 2-dimensional grids.

Proposition 5. *It follows from the application of Chan's results ([Ch1], [Ch4]) for 2-dimensional grids that: If $\lceil \log x \rceil + \lceil \log(yz) \rceil \le \lceil \log(xyz) \rceil$ with $x, y, z \in \mathbb{N}$ holds, then the $x \times y \times z$ grid is embeddable into its optimal hypercube of dimension $\lceil \log(xyz) \rceil$ with dilation at most 2.*

Furthermore, Chan showed that 7 is an upper bound for the dilation of embedding 3-dimensional grids ([Ch2]). Sudborough improved the upper bound up to 5 ([Sud]). The basic idea of the commonly used methods of embedding grids into their optimal hypercubes can be described as follows:

If the grid G should be embedded into its optimal hypercube H then embed G into a grid G', which is a subgraph of H. The dilation of the embedding of G into H equals the dilation of the embedding of G into G' because one can easily embed the grid G' into H with dilation 1 using the *GrayCode* which is defined as follows:

Definition 6. Let $m, s \in \mathbb{N}$ and $1 \leq m \leq 2^s$. Represent the number $m - 1$ as the binary string $a_{s-1} a_{s-2} \cdots a_1 a_0$ of length s. The $GrayCode(m, s)$ is defined as:

$$GrayCode(m, s) := a_{s-1} \circ (a_{s-1} \oplus a_{s-2}) \circ (a_{s-2} \oplus a_{s-3}) \circ \ldots \circ (a_1 \oplus a_0),$$

where $\oplus$ is the modulo 2 addition and $\circ$ represents the concatenation of the bits.

Under this technique one does not consider the edges of the hypercube that do not simulate the edges of G'. We will find in this paper another embedding in order to eliminate this disadvantage of the mentioned technique.

Ho and Johnsson ([HJ]) explores embeddings of 3-dimensional grids into optimal hypercubes by graph decomposition, which implies a more widely applicable embedding technique than the above mentioned methods. The following is a consequence of this technique:

Proposition 7. *If a $x_i \times y_i \times z_i$ grid is embeddable into its optimal hypercube $Q(n_i)$ with dilation at most 2, $i \in \{1, 2\}$, then there exists also an embedding of the $G = (x_1 \cdot x_2) \times (y_1 \cdot y_2) \times (z_1 \cdot z_2)$ grid into the $Q(n_1 + n_2)$ hypercube with dilation at most 2.*

Notice in the case that the dimension of the optimal hypercube for G is equal to $n_1 + n_2$, we obtain the requested embedding.

One of our main results is Theorem 13 which implies a widely applicable and easily computable embedding technique, but it still may not be applied to all 3-dimensional grids. Combining our new technique with all known methods of embedding grids into their optimal hypercubes yields embeddings with dilation at most 2 of 97 % of all 3-dimensional grids contained in a $512 \times 512 \times 512$ grid.

Moreover, we reduce the mentioned open problem to a consideration of a particular class of grids. The former authors were unable to embed all 3-dimensional grids up to 128 nodes with dilation at most 2. There was the conjecture that the $5 \times 5 \times 5$ grid cannot be embedded in $Q(7)$ with dilation 2. Examining this conjecture, we developed embeddings so that we can prove that every 3-dimensional grid up to $2^9 - 18$ nodes is embeddable with dilation at most 2. In other words we disprove the conjecture about the $5 \times 5 \times 5$ grid.

This paper is organized in the following manner. In Sect. 2 the $GrayCode$ is analyzed, followed by Sect. 3 that presents Theorem 11 and the basic techniques for our new embedding. Since the usage of Theorem 11 is too limited, we make a simple observation to extend the applicability of the embedding and state this in Theorem 13. In Sect. 4 we present a conclusion of our results. Furthermore we define a new class of grids and reduce the problem of finding embeddings for all 3-dimensional grids into their optimal hypercubes with dilation 2 to this class.

2   Some Properties of the *GrayCode*

As mentioned above, by using the common methods of embedding the grid G into its optimal hypercube H one has to take a grid G', with G' is a subgraph of H, and then embed G into G'. One can easily embed the grid G' into H with dilation 1 using the *GrayCode*. The following function g embeds the $x \times y \times z$ grid which is a subgraph of its optimal hypercube into $Q(\lceil \log(xyz) \rceil)$ with dilation 1. For all a, b, c with $1 \le a \le x$, $1 \le b \le y$, $1 \le c \le z$ we define:

$$g(a, b, c) := GrayCode(a, \lceil \log x \rceil) \circ GrayCode(b, \lceil \log y \rceil) \circ GrayCode(c, \lceil \log z \rceil)$$

Notice that the technique of embedding G into H does not consider the edges of the hypercube which do not simulate the edges of G'. To eliminate this disadvantage we analyse the structure of the hypercube in accordance with the *GrayCode* mapping. Figure 1 shows a classification of the hypercube edges. For simplicity we show in the figure only a 1-dimensional cross-section of a grid. The numbers represent the coordinates of the grid nodes, while the binary strings correspond to the hypercube nodes in accordance with the *GrayCode*. The following corollaries list some observations of the *GrayCode* which imply relations between some grid node images of G'. Therefore, we consider the existence of the "unused hypercube edges" (see Fig. 1). Later we will use these facts to construct new embeddings for many 3-dimensional grids.

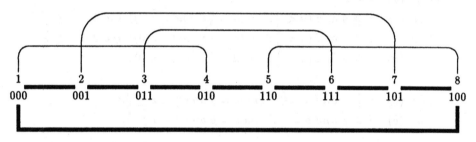

━━━━━━ hypercube edges that simulate grid edges

⌐‾‾‾‾⌐ unused hypercube edges

Fig. 1. Embedding into $Q(3)$

Corollary 8. *For all $m, s \in \mathbb{N}$ with $k \in \mathbb{N}_0$ and $m, m + 2^k \le 2^s$:*

$$Bitdiff(GrayCode(m, s), GrayCode(m + 2^k, s)) = \begin{cases} 1 & if \ k = 0 \\ 2 & if \ k > 0 \end{cases}$$

Corollary 9. *For all $m, s \in \mathbb{N}$ with $1 \le m, 2^s - m + 1 \le 2^s$:*

$$Bitdiff(GrayCode(m, s), GrayCode(2^s - m + 1, s)) = 1$$

Corollary 10. *For all $m, s \in \mathbb{N}$, $s > 1$ with $1 \le m, 3 \cdot 2^{s-1} - m + 1 \le 2^s$:*

$$Bitdiff(GrayCode(m, s), GrayCode(3 \cdot 2^{s-1} - m + 1, s)) = 1$$

3 Dilation 2 Embedding

Now, we will introduce a Theorem which lists some sufficient conditions so that we can guarantee the existence of easily computable embeddings with dilation at most 2.

Theorem 11. Let $G = x \times y \times z$ be a 3-dimensional grid. Let furthermore $y' = 2^{\lceil \log y \rceil}$ or $y' = 2^{\lfloor \log y \rfloor}$ and $x' = 2^{\lfloor \log x \rfloor}$, $z' = 2^{\lceil \log(xyz) \rceil}/(x'y')$, $d = \lfloor (yz)/y' \rfloor$, $r = (yz) \pmod{y'}$.
If $((I)$ and $(II))$ or $((I)(c)$ and $(II'))$ holds for G then the grid G is embeddable into its optimal hypercube with dilation at most 2.

(I) (a) $y = y'$

    or (b) $y = y' + 1$

    or (c) $y = y' - 1$

(II) (a) $x = x'$

    or (b) $z' \geq 4$ and $(d < \frac{3}{4}z'$ or $(d = \frac{3}{4}z'$ and $r = 0))$ and $4 < x \leq \frac{5}{4}x'$

    or (c) $z' \geq 8$ and $(d < \frac{5}{8}z'$ or $(d = \frac{5}{8}z'$ and $r = 0))$ and $2 < x \leq \frac{3}{2}x'$

    or (d) $z' \geq 2$ and $(d < \frac{1}{2}z'$ or $(d = \frac{1}{2}z'$ and $r = 0))$

(II') (a) $z' \geq 8$ and $d = \frac{3}{4}z'$ and $r = y$ and $8 < x \leq \frac{9}{8}x'$

    or (b) $z' \geq 8$ and $d = \frac{5}{8}z'$ and $r = y$ and $4 < x \leq \frac{5}{4}x'$

If $(I)(a)$ and $(II)(a)$ is valid for G then G is a subgraph of $Q(\lceil \log(xyz) \rceil)$. $\square$

Proof: To prove Theorem 11 we specify an embedding h which results in a concatenation of three binary strings. Therefore we define three functions f_Z, f_S and f_G which map grid nodes to the natural numbers. For an arbitrary 3-dimensional grid $G = x \times y \times z$ that satisfies the condition $((I)$ and $(II))$ or $((I)(c)$ and $(II'))$ of Theorem 11, we define the embedding h for all a, b, c with $1 \leq a \leq x$, $1 \leq b \leq y$, $1 \leq c \leq z$ as follows:

$$h(a, b, c) := GrayCode(f_Z(a, b, c), \log x') \circ GrayCode(f_S(a, b, c), \log y') \circ$$
$$GrayCode(f_G(a, b, c), \log z')$$

The function h maps the nodes of the grid G to the nodes of the hypercube $H = Q(\lceil \log(xyz) \rceil)$. The construction of the functions $f_Z(a, b, c)$, $f_S(a, b, c)$ and $f_G(a, b, c)$, for all a, b, c with $1 \leq a \leq x$, $1 \leq b \leq y$, $1 \leq c \leq z$, completes the definition of the embedding h. Defining these functions we utilize the Corollaries 8, 9 and 10.

If $(I)(a)$ holds then define:

$$f_S(a, b, c) := b$$

If (I)(b) holds then define:

$$f_S(a, b, c) := (c - \lfloor (y(c-1) + b - 1)/(2y') \rfloor + b - 2) \pmod{y'} + 1$$

If (I)(c) holds then define:

$$f_S(a, b, c) := (b - c) \pmod{y'} + 1$$

If (II)(a) holds then define:

$$f_G(a, b, c) := \lfloor (y(c-1) + b - 1)/y' \rfloor + 1$$
$$f_Z(a, b, c) := a$$

If (II)(b) holds then define:

$$f_G(a, b, c) := \begin{cases} \lfloor (y(c-1) + b - 1)/y' \rfloor + 1 & \text{if } a \leq x' \\ z' - f_G(1, b, c) + 1 & \text{if } a > x' \text{ and } f_G(1, b, c) \leq \frac{1}{4}z' \\ \frac{1}{2}z' + f_G(1, b, c) & \text{if } a > x' \text{ and } \frac{z'}{4} < f_G(1, b, c) \leq \frac{z'}{2} \\ \frac{3}{2}z' - f_G(1, b, c) + 1 & \text{if } a > x' \text{ and } f_G(1, b, c) > \frac{1}{2}z' \end{cases}$$

$$f_Z(a, b, c) := \begin{cases} a & \text{if } a \leq x' \\ a - x' & \text{if } a > x' \text{ and } \qquad f_G(1, b, c) \leq \frac{1}{4}z' \\ 2x' + 1 - a & \text{if } a > x' \text{ and } \frac{1}{4}z' < f_G(1, b, c) \leq \frac{1}{2}z' \\ a - \frac{1}{2}x' & \text{if } a > x' \text{ and } \qquad f_G(1, b, c) > \frac{1}{2}z' \end{cases}$$

If (II)(c) holds then define:

$$f_G(a, b, c) := \begin{cases} \lfloor (y(c-1) + b - 1)/y' \rfloor + 1 & \text{if } a \leq x' \\ z' - f_G(1, b, c) + 1 & \text{if } a > x' \text{ and } f_G(1, b, c) \leq \frac{3}{8}z' \\ \frac{1}{4}z' + f_G(1, b, c) & \text{if } a > x' \text{ and } f_G(1, b, c) > \frac{3}{8}z' \end{cases}$$

$$f_Z(a, b, c) := \begin{cases} a & \text{if } a \leq x' \\ a - x' & \text{if } a > x' \text{ and } f_G(1, b, c) \leq \frac{3}{8}z' \\ 2x' + 1 - a & \text{if } a > x' \text{ and } f_G(1, b, c) > \frac{3}{8}z' \end{cases}$$

If (II)(d) holds then define:

$$f_G(a, b, c) := \begin{cases} \lfloor (y(c-1) + b - 1)/y' \rfloor + 1 & \text{if } a \leq x' \\ z' - f_G(1, b, c) + 1 & \text{if } a > x' \end{cases}$$

$$f_Z(a, b, c) := \begin{cases} a & \text{if } a \leq x' \\ 2x' + 1 - a & \text{if } a > x' \end{cases}$$

If (I)(c) and (II')(a) holds then define f_S, f_G and f_Z as follows:
For the grid nodes (a, b, c) with $c < z$ take the same functions f_S, f_G and f_Z as in the case (I)(c) and (II)(b). For the grid nodes (a, b, z) we use the following functions:

$$f_S(a, b, z) := f_S(a, b, z - 1)$$

$$f_G(a, b, z) := \begin{cases} \frac{3}{4}z' + 1 & \text{if} \quad a \leq \frac{1}{8}x' \\ \frac{3}{4}z' + 2 & \text{if} \ \frac{1}{8}x' < a \leq \frac{3}{8}x' \\ \frac{3}{4}z' + 1 & \text{if} \ \frac{3}{8}x' < a \leq \frac{5}{8}x' \\ \frac{3}{4}z' + 2 & \text{if} \ \frac{5}{8}x' < a \leq \frac{7}{8}x' \\ \frac{3}{4}z' + 1 & \text{if} \ \frac{7}{8}x' < a \end{cases}$$

$$f_Z(a, b, z) := \begin{cases} \frac{1}{4}x' + 1 - a & \text{if} \quad a \leq \frac{1}{8}x' \\ a & \text{if} \ \frac{1}{8}x' < a \leq \frac{1}{2}x' \\ \frac{5}{4}x' + 1 - a & \text{if} \ \frac{1}{2}x' < a \leq \frac{5}{8}x' \\ a & \text{if} \ \frac{5}{8}x' < a \leq \frac{7}{8}x' \\ \frac{7}{4}x' + 1 - a & \text{if} \ \frac{7}{8}x' < a \leq x' \\ a - \frac{3}{4}x' & \text{if} \ x' < a \end{cases}$$

If (I)(c) and (II')(b) holds then define f_S, f_G and f_Z as follows:
For the grid nodes (a, b, c) with $c < z$ we take the same functions f_S, f_G and f_Z as in case (I)(c) and (II)(c). For the grid nodes (a, b, z) we use the following functions:

$$f_S(a, b, z) := f_S(a, b, z - 1)$$

$$f_G(a, b, z) := \begin{cases} \frac{5}{8}z' + 1 & \text{if} \quad a \leq \frac{1}{4}x' \\ \frac{5}{8}z' + 2 & \text{if} \ \frac{1}{4}x' < a \leq \frac{3}{4}x' \\ \frac{5}{8}z' + 1 & \text{if} \ \frac{3}{4}x' < a \leq x' \\ \frac{7}{8}z' + 1 & \text{if} \ x' < a \end{cases}$$

$$f_Z(a, b, z) := \begin{cases} \frac{1}{2}x' + 1 - a & \text{if} \quad a \leq \frac{1}{4}x' \\ a & \text{if} \ \frac{1}{4}x' < a \leq \frac{3}{4}x' \\ \frac{3}{2}x' + 1 - a & \text{if} \ \frac{3}{4}x' < a \leq x' \\ a - \frac{1}{2}x' & \text{if} \ x' < a \end{cases}$$

If more than one condition of (I), (II) or (II') holds for various orderings of x, y and z, then one has to choose one of them to guarantee that the functions f_Z, f_S and f_G are not ambiguous.

In the Appendix we show the main idea of the proof that h is an injective mapping and one gets always the dilation at most 2 by embedding G into

$Q(\lceil \log(xyz) \rceil)$ using h. The full proof of Theorem 11 would stretch the page limit of this paper. We give in the Appendix the proof of one case as an example. □

Figure 2 shows the mappings f_G, f_S and f_Z for the $3 \times 3 \times 3$ grid. We represent the grid as three horizontally arranged 3×3 grids. Notice that $x' = 2$ and let us set $y' = 2$. Then $z' = 8$, $d = 4$ and $r = 1$, and so (I)(b) and (II)(c) hold. Now it is easy to see that the function h defined above delivers an embedding of the $3 \times 3 \times 3$ grid into its optimal hypercube $Q(5)$ with dilation 2.

1	1	2	2	3	3	4	4	5
1	1	2	2	3	3	4	4	5
8	8	7	7	6	6	6	6	7

(a) The mapping f_G

1	2	1	2	2	1	2	1	1
1	2	1	2	2	1	2	1	1
1	2	1	2	2	1	2	1	1

(b) The mapping f_S

1	1	1	1	1	1	1	1	1
2	2	2	2	2	2	2	2	2
1	1	1	1	1	1	2	2	2

(c) The mapping f_Z

Fig. 2. The mappings f_G, f_S and f_Z for the $3 \times 3 \times 3$ grid

Since the condition (I) of Theorem 11 is too strong, the usage of it is very limited. To extend the applicability of the embedding we make the following simple observation which is a special case of Proposition 7.

Lemma 12. *If the 3-dimensional $x \times y \times z$ grid is embeddable into its optimal hypercube with dilation 2, then the $(2^{i_1} \cdot x) \times (2^{i_2} \cdot y) \times (2^{i_3} \cdot z)$ grid with $i_j \in \mathbb{N}_0$ for $j \in \{1, 2, 3\}$ is also embeddable into its optimal hypercube with dilation 2.*

Proof: Let g be the function which embeds the $x \times y \times z$ grid G into its optimal hypercube of dimension $\lceil \log(xyz) \rceil$ with dilation 2, then the following function embeds the $(2 \cdot x) \times y \times z$ grid G' into its optimal hypercube of dimension $1 + \lceil \log(xyz) \rceil$. Define for all a, b, c with $1 \le a \le 2 \cdot x$, $1 \le b \le y$, $1 \le c \le z$

$$g'(a, b, c) := \begin{cases} 0 \circ g(a, b, c) & \text{if } a \le x \\ 1 \circ g(2 \cdot x + 1 - a, b, c) & \text{if } a > x \end{cases}$$

Notice that g' embeds G' into its optimal hypercube with dilation 2. Apply now the function g' $(i_1 - 1)$ times to the first dimension of the grid, i_2 times to the second and i_3 times to the third dimension, respectively. As a result we get an embedding of the $(2^{i_1} \cdot x) \times (2^{i_2} \cdot y) \times (2^{i_3} \cdot z)$ grid into its optimal hypercube of dimension $i_1 + i_2 + i_3 + \lceil \log(xyz) \rceil$ with dilation 2. $\qquad \square$

Applying Lemma 12 to Theorem 11 we are able to modify the restrictive conditions under (I).

Theorem 13. *Let $G = x \times y \times z$ be a 3-dimensional grid. Let furthermore $y' = 2^{\lceil \log y \rceil}$ or $y' = 2^{\lfloor \log y \rfloor}$ and $x' = 2^{\lfloor \log x \rfloor}$, $z' = 2^{\lceil \log(xyz) \rceil} / (x'y')$, $d = \lfloor (yz)/y' \rfloor$, $r = (yz) \pmod{y'}$.*
If $((I)$ and $(II))$ or $((I)(c)$ and $(II'))$ holds for G then the grid G is embeddable into its optimal hypercube with dilation at most 2.

$$(I) \qquad (a) \quad y = 2^j \text{ for any } j \in \mathbb{N}$$
$$or \qquad (b) \quad y = 2^k \cdot (2^j + 1) \text{ for any } j \in \mathbb{N}, k \in \mathbb{N}_0$$
$$or \qquad (c) \quad y = 2^k \cdot (2^j - 1) \text{ for any } j \in \mathbb{N} \text{ with } j > 1, k \in \mathbb{N}_0$$

$$(II), (II') \qquad \text{see Theorem 11}$$

Proof: Apply Lemma 12 to Theorem 11 $\qquad \square$

It is obvious that under certain conditions one can use an embedding for a grid G in order to embed a grid G' which is a subgraph of G.

Corollary 14. *Let the $x \times y \times z$ grid be embeddable into its optimal hypercube of dimension $n = \lceil \log(xyz) \rceil$ with dilation 2. If $(x - i_1) \cdot (y - i_2) \cdot (z - i_3) > 2^{n-1}$ with $i_1, i_2, i_3 \in \mathbb{N}_0$ holds, then the $(x - i_1) \times (y - i_2) \times (z - i_3)$ grid is also embeddable into its optimal hypercube of dimension n with dilation 2.*

For example consider the $3 \times 3 \times 3$ grid which is embeddable into $Q(5)$ with dilation 2. Applying Lemma 12 it follows that the $3 \times 3 \times 6$ grid, the $3 \times 6 \times 6$ grid, the $6 \times 6 \times 6$ grid, etc. are embeddable into their optimal hypercubes with equal dilation. It follows from Corollary 14 that for the $3 \times 3 \times 5$ grid, the $3 \times 5 \times 6$ grid, the $5 \times 6 \times 6$ grid, etc. there exist embeddings into their optimal hypercubes with dilation 2. Notice that all these embeddings are reducible only to the embedding of the $3 \times 3 \times 3$ grid.

4 Conclusion

By using our new technique with respect to the Propositions 4, 5 and Corollary 14 one only has to embed the $3 \times 3 \times 7$ and the $5 \times 5 \times 5$ grid to ensure that all 3-dimensional grids up to 2^7 nodes are embeddable into their optimal hypercubes with dilation at most 2. None of the known methods can be used to embed one of these two grids with dilation 2. Based on our embedding function we have developed an algorithm which also delivers embeddings with dilation 2 for these two grids (for example we show in Fig. 4 the embedding of the $5 \times 5 \times 5$ grid). By finding embeddings for three more grids (the $3 \times 5 \times 17$, the $3 \times 9 \times 9$ and the $5 \times 7 \times 7$ grid) one can guarantee that all 3-dimensional grids up to 2^8 nodes are embeddable into their optimal hypercubes with dilation at most 2. Because the known methods fail again, we have used the mentioned algorithm that works for these grids too. The resulting embedding for the $5 \times 7 \times 7$ grid is shown as an example in Fig. 4. As a conclusion of the above we can describe the properties of all the grids that one has to embed to give a positive answer to the question "Is it possible to embed every 3-dimensional grid into its optimal hypercube with dilation at most 2?". We summarize these grids in the set $\mathcal{P}$:

$$\mathcal{P} := \{x \times y \times z \mid 1 < x \le y \le z \ \wedge \ xyz \ (\bmod \ 2) \not\equiv 0 \ \wedge \ xy(z+1) > 2^{\lceil \log(xyz) \rceil}\}$$

After applying all known embeddings it is clear that every 3-dimensional grid up to $2^9 - 18$ nodes is embeddable with dilation at most 2, because the $5 \times 9 \times 11$ grid is the smallest grid in $\mathcal{P}$ we cannot yet embed with dilation 2. For the $3 \times 11 \times 15$ grid, that is an element of $\mathcal{P}$ and has just as many nodes as the $5 \times 9 \times 11$ grid, the algorithm delivers an embedding with dilation 2. Since the algorithm is not implemented yet we are not able to discuss its behaviour on any given grid of $\mathcal{P}$.

Notice that every candidate to give a negative answer to the question above is an element of $\mathcal{P}$ (consider for example the $55 \times 123 \times 155$ grid with $2^{20} - 1$ nodes that has to be embedded into the hypercube with $1\ 048\ 576 = 2^{20}$ nodes). So the problem of finding embeddings for all 3-dimensional grids into their optimal hypercubes with dilation 2 is now reduced to the class $\mathcal{P}$.

Table 1 shows the number of 3-dimensional grids in comparison with the number of grids in $\mathcal{P}$. We denote by $\mathcal{P}'$ the set of grids in $\mathcal{P}$, for which no embeddings with dilation at most 2 are known. In particular Table 1 shows that there exist exactly 120 distinct 3-dimensional grids which are contained in an $8 \times 8 \times 8$ grid. Notice that only 4 of them (the $3 \times 3 \times 3$, the $3 \times 3 \times 7$, the $5 \times 5 \times 5$ and the $5 \times 7 \times 7$ grid) are elements of $\mathcal{P}$ and for all of them an embedding with dilation 2 is known. One can deduce the embeddings for all 120 grids by using the embeddings of these 4 grids or the embeddings resulting from Propositions 4, 5.

If we consider all 22 500 864 different grids contained in a $512 \times 512 \times 512$ grid, we could prove by combining our new technique with all known methods of embedding grids into their optimal hypercubes that 97 % of all these grids are embeddable into their optimal hypercubes with dilation at most 2. Notice that only 11 752 of them are elements of $\mathcal{P}$.

```
431 421 441 111 112        132 122 142 113 123        143 133 144 114 424

131 121 141 211 212        232 222 242 213 223        243 233 244 214 324

231 221 241 311 312        332 322 342 313 323        343 333 344 314 224

331 321 341 411 412        432 422 442 413 423        443 433 444 414 124

238 338 348 318 317        337 327 347 417 427        447 437 448 418 128



134 135 445 415 425        136 146 416 426 428

234 235 345 315 325        236 246 316 326 328

334 335 245 215 225        336 346 216 226 228

434 435 145 115 125        436 446 116 126 127

438 138 148 218 118        137 147 117 227 217
```

Fig. 3. Embedding of the $5 \times 5 \times 5$ grid into $Q(7)$. First number of the triplet is equal to f_Z, second number to f_S, third number to f_G.

```
141 161 151 181 171 421 411    131 162 152 482 472 422 112    132 142 153 183 173 123 413
241 261 251 281 271 321 311    231 262 252 382 372 322 412    232 242 253 283 273 223 313
341 361 351 381 371 221 211    331 362 352 282 272 222 312    332 342 353 383 373 323 213
441 461 451 481 471 121 111    431 462 452 182 172 122 212    432 442 453 483 473 423 113
148 168 158 188 178 128 118    138 167 157 177 277 127 217    137 147 457 487 477 427 117


143 133 163 174 184 124 114    144 134 164 154 185 125 115    155 145 135 165 275 226 216
243 233 263 274 284 224 214    244 234 264 254 285 225 215    255 245 235 265 175 126 116
343 333 363 374 384 324 314    344 334 364 354 385 325 415    355 345 335 365 475 426 416
443 433 463 474 484 424 414    444 434 464 454 485 425 315    455 445 435 465 375 326 316
447 437 467 478 488 428 418    448 438 468 458 388 328 218    358 348 338 368 278 227 317


156 146 136 266 276 286 287
256 246 236 166 176 186 187
356 346 336 466 476 486 417
456 446 436 366 376 386 387
357 347 337 367 378 377 327
```

Fig. 4. Embedding of the $5 \times 7 \times 7$ grid into $Q(8)$. First number of the triplet is equal to f_Z, second number to f_S, third number to f_G.

Table 1. Analysis of 3-dimensional grids according to the set $\mathcal{P}$.

$G = 2^n \times 2^n \times 2^n$	# of grids contained in G	# of grids in $\mathcal{P}$	# of grids in $\mathcal{P}'$
$n = 2$	20	1	0
$n = 3$	120	4	0
$n = 4$	816	11	5
$n = 5$	5 984	41	32
$n = 6$	45 760	177	163
$n = 7$	357 760	710	675
$n = 8$	2 829 056	2 895	2 831
$n = 9$	22 500 864	11 752	11 562

5 Appendix

To the proof of Theorem 11

One has to show that h is an injective function and that the dilation of the embedding h is at most 2.

That h is an injective function is easy to prove by assuming the opposite and then deriving a contradiction. For brevity we show the proof that one gets always dilation at most 2 by embedding G into $Q(\lceil \log(xyz) \rceil)$ using h only for the case that (I)(b) and (II)(b) holds. The other cases are similar. Using h one can compute the distance in the hypercube between the images of any pair of grid nodes K and K' as follows:

$$dist(K, K') := Bitdiff(S) + Bitdiff(G) + Bitdiff(Z) \quad \text{with}$$
$$Bitdiff(S) := Bitdiff(GrayCode(f_S(K), \log y'), GrayCode(f_S(K'), \log y'))$$
$$Bitdiff(G) := Bitdiff(GrayCode(f_G(K), \log z'), GrayCode(f_G(K'), \log z'))$$
$$Bitdiff(Z) := Bitdiff(GrayCode(f_Z(K), \log x'), GrayCode(f_Z(K'), \log x'))$$

Notice that the dilation of the embedding h equals the maximum of $dist(K, K')$ over all grid edges $\{K, K'\}$. So we have to prove that for every grid edge $\{K, K'\}$ the following holds: $dist(K, K') \leq 2$. Let K be an arbitrary grid node with the coordinates (a, b, c). Since K' is a grid neighbour of K the following three cases are relevant:

Case 1. Let K' be the grid node with the coordinates $(a + 1, b, c)$.
Case 2. Let K' be the grid node with the coordinates $(a, b + 1, c)$.
Case 3. Let K' be the grid node with the coordinates $(a, b, c + 1)$.

Ad Case 1:

1.1 $a < x'$ is valid, then holds:

$$
\begin{aligned}
Bitdiff(S) \quad &= 0 \ \text{notice:} \ f_S \ \text{does not depend on} \ a \\
f_G(a,b,c) \quad &= \lfloor (y(c-1)+b-1)/y' \rfloor + 1 \\
f_G(a+1,b,c) &= \lfloor (y(c-1)+b-1)/y' \rfloor + 1 \\
&\Rightarrow Bitdiff(G) = 0 \\
f_Z(a,b,c) \quad &= a \\
f_Z(a+1,b,c) &= a+1 \Rightarrow Bitdiff(Z) = 1 \ \text{see Corollary 8}
\end{aligned}
$$

1.2 $a = x'$ is valid.

    1.2.1 $f_G(1,b,c) \leq \frac{1}{4}z'$ is valid, then holds:

$$
\begin{aligned}
Bitdiff(S) \quad &= 0 \ \text{see above} \\
f_G(a,b,c) \quad &= \lfloor (y(c-1)+b-1)/y' \rfloor + 1 = f_G(1,b,c) \\
f_G(a+1,b,c) &= z' - f_G(1,b,c) + 1 \\
&\Rightarrow Bitdiff(G) = 1 \ \text{see Corollary 9} \\
f_Z(a,b,c) \quad &= a = x' \\
f_Z(a+1,b,c) &= a+1-x' = 1 \\
&\Rightarrow Bitdiff(Z) = 1 \ \text{see Corollary 9}
\end{aligned}
$$

    1.2.2 $\frac{1}{4}z' < f_G(1,b,c) \leq \frac{1}{2}z'$ is valid, then holds:

$$
\begin{aligned}
Bitdiff(S) \quad &= 0 \ \text{see above} \\
f_G(a,b,c) \quad &= \lfloor (y(c-1)+b-1)/y' \rfloor + 1 = f_G(1,b,c) \\
f_G(a+1,b,c) &= \tfrac{1}{2}z' + f_G(1,b,c) \\
&\Rightarrow Bitdiff(G) = 2 \ \text{see Corollary 8} \\
f_Z(a,b,c) \quad &= a = x' \\
f_Z(a+1,b,c) &= 2x' + 1 - (a+1) = x' \Rightarrow Bitdiff(Z) = 0
\end{aligned}
$$

    1.2.3 $f_G(1,b,c) > \frac{1}{2}z'$ is valid, then holds:

$$
\begin{aligned}
Bitdiff(S) \quad &= 0 \ \text{see above} \\
f_G(a,b,c) \quad &= \lfloor (y(c-1)+b-1)/y' \rfloor + 1 = f_G(1,b,c) \\
f_G(a+1,b,c) &= \tfrac{3}{2}z' - f_G(1,b,c) + 1 \\
&\Rightarrow Bitdiff(G) = 1 \ \text{see Corollary 10} \\
f_Z(a,b,c) \quad &= a = x' \\
f_Z(a+1,b,c) &= a+1-\tfrac{1}{2}x' = \tfrac{1}{2}x' + 1 \\
&\Rightarrow Bitdiff(Z) = 1 \ \text{see Corollary 10}
\end{aligned}
$$

1.3 $a > x'$ is valid.

    1.3.1 $f_G(1,b,c) \leq \frac{1}{4}z'$ is valid, then holds:

$$
\begin{aligned}
Bitdiff(S) \quad &= 0 \ \text{see above} \\
f_G(a,b,c) \quad &= z' - f_G(1,b,c) + 1
\end{aligned}
$$

$$f_G(a+1,b,c) = z' - f_G(1,b,c) + 1 \;\Rightarrow\; Bitdiff(G) = 0$$
$$f_Z(a,b,c) \quad\;\; = a - x'$$
$$f_Z(a+1,b,c) = a + 1 - x' \;\Rightarrow\; Bitdiff(Z) = 1 \text{ see Corollary 8}$$

1.3.2 $\frac{1}{4}z' < f_G(1,b,c) \leq \frac{1}{2}z'$ is valid, then holds:

$$Bitdiff(S) \quad\;\; = 0 \text{ see above}$$
$$f_G(a,b,c) \quad\;\; = \tfrac{1}{2}z' + f_G(1,b,c)$$
$$f_G(a+1,b,c) = \tfrac{1}{2}z' + f_G(1,b,c) \;\Rightarrow\; Bitdiff(G) = 0$$
$$f_Z(a,b,c) \quad\;\; = 2x' + 1 - a$$
$$f_Z(a+1,b,c) = 2x' - a \;\Rightarrow\; Bitdiff(Z) = 1 \text{ see Corollary 8}$$

1.3.3 $f_G(1,b,c) > \frac{1}{2}z'$ is valid, then holds:

$$Bitdiff(S) \quad\;\; = 0 \text{ see above}$$
$$f_G(a,b,c) \quad\;\; = \tfrac{3}{2}z' - f_G(1,b,c) + 1$$
$$f_G(a+1,b,c) = \tfrac{3}{2}z' - f_G(1,b,c) + 1 \;\Rightarrow\; Bitdiff(G) = 0$$
$$f_Z(a,b,c) \quad\;\; = a - \tfrac{1}{2}x'$$
$$f_Z(a+1,b,c) = a + 1 - \tfrac{1}{2}x' \;\Rightarrow\; Bitdiff(Z) = 1 \text{ see Corollary 8}$$

Notice that Corollaries 8, 9 and 10 are the main tools of the proof. We have only shown the proof for Case 1 here. In the other cases the technique is quite similar. □

References

[BMS] Bettayeb, S., Miller, Z., Sudborough, I.H.: Embedding Grids into Hypercubes, Proc. of Aegean Workshop on Computing, Springer Verlag's Lecture Notes in Computer Science, **Vol. 319**, 210–211, 1988.

[Ch1] Chan, M.Y.: Dilation 2-embedding of grids into hypercubes, in Proc. International Conference on Parallel Processing, St. Charles, IL, 295–298, August 1988.

[Ch2] Chan, M.Y.: Embedding of 3-dimensional grids into hypercubes, in Proc. Fourth Conference on Hypercubes, Concurrent Computers, and Applications, Monterey, CA, March 1989.

[Ch3] Chan, M.Y.: Embedding of d-dimensional grids into optimal hypercubes, in Proc. 1989 ACM Symposium on Parallel Algorithms and Architectures, Santa Fe, NM, 52–57, June 1989.

[Ch4] Chan, M.Y.: Embedding of grids into optimal hypercubes, SIAM J. Computing, **Vol. 20**, No 5, 834–864, October 1991.

[HJ] Ho, C.T., Johnsson, S.L.: Embedding Meshes in Boolean Cubes by Graph Decomposition, Journal of Parallel an Distributed Computing 8, 325–339, 1990.

[MS] Monien, B., Sudborough, I.H.: Embedding one Interconnection Network in Another, Springer Verlag, Computing Suppl. **7**, 257–282, 1990.

[SB] Scott, D.S., Brandenburg, J.: Minimal mesh embeddings in binary hypercubes, IEEE Transaction on computers, **Vol. 37**, No. 10, October 1988.

[Sud] Sudborough, I.H.: Embedding k-D Meshes into Optimum Hypercubes with Dilation 2k-1, University of Texas at Dallas, to appear.

Distributed Cyclic Reference Counting *

Frank Dehne [†]
School of Computer Science
Carleton University
Ottawa, Canada K1S 5B6
dehne@scs.carleton.ca

Rafael D. Lins [‡]
Departamento de Informática
Univ. Federal de Pernambuco
50732-970 Recife, Pe, Brasil
rdl@di.ufpe.br

Abstract

We present a distributed cyclic reference counting algorithm which incorporates both, the correct management of cyclic data structures and the improvement of lazy mark-scan. The algorithm allows processors to run local mark-scan simultaneously without any need of synchronisation between phases of different local mark-scans either on the same processor or on different processors.

1 Introduction

In distributed memory multiprocessors, each processor is responsible for reclaiming unused structures residing in its local memory (distributed garbage collection). As in the case of uni-processors, the algorithms are usually based on (global) mark-scan or on reference counting [5].

A number of algorithms use *(global) mark-scan* in distributed architectures [11, 1, 12]. The major disadvantage of these methods is that, as in the sequential case, the application is suspended during the garbage collection phase. One attempt [21] to improve this uses dual processors on each local memory and transfers objects with non-local references. This can however create very high communication overhead, depending on the size of the objects being transfered, and it may be unable to reclaim large cyclic structures that span over several processors.

Reference counting has the advantage that it is performed in small steps interleaved with application computation and, hence, does not need to suspend the application. The major drawback of standard reference counting [6] is its inability to reclaim cyclic data structures. For the sequential domain, Friedman & Wise [8], Bobrow [4], and Hughes [13] have solved this problem in the context of implementing Lisp and functional languages such as Miranda. A general uniprocessor algorithm for cyclic reference counting with local mark-scan was presented in [19] and substantially improved in [17].

Two algorithms made standard reference counting suitable for use in loosely-coupled multiprocessor architectures: weighted reference counting [3, 24] and generational reference counting [9]. However , as in sequential standard reference counting, these both algorithms are also not able to reclaim cyclic data structures. A multiprocessor algorithm that merges weighted reference counting with Lins' cyclic reference counting [17] was described in [18]. This method can reclaim cyclic data structures. Plainfossé and Shapiro [20] pointed out that this algorithm is simple but does not allow several processors to invoke mark-scan simultaneously. A first attempt to solve this problem is presented in [14]. However, this new algorithm needs global synchronisation between the phases of the local mark scan procedures on the same processor as well as on different processors.

In this paper, a distributed cyclic reference counting algorithm is presented. This algorithm allows processors to run local mark-scan simultaneously without any need of synchronisation between phases of different local mark-scans either on the same processor or on different processors. It incorporates both, the correct management of cyclic data structures [19] and the improvement of lazy mark-scan [17]. As shown in [17], lazy mark-scan evaluation can considerably reduce the garbage collection overhead by avoiding unnecessary local mark-scans.

We concentrate on the distributed computing issues of the algorithm without taki ng into account the choice of a suitable communication protocol as described in [16]. As an alternative, the algorithm presented can easily be modified to work with weighted reference conting [3, 24].

The remainder of this paper is organized as follows. In Section 2 we present our algorithm and in Section 3 we prove its correctness.

---

*This work was done while the first author was visiting the Universidade Federal de Pernambuco, Brasil.
[†]Research partially supported by the Natural Sciences and Engineering Research Council of Canada.
[‡]Research partially supported by CNPq research grant no. 40.9110/88.4.

2 The Algorithm

For reference counting with local mark-scan, the basic interface between the application and the garbage collection consists of three procedures New(R), Copy(R,<S,T>), and Delete(<R,S>) used by the application to allocate a new cell, copy a new pointer, and delete a pointer, respectively. Unused cells, i.e. cells which can not be referenced any more by the application, are organised in a *free list*.

Every cell S has a reference couter RC(S) which counts the number of pointers to S as well as a colour which could be yellow, green, black, red, pink , or blue. All cells in the free list are coloured yellow.

The garbage collection code uses three data structures. A control heap CH is used for implementing the lazy evaluation scheme. It is a data structure created locally on each processor. mr-Q(S) denotes a queue of mark-red(S) processes attempting to mark red a cell S , and rec-Q(S) denotes a queue of reclaim(S) processes, where procedures mark-red(S) and reclaim(S) are described below. The mr-Q(S) and rec-Q(S) for all cells S on one processor can be easily implemented as one data structure, each.

A call of Delete(<R,S>) spawns an independent process $reclaim_p(S)$ on the same processor which attempts to recover unused cells. The index p denotes a process number given to the reclaim process, and it will later be used to determine if this process has terminated. For the remainder, all indices attached to procedure calls refer to such process numbers.

> **New(R)** =
> select a new cell S from the free-list
> RC(S):=1
> colour(S):=green
> mr-Q(S):=nil
> rec-Q(S):=nil
> create a pointer from R to S

> **Copy(R,<S,T>)** =
> create a pointer from R to T
> RC(T):= RC(T)+1

> **Delete(<R,S>)** =
> remove the pointer <R,S>
> RC(T):= RC(T)-1
> spawn a new process $reclaim_p(S)$
> for some new process number p

Procedure $reclaim_p(S)$ checks if the reference counter of S is zero, in which case S can be immediately linked to the free list. Otherwise, a reference to S is added to the control heap CH (if not already ther e) for later evaluation on whether S is unreachable by the application. Sons(S) refers to all cells T such that there exists a pointer from S to T.

> **$reclaim_p(S)$** =
> if colour(S) is green or black then
>   if RC(S)=0 then
>   for all cells T in Sons(S) do
>   Delete(<S,T>)
>   colour(S):=yellow
>   link S to the free-list
>   else
>   if colour(S) is green then
>   colour(S):=black
>   add S to the control heap CH
> else
>   add p to rec-Q(S)

Control Heap CH

Continuously select the *next best* S from CH and do:
   remove S from CH
   if colour(S) is black then
      spawn a process mark-red$_{p_1}$(S)
         for some new process number p_1
      suspend until no active process with number p_1 exists
      spawn a process scan$_{p_2}$(S)
         for some new process number p_2
      suspend until no active process with number p_2 exists
      spawn a process collect-blue(S)

When a cell S is taken from the control heap CH, there are three processes started at S: mark-red$_{p_1}$(S), scan$_{p_2}$(S), and collect-blue(S). As these processes spawn other subprocesses, the process numbers are used to determine their termination.

Procedure mark-red$_{p_1}$(S) paints red S and all cells in the subgraph of cells reachable from S. It also decrements the reference counters of the cells visited, so that fianl reference counts are associated only with pointers from outside the subgraph (external references).

mark-red$_p$(S) =
   if colour(S) is green or black then
      colour(S):=red
      for all cells T in Sons(S) do
         mark-red(T)
      for all cells T in Sons(S) do
         RC(T):=RC(T)-1
   else
      add p to mr-Q(S)

scan$_p$(S) searches the subgraph now painted red and repaints green all cells reachable from external references. All other cells are painted blue.

scan$_p$(S) =
   if colour(S) is red then
      if RC(S)>0 then
         colour(s) := pink
         spawn a process scan-green$_p$(S)
      else
         colour(S):=blue
         for all cells T in Sons(S) do
            spawn a process scan$_p$(T)

scan-green$_p$(S) =
   for all cells T in Sons(S) do
      RC(T):= RC(T)+1
      colour-green(T)
   for all cells T in Sons(S) do
      if colour(T) is red or blue then
         spawn a process scan-green$_p$(T)

```
colour-green(S) =
    if mr-Q(S) not nil then
        colour(S) := red
        take q from mr-Q(S)
        spawn a process mark-red_q(S)
    else
        colour(S) := green
        if rec-Q(S) not nil then
            take q from rec-Q(S)
            spawn a process reclaim_q(S)
```

Procedure collect-blue(S) links to the free list all cells that are blue after the previous $scan_p(S)$ has terminated.

```
collect-blue(S) =
    if colour(S)=pink then
        colour-green(S)
    if colour(S)=blue then
        colour(S):=yellow
        Temp := Sons(S)
        link S to free-list
        for all cells T in Temp do
            spawn a process collect-blue(T)
```

3 Proof of Correctness

We now show the correctness of our distributed reference counting algorithm. In particular, we show that (a) if some cell S is collected for the free list, then S is not in use and (b) every unused cell is collected for the free list.

Lemma 1 *If colour(S)=green then RC(S) has the actual value.*

Proof. When S is created by a New(R), then S is coloured green and RC(S) has the actual value. The only procedures possibly changing RC(S) are Copy, Delete, mark-red, and scan-green. Obviously, changes made by Copy and Delete are always correct. A mark-red(S) changes colour(S) to red and, hence, the assumption of this lemma does not hold for S. Assume that S is coloured green by colour-green(S) called within scan-green(S). We observe that scan and scan-green can only reach nodes which have previously been reached by a mark-red started at some deleted pointer, and that this entire mark-red process has been terminated. Hence, RC(S) has first been decremented by one in a mark-red(S), and is now being incremented by one in a scan-green(S). Assuming, by induction, that RC(S) was correct before mark-red(S) changed it, it has now again the actual value. □

Lemma 2 *If colour(S) is not green then in finite time either*

(a) *colour(S) will change to green (if S is in use) or*

(b) *colour(S) will change to yellow (if S is not in use) and S will be collected for the free list.*

Proof. (a) Assume S is in use. If colour(S)=red then a scan process will colour it blue, green, or pink by using the same path as the mark-red process which coloured S red. If colour(S)=blue then it lies on a cycle and is reachable from a cell that is in use. Hence, a subsequent scan-green process will colour S green. If colour(S)=black then it is on the control heap CH, and after finite time it will be taken from CH and coloured red. As shown above, all red cells (in use) are eventually coloured green. If colour(S)=pink and S is part of a cycle, then the scan-green process called by the scan that coloured S pink will colour S green. If colour(S)=pink and S is not part of a cycle, the collect-blue phase following the scan that marked S pink will colour S green.

(b) Assume S is not in use. If colour(S)=blue then the collect-blue process following the scan that made S blue will change it to yellow and collect it for the free list. If colour(S)=red then the subsequent scan process will change it to blue, and from there it will be changed to yellow as indicated above. If colour(S)=black then it is on the control heap CH, and after finite time it will be taken from CH and coloured red. As shown above, all red cells (not in use) are eventually coloured yellow and collected for the free list. If colour(S)=pink and S is not in use, then the mark-red originated at the pointer whose deletion made S useless has not yet reached S. However, it is reachable from that pointer and will therefore be set to red by that mark-red process. □

Lemma 3 *If colour(S)=black then RC(S) has the actual value.*

Proof. The only procedure to change colour(S) to black is reclaim, and reclaim changes colour(S) from green to black. Hence, by Lemma 1, RC(S) has the actual value. For black cells, only Copy and Delete can change their reference counters, and both procedures obviously keep the reference counter at the actual value. □

Theorem 1 *If S is collected for the free list, then S is not in use.*

Proof. S can be collected either by reclaim or by collect-blue. Reclaim deletes S only if colour(S) is green or black and RC(S)=0. Hence, it follows from Lemma 1 and Lemma 3 that S is not in use. S is collected by collect-blue only if colour(S)=blue and collect-blue(S) is called. This implies that the previous mark-red and scan phases for the same deleted pointer have terminated (see Control Heap CH). Thus, the previous scan marked S blue and could not repaint it green. Therefore, S is part of an unreachable cycle. □

Theorem 2 *Every cell not in use is collected for the free list.*

Proof. Assume that cell S is not in use. If colour(S) is not green then, by Lemma 2, after finite time its colour will be set to yellow and S will be collected for the free list. Assume that colour(S)=green, then S is reachable from a cell T for which a reclaim(T) is still in process. Hence, this reclaim(T) process will change colour(S). □

4 Conclusion

We presented a distributed cyclic reference counting algorithm which solves the previously open problem of performing cyclic reference counting which recovers unused cyclic structure and at the same time allowing multiple such processes without synchronisations between these processes.

References

[1] K.A.M.Ali. *Object-oriented storage management and garbage collection in distributed processing systems.* PhD thesis, Royal Institute of Technology, Stockholm, December 1984.

[2] M.Ben-Ari. Algorithms for on-the-fly garbage collection. *ACM Transactions on Programming Languages and Systems*, 6(3):333–344, July 1984.

[3] D.I.Bevan. Distributed garbage collection using reference counting. In *PARLE Parallel Architectures and Languages Europe*, pages 176–187. Springer Verlag, LNCS 259, June 1987.

[4] D.G. Bobrow. Managing reentrant structures using reference counts. *ACM Transactions on Programming Languages and Systems*, 2(3): 269–273, March 1980.

[5] J.Cohen. Garbage collection of linked data structures. *ACM Computing Surveys*, 13(3):341-367, September 1981.

[6] G.E. Collins A method for overlapping and erasure of lists *Communications of the ACM*, 3(12): 655-657, 1960.

[7] E.W.Dijkstra, L.Lamport, A.J.Martin, C.S.Scholten & E.M.F.Steffens. On-the-fly garbage collection: an exercise in cooperation. *Communications of ACM*, 21(11):966-975, November 1978.

[8] D.P.Friedman and D.S.Wise. Reference counting can manage the circular environment of mutual recursion. *Information Processing Letters*, 8(1):921-930, January 1979.

[9] B.Goldberg. Generational reference counting: A reduced-communication distributed storage reclamation scheme. In *Proceedings of SIGPLAN'89 Conference on Programming Languages Design and Implementation*, pages 313-321. ACM Press, June 1989.

[10] D.Gries. An exercise in proving parallel programs correct. *Communications of ACM*, 20(12):921–930, December 1977.

[11] P.Hudak and R.M.Keller. Garbage collection and task deletion in distributed applicative processing systems. In *Proceedings of 1986 ACM Conference on Lisp and Functional Programming*, pages 168–178, Pittsburg, August 1982.

[12] R.J.M.Hughes. A distributed garbage collection algorithm. In J. P. Jouannaud (Ed.), *Functional Programming Languages and Computer Architecture*, Springer-Verlag, LNCS 201, pages 256–272, 1985.

[13] R.J.M.Hughes. Managing reduction graphs with reference counts. Departmental Research Report CSC/87/R2, University of Glasgow, March 1987.

[14] R. Jones and R.D.Lins Cyclic weighted reference counting without delay In *In PARLE'93 Parallel Architectures and Languages Europe*, Springer Verlag, LNCS 694, Arndt Bode and Mike Reeve and Gottfried Wolf Editors, pp 512–515, June 1993.

[15] H.T.Kung and S.W.Song. An efficient parallel garbage collection system and its correctness proof. In *Proc. IEEE Symposium on Foundations of Computer Science*, pa ges 120–131, 1977.

[16] C-W.Lermen and D.Maurer. A protocol for distributed reference counting. In *Proceedings of 1986 ACM Conference on Lisp and Functional Programming*, pages 343–350, Cambridge, Massachusetts, August 1986.

[17] R.D.Lins. Cyclic reference counting with lazy mark-scan. *Information Processin Letters* 44:215–220, 1992.

[18] R.D.Lins and R. Jones Cyclic weighted reference counting In K.Boyanov (Ed.), Parallel and Distributed Processing'93 (WP&DP'93), Bulgarian Academy of Sciences, Sofia, 1993, pages 369–382, to be published by North Holland.

[19] A.D.Martinez, R.Wachenchauzer and R.D.Lins. Cyclic reference counting with local mark-scan. *Information Processing Letters*, 34:31–35, 1990.

[20] D.Plainfossé and M. Shapiro. Experience with a fault-tolerant garbage collector in a distributed Lisp system. in Y. Bekkers and J.Cohen (Eds.) Proceedings of *Memory Management - International Workshop*, St. Malo, France, 1992, volume LNCS 637, pages 116–133. Springer-Verlag, 1992.

[21] M.Shapiro, O.Gruber and D.Plainfossé. A garbage detection protocol for a realistic distributed object-support system. Technical Report 1320, Rapports de Recherche, INRIA-Rocqencourt, November 1990.

[22] G.L.Steele. Multiprocessing compactifying garbage collection. *Communications of ACM*, 18(09):495–508, September 1975.

[23] D.A. Turner. Miranda: a non-strict functional language with polymorphic types. In J. P. Jouannaud (Ed.), *Functional Programming Languages and Computer Architecture*, Springer-Verlag, LNCS 201, pages 1–16, 1985.

[24] P.Watson and I.Watson. An efficient garbage collection scheme for parallel computer architectures. In *In PARLE'87 Parallel Architectures and Languages Europe*, Springer Verlag, LNCS 259, pages 432–443, June 1987.

Efficient reconstruction of the causal relationship in distributed systems [*]

P. Baldy[1], H. Dicky[1], R. Medina[1], M. Morvan[2], J.F. Vilarem[1]

[1] Département d'Informatique Fondamentale, LIRMM, Université Montpellier II —
CNRS UMR C09928, 161 rue Ada, 34393 Montpellier Cedex 5, FRANCE.
[2] LIP-IMAG, CNRS URA 1398, Ecole Normale Supérieure de Lyon, 46 allée d'Italie,
69364 Lyon Cedex 07, FRANCE.

Abstract. This paper analyzes computation of causality relationships in distributed systems. We give a formal framework based upon partial orders for this study. The initial algorithm dealing with reconstruction of the causal past of given events is greatly improved. We show a greedy algorithm whose complexity is $O(S^2)$ where S is the number of processes involved in the system. Furthermore, distributed implementation of our algorithm is optimal since it uses only messages.

1 Introduction

The importance of causality relations between two events in distributed programs has been shown in [1, 4, 9, 8, 15]. The causal order is used to analyze, monitor and debug distributed programs, manage replicated data, and for resource allocation problems. Lamport in [10] termed this relation "happened before":

> a *"happened before"* b **iff:**
>   - *a and b are in the same process and a comes before b*
>   - *or a is the sending message and b is its receipt*
>   - *or ∃ c: a "happened before" c and c "happened before" b*

The causality relation is a partial order. Upon this principle, Lamport provided an algorithm that calculates an extension of this order. His algorithm computes the height of each event during the observed distributed computation, where the height of an event e (noted $heightL(e)$) is the length of the longest path ending with e. The height is the greatest height of the predecessors of e plus one. It is easy to see that if a "happened before" b then $height(a) < height(b)$. This extension of the causality order is sufficient for some distributed problems but does not give total information on the causality order, particularly the concurrency relation required in debugging activity for instance .

Lamport's algorithm was generalized independently by Fidge in [6] and Mattern in [11]. This generalization allows exact coding of the causality order by

---

[*] This work was partially supported by the french PRC C3

associating to each event e, a vector named $VectorTime(e)$. The ith component represents the height restricted to events of process i, i.e. the number of predecessors of e on process i, possibly including e; its computation is the same type as the previous one. But the problem of this algorithm is that each emission must overload its message with its $VectorTime$; when the number of processes is high this may induce a congestion of the communication channels and slow down the distributed execution.

Indeed, only knowledge of the causality relation between some particular events is useful. In other words, given an event e, the aim is to respond in constant time for any other event x to the question: Does x belong to the causal past of e ? Thus the idea of reconstructing the $VectorTime$ of an event only when necessary appeared. In [7], Fowler and Zwaenepoel provided an algorithm in $O(min(N, S!))$ where N is the number of events and S the number of processes.

In this paper, we propose an algorithm that can be implemented sequentially in a distributed environment with complexity in $O(min(N, S^2))$ but with only S messages.

In Sect. 2, we present our modelisation of a distributed system. In Sect. 3, we expose our algorithm. In Sect. 4, we propose and analyse an implementation of this algorithm.

2 A Distributed System Model

Following [10], we model a distributed system as S processes or sites : $P_1, \ldots, P_S$. An observed distributed computation consists of a set of events $E = \{e_1, \ldots, e_n\}$. There are three types of events: *send*, *receive* and *internal* events. Each event e occurs in a single process denoted by $site(e)$.

2.1 Distributed Computation and Causality Relation

A distributed computation is represented by a binary relation $(E, \rightarrow)$, derived from the "happened before" relation in [10]:

1. The set E_i of events occurring in a process i is totally ordered, i.e. the restriction of $(E, \rightarrow)$ to E_i is a total order. We denote by $(E_i, <_i)$ this relation and call it the *chain i*.
2. Given x and y occurring on differents sites, $x \rightarrow y$ holds **iff** x is a send event, and y is a corresponding receive event.
3. For the sake of completeness, we add S elements to E, $\perp_1, \ldots, \perp_S$ such that : $\forall i \in 1..S, \perp_i \in E_i$ and $\forall e_i \in E_i : \perp_i <_i e_i$
4. There is no cycle in $\rightarrow$.

The *causality* relation is the reflexive and transitive closure of $\rightarrow$ in E. We denote by $(E, \prec)$ this order.

The causality relation is not computed by using a general transitive closure algorithm ([5] or [13] for instance), firstly because of the distributed nature of the data structure, and secondly because of the complexity of such algorithms ($O(N^{2.376})$ for [5]).

Indeed, causality is either coded dynamically [6, 11], or partially reconstructed on-query [7, 2, 15]. Both methods rely on the following relation which is reminiscent of the "vector time", as defined in [11, 6]. This relation uses a decomposition of $(E, \prec)$ along the chains as an equivalent order.

To the event x, we associate the vector [3] $CausalEvents(x)$ of size S such that:

$$\forall i \in 1..S : CausalEvents(x)[i] = \max_{\leq_i} (\{y \in E_i \ : \ y \prec x\} \cup \{\perp_i\}) \ . \quad (1)$$

The following results can be easily checked:

$$x, y \in E : x \prec y \text{ iff } \forall i \in 1..S : CausalEvents(x)[i] \leq_i CausalEvents(y)[i] \quad (2)$$

$$x \in E_i, y \in E : x \prec y \text{ iff } x \leq_i CausalEvents(y)[i] \quad [11, 3] \ . \quad (3)$$

2.2 The Main Question

Given an event e of a distributed computation $(E, \rightarrow)$, our aim is to compute the set $\{y \in E \ : \ y \prec e\}$, i.e. to know the causal past of e.

A first answer was given by Fidge and Mattern [6, 11]. They compute on-the-fly an embedding χ from $(E, \prec)$ in the product order $(\mathbb{IN}^S, <)$. Each event being represented by its height on its chain. Their algorithm stores a size S vector with each event. A send event e must overload its message with $CausalEvents(e)$. With this method $y \prec e \Leftrightarrow \chi(y) < \chi(e)$.

Another approach to answer this question is to reconstruct the causal past of a given event on-query [7, 2, 15]. The principle of this method is to partially search $(E, \rightarrow)$ in order to compute the $CausalEvents$ vector of a given event. With this method, $y \in E_i, y \prec e \Leftrightarrow y \leq_i CausalEvents(e)[i]$ from (3). The algorithm shown in this paper uses this approach.

2.3 Hypothesis of the Reconstruction Algorithms

The *DirectDependency* Vector. This vector is essential in reconstruction algorithms: it is used to compute the $CausalEvents$ vector of an event e from the knowledge of its direct most recent predecessors with respect to each chain.

On each process, knowledge of the "most recent" predecessors of each event may be maintained, with respect to each chain $(E_i, <_i), i \in 1..S$. In fact, the S most recent events — one per chain — related to an event e, are given by the "direct dependencies" vector $DirectDependency(e)$ in the following way:

Given e an event occurring on process i, and for any j in $1..S$ we have

$$DirectDependency(e)[j] = \max_{<_j} (\{y \in E_j : y \rightarrow z \wedge z \in E_i \wedge z \leq_i e\} \cup \{\perp_j\})$$

$$(4)$$

---

[3] In this paper, we use the same notation for a vector and for the set of components of this vector where this notation is not ambiguous

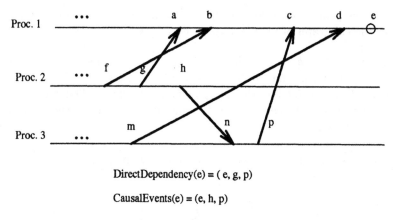

DirectDependency(e) = (e, g, p)

CausalEvents(e) = (e, h, p)

Fig. 1. *DirectDependency* and *CausalEvents* vectors

Complexity Hypothesis. The complexity of the algorithm presented in this paper is analyzed, based on the fact that, given a site i and an event $e \in E_i$, we compute in constant time (notation $O(1)$) :

1. $\forall y \in E_i$, the Boolean : $(y <_i e)$
2. $DirectDependency(e)$

This is shown to be a reasonable hypothesis in Sect. 4.

3 Analysis Using Graphs: A Greedy Algorithm

3.1 Analysis of the Reconstruction Problem Using Graphs

In this section, we show and prove some results required for our algorithm. All results shown in this section are given with a multi-send/multi-receive events model.

We define the binary relation Γ upon $E \cup \{\perp_1, \ldots, \perp_S\}$, $e\ \Gamma\ e'$ standing for $e' \in DirectDependency(e)$. We show that the associated graph (E, Γ) is cycle-free. For a given event $e \in E$, we define the set $\Gamma(e)$ as the set of successors of e in (E, Γ), and the set $\Gamma^*(e)$ as the set of events $y \in E$ such that there is a path from e to y in (E, Γ). We denote by $G(e)$ the subgraph [4] of (E, Γ) induced by $CausalEvents(e)$.

Our algorithm is based on particular properties of $G(e)$, namely :

1. $G(e)$ is connected, cycle-free and e is the minimum element in $G(e)$.
2. $G(e)$ can be computed using a partial search in the subgraph of (E, Γ) induced by $\Gamma^*(e)$, observing that :

$$\forall i \in 1..S, CausalEvents(e)[i] = \max_{<_i}(\{y \in E_i : y \in \Gamma^*(e)\}) \qquad (5)$$

---

[4] Let E be the set of edges and V be the set of vertices of a graph G. The subgraph of G induced by $V' \subseteq V$ is $G' = (V', E')$ where $E' = E \cap V' \times V'$.

With these properties, different algorithms may be used in order to partially search in the subgraph of (E, Γ) induced by $\Gamma^*(e)$.

1. *Depth-First Search* [7, 15] : the complexity is in $O(S!)$ since we can restrict the search to paths that are less or equal to S in size.
2. *Breadth-First Search* [2] : the complexity is in $O(S^3)$. The algorithm proceeds in S steps. At each step i, the algorithm computes the set of events $x \in E$ such that :
   (a) x is maximal in the chain $site(x)$.
   (b) there is a path, of length $i - 1$, from e to x in (E, Γ).

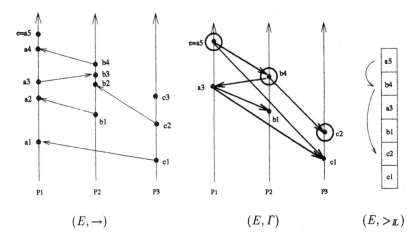

$(E, \rightarrow)$ (E, Γ) $(E, >_{I\!L})$

Fig. 2. Graphs associated to the greedy search algorithm.

3. *Greedy Search* : It uses $>_{I\!L}$ a linear extension of (E, Γ) - for instance, Lamport's clocks could be used to obtain such a linear extension. The subgraph induced by $\Gamma^*(e)$ is searched according to the order defined by $>_{I\!L}$ – see Fig. 2. The algorithm proceeds in S step. At each step, the greedily selected event is the first - in order $>_{I\!L}$ - that does not belong to a site still having an element of $G(e)$ [5] .

This greedy algorithm is based on the following results :

Lemma 1. *Given an event $e \in E$: For each process $i \neq site(e)$, let $x \in G(e) \cap E_i$ -i.e. $x = CausalEvents(e)[i]$. There is a process $k \neq i$ and $y \in G(e) \cap E_k$ such that : $y \Gamma x$.*

Proof. Clearly since $x = \max_{\leq_i} (u \in E_i : u \prec e)$, x is a send event. Thus a receive event z on a site k corresponding to the send event x exists. z belongs to

---
[5] We use the same term $G(e)$ for a graph and for its set of vertices where it is not ambiguous.

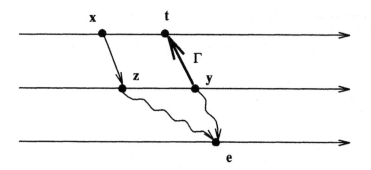

Fig. 3.

a path from x to e in $(E, \rightarrow)$. Let $y \in G(e) \cap E_k$ and $t = DirectDependency(y)[i]$, see Fig. 3. Let us prove that $x = t$.

1. $x \leq_i t$: From (4) $t = \max_{\leq_i} \{u \in E_i : u \rightarrow w \wedge w \in E_k \wedge w \leq_k y\}$. On process k, we have $z \prec e$ and $y = \max_{\leq_k} (u \in E_k : u \prec e)$ thus $x \rightarrow z \wedge z \leq_k y$. The definition of t achieves this point.
2. $x \geq_i t$ since $x = \max_{\leq_i} (u \in E_i : u \prec e)$, and there is a path from t to e.

□.

Lemma 2. *Given an event $e \in E$ and $x \in G(e)$ - i.e. x is a component of the vector $CausalEvents(e)$. There exists a sequence $y_0 = e, y_1, \ldots, y_k = x - k \geq 0$ — of elements in $G(e)$ such that : $\forall i \in 1..k : y_{i-1} \, \Gamma \, y_i$.*

The proof is a simple iteration of Lemma 1 giving a path from e to x in $G(e)$ which contains different components of $CausalEvents(e)$. This iteration stops when e is reached. (The acyclicity of $(E, \prec)$ proves the termination of this computation).

Proposition 3. *The subgraph $G(e)$ of (E, Γ) induced by $CausalEvents(e)$ is connected, cycle-free and e is the minimum element in $G(e)$.*

Proof. (E, Γ^{-1}) is a subgraph of the order $(E, \prec)$, hence (E, Γ) is acyclic. This proves $G(e)$ is acyclic. From Lemma 2, there exists a path in $G(e)$ from e to each vertex of $G(e)$. This proves $G(e)$ is connected and e is the minimum element in $G(e)$.

□.

3.2 A Greedy Algorithm

Let $(E, <)$ be a total order compatible with $(E, \prec)$. Since (E, Γ^{-1}) is a partial graph of the order $(E, \prec)$, $(E, >)$ is also compatible with (E, Γ) and may be used for a greedy search of $CausalEvents(e)$ in the subgraph of (E, Γ) induced by $\Gamma^*(e)$.

This leads to the following algorithm that, given an event e, computes the set $CausalEvents(e)$:

Reconstruction_greedy_search (e : event)

```
    Solution := {e};
    /* The set Visited holds the processes visited during the search */
    Visited := site(e) ;
    Candidates := Γ(e) − {e} ;
    i := 1 ;
    while i ≤ S do
        Candidates := {Γ(x) ; x ∈ Solution} − ∪_{j∈Visited}(E_j) ;
        x := max_<(Candidates) ;
        Solution := Solution ∪ {x} ;
        Visited := Visited ∪ site( x ) ;
        i := i + 1 ;
    endwhile
    return( Solution) ;
end Reconstruction_greedy_search
```

Proof. Let the invariants be:

1. *Solution* contains the i greatest — in the total order $(E, <)$ — vertices of $G(e)$
2. *Visited* contains the sites of these vertices
3. $Candidates = \{\Gamma(x) ; x \in Solution\} - \bigcup_{j \in Visited}(E_j)$
4. *Candidates* contains at least $S - i$ elements.

The invariants are initially verified. Let us call $Solution'$, i', $Candidates'$, $Visited'$ the values of $Solution, i, Candidates, Visited$ at the end of an iteration. Suppose that *Solution* contains $y_1 = e > y_2 > \cdots > y_i$ which are the i greatest (in $<$) vertices of $G(e)$. Let $x = \max_<(Candidates')$. This means:

1. $\exists j \in 1..i : y_j \;\Gamma\; x$
2. $\forall z \in Candidates' : z \leq x$

Let t be a vertex of $G(e)$ which is not in *Solution*. From the connexity of $G(e)$ we know that there exists a path from e to t in $G(e)$. Let z be the first element in this path which is not in *Solution*. We have $z \in Candidates'$, thus, from 2., $z \leq x$. Since there is a path from z to t in $G(e)$, we have $t \prec z$, thus $t \leq z$. Consequently $t \leq x$.

Therefore $\forall t \in G(e) - Solution : t \leq x$. But, from $CausalEvents$ definition, the element t_x of $G(e) - Solution$ which occurs on site(x) is such that $x \leq t_x$. So x is the maximum vertex in $G(e) - Solution$.

This proves $Solution' = Solution \cup \{x\}$ contains $y_1 = e > y_2 > \cdots > y_i > y_{i'} = x$ which are the $i' = i + 1$ greatest (in $<$) vertices of $G(e)$, which proves the first invariant. The two others come immediately.

□.

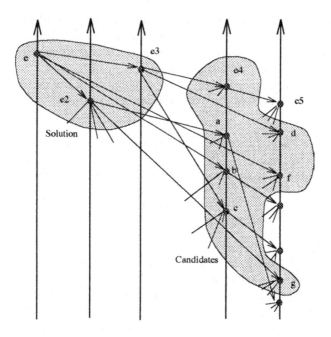

Fig. 4. Search in Γ

In Fig. 4, we show an example where $i = 3$, *Solution* contains the 3 maximum vertices e, e_2, e_3, *Candidates'* is $\{a, b, c, d, f, g, e_4\}$, and $e_4 > d > a > b > f > c > g$ for example. Thus e_4 is chosen as the next vertex in $G(e)$.

4 Algorithms and Complexity

4.1 Data Structures

In order to implement this algorithm, we represent each element with the index of its chain and its height on the chain. We code it with the type *Event* : **record** { LocalHeight : **integer**; Site : 1..S }.

The arbitrary fixed total order used for the greedy search is the order given by the Lamport mechanism [10]. It uses S integer "local clocks" t_i timestamping events on site i, and a global synchronization mechanism between the different local clocks, (see [10]). As a result, an event e is "timestamped" $height(e)$, where $height(e)$ is the global height of e in $(E, \prec)$ (length of the longest path ending with e in $(E, \prec)$). In order to separate timestamped events with the same height, the total order is the lexicographic order defined upon $(height(e) , site(e))$. We denote this total order by $<_{I\!L}$.

We also need direct access to the vector $DirectDependency(e)$ using the representation element e as a key. In order to combine comparability in $(E , <_{I\!L})$ and direct access, we code each component e_j of $DirectDependency(e)$ with the pair $(height_j(e_j), height(e_j))$.

We use the data structure $DirectDependency$: **array 1..S of record** { Local-Height, Height : integer }. We then represent the vector $CausalEvents$ in the same way as the vector $DirectDependency$. So, if $x \in E_k$ and $y \in E$, we have $x \prec y \Leftrightarrow x.LocalHeight \leq CausalEvents(y)[k].LocalHeight$.

The $DirectDependency$ vector and the global clock are usually maintained "on the fly" in each process using the following mechanism:

On process i :

1. Initially, we assign $(0,0)$ to each component of $DirectDependency(\perp_i)$, and 0 to the local clock t_i.

2. For each internal or send event e, let x be e's predecessor on process i :

$$t_i := t_i + 1 \; ; DirectDependency(e)[i] := (e.LocalHeight \; , \; t_i);$$
$$\forall k \neq i, DirectDependency(e)[k] := DirectDependency(x)[k] \; ;$$

3. For each send event e, the messages are timestamped $(e.LocalHeight, t_i)$.

4. For each receive event e, let D be the set of sending processes in case of multi-receive events model. For each j in D let $(h_j \; , \; t_j)$ be the timestamp of the received message from process j (wlog we assume that only one message can be received at a time from a given process). Let us denote the value of $DirectDependency(x)[j]$ by $(previous\_lh_j \; , \; previous\_t_j)$. We sequentially achieve the following assignments:

$$t_i := max(t_i, max(t_j \; ; \; j \in D)) + 1 \; ;$$
$$DirectDependency(e)[i] := (e.LocalHeight \; , \; t_i) \; ;$$
$$\forall j \in D : DirectDependency(e)[j] :=$$
$$\qquad (\textbf{if } previous\_t_j < t_j$$
$$\qquad\qquad \textbf{then } (h_j \; , \; t_j)$$
$$\qquad\qquad \textbf{else } (previous\_h_j \; , \; previous\_t_j) \;)$$
$$\qquad\qquad \text{/* The else case is necessary */}$$
$$\qquad\qquad \text{/* when communications occur */}$$
$$\qquad\qquad \text{/* in a non FIFO mode */}$$
$$\qquad).$$
$$\forall j \notin D, j \neq i, DirectDependency(e)[j] \text{ keeps the same value.}$$

This mechanism is related to the algorithm shown by Mattern in [11].

Let us now give the algorithm which computes the $CausalEvents$ vector.

4.2   A Centralized Algorithm Computing CausalEvents

Reconstruction_greedy_search (e : Event)

```
    /* The type Event is : record { LocalHeight : integer ; Site : 1..S } */
    /* The CausalEvents vector holds the result : */
    /* The CausalEvents and DirectDependency vectors */
    /* are both arrays 1..S such that : */
    /* The jth component type is record { LocalHeight, Height : integer } */


    CausalEvents := DirectDependency(e) ;
    /* The list ToVisit holds the processes to be visited during the search */
    ToVisit := 1..S - e.Site ;
    i := 1;


    while i ≤ S do
        FindNextComponent( CausalEvents, ToVisit, x, site );
        /* The event x and site are computed */
        /* and returned by FindNextComponent */


        Merge( CausalEvents, DirectDependency( x ), ToVisit );
        /* CausalEvents is modified by the procedure Merge */


        ToVisit := ToVisit - site ;
        i := i + 1 ;
    endwhile
end Reconstruction_greedy_search
```

Procedure FindNextComponent(CausalEvents,ToVisit,x,site);

```
    /* Given CausalEvents and ToVisit returns x and site, */
    /* CausalEvents contains exactly the elements of Solution —as described */
    /* in Sect. 3.2 — */
    /* and the elements of Candidates which are maximum on their chain, */
    /* and belong to ToVisit (see also Merge) */
    /* x = max<L(CausalEvents[j] ; j ∈ ToVisit) */


    site := first_element_in( ToVisit ) ;
    x := CausalEvents[ site ] ;
    foreach site p in ToVisit - Site do
        if ( CausalEvents[ p ].Height > x.Height )
            then x := CausalEvents[ p ] ; site := p ;
        endif
    endforeach
end FindNextComponent
```

Procedure Merge(`CausalEvents, DirectDependency, ToVisit`);
    /* Returns in CausalEvents the merging of maximum ($<_{\it I\!L}$) components */
    /* between CausalEvents and DirectDependency */
    /* The only concerned components are those in ToVisit */
    foreach site p in ToVisit do
        if (`DirectDependency[p].Height` > `CausalEvents[p].Height`)
          then `CausalEvents[ p ] := DirectDependency[p]`
      endif
    endforeach
end Merge

The algorithm complexity is clearly in $O(S^2)$ since each instruction in each iteration is computed in constant time.

4.3   A Distributed Algorithm Computing *CausalEvents*

A distributed implementation is straightforward: Each intermediate process sequentially runs the *Merge*, then the *FindNextComponent* procedures. The *FindNextComponent* procedure returns the index of the next process where the message (x, *ToVisit*, *CausalEvents*, *InitiatingSite*) is sent. The last process sends *CausalEvents* to the site *InitiatingSite*.

Initiating Process
    Reconstruction_greedy_search (e : **Event**)
        `CausalEvents := DirectDependency(e)` ;
        `ToVisit := 1..S - e.Site` ;
        `FindNextComponent( CausalEvents, ToVisit, x, Nextsite )`;
        /* Event x and Nextsite are computed and returned by FindNextComponent */
        `send ( x, ToVisit, CausalEvents, e.site) to Nextsite;`
        `wait until receipt of CausalEvents` ;
    end Reconstruction_greedy_search

Intermediate Process P
    upon receipt of (`x, ToVisit, CausalEvents, InitiatingSite`) **do**
        `Merge( CausalEvents, DirectDependency( x ), ToVisit )`;
        /* CausalEvents is modified by the procedure Merge */
        `ToVisit := ToVisit - P` ;
        if `ToVisit` = $\emptyset$ **then send** (`CausalEvents`) **to** `InitiatingSite`;
          /* Process P is the last visited process of the distributed computation */
        else
          `FindNextComponent( CausalEvents, ToVisit, x, Nextsite )`;
          /* FindNextComponent computes and returns event x and *Nextsite* */
          `send ( x, ToVisit, CausalEvents, e.site) to Nextsite;`
        endif
    end

This distributed version has global complexity in $O(S^2)$, and requires only S messages, which makes it attractive. Note that a simple test may reduce the worst case complexity to $O(min(N, S^2))$.

5 Conclusion and Prospects

In order to describe distributed computations, we have proposed a formal model using graphs and partially ordered sets. This allowed us to prove some properties of distributed computations.

With this framework, we exhibited a greedy algorithm computing the causal past for one event. The complexity of this algorithm is in $O(min(N, S^2)$ where S is the number of sites. Furthermore, a distributed version is efficient as it needs only S messages to proceed.

One problem remains: this algorithm needs a great amount of maintained data. In fact, each event is associated with a vector of size S, which makes the secondary memory space grow continuously. This drawback, which is not inherent to our algorithm, may be solved using some mechanisms that discard unnecessary data [12].

References

1. Adam, M., Hurfin, M., Raynal, M., Plouzeau N.: Distributed debugging techniques. Research Report 1459 INRIA, Rennes France (1991)
2. Baldy, P., Dicky, H., Medina, R., Morvan, M., Vilarem, J.-F.: Efficient reconstruction of the causal relationship in distributed computations. Research Report LIRMM 92-013, LIRMM, Montpellier France (1992)
3. Charron-Bost, B.: Mesures de la concurrence et du parallélisme des calculs répartis. PhD thesis, Universit'e Paris VII, Paris France (1992)
4. Cooper, R., Marzullo, K.: Consistent detection of global predicates. In Proc. ACM/ONR Workshop on Parallel and Distributed Debugging Santa-Cruz, California (1991) 163–173
5. Coppersmith, D., Winograd, S.: Matrix multiplication via arithmetic progressions. Proc. 19th Annual Symposium on the Theory of Computation (1987) 1–6
6. Fidge, C.J.: Timestamps in message-passing systems that preserve the partial ordering. Proc. 11th Australian Computer Science Conference, University of Queensland Australia (1988) 55–66
7. Fowler, J., Zwaenepoel, W.: Causal distributed breakpoints. Proc. 10th Int. Conference on Distributed Computing Systems, Paris France (1990) 134–141
8. Hseush, W., Kaiser, G.E.: Modeling concurrency in parallel debugging. ACM SIGPLAN Notices 25(3) (1990) 11–20
9. Haban, D., Weigel, W.: Global events and global breakpoints in distributed systems. Proc. 21th Annual Hawaii Int. Conference on System Sciences (1988) 166–175
10. Lamport, L.: Time, clocks and the ordering of events in a distributed system. Communications of the ACM, 21(3) (1978) 558–565
11. Mattern, F.: Virtual time and global states of distributed systems. In M. Cosnard et al. editor, Parallel and Distributed Algorithms (1989) Elsevier / North-Holland 215–226

12. Medina, R.: Incremental garbage collection for causal relationship computation in distributed systems. Proc. 5th IEEE Symposium on Parallel and Distributed Processing (1993) 650-655
13. Morvan, M.: Algoritmes linéaires et invariants d'ordres. PhD thesis, Université de Montpellier II, Montpellier France (1991)
14. Raynal, M., Schiper, A., Toueg, S.: The causal ordering abstraction and a simple way to implement it. Information Processing Letters **39** (1991) 343-350
15. Schwarz, R., Mattern, F.: Detecting causal relationships in distributed computations: in search of the holy-grail. To appear in Distributed Computing.

Scalable Parallel Computational Geometry *
– *Summary* –

Frank Dehne [†]
School of Computer Science
Carleton University
Ottawa, Canada K1S 5B6
dehne@scs.carleton.ca

Parallel Computational Geometry is concerned with solving some given geometric problem of size n on a parallel computer with p processors (e.g., a PRAM, mesh, or hypercube multiprocessor) in time $T_{parallel}$. Let $T_{sequential}$ and $S_p = \frac{T_{sequential}}{T_{parallel}}$ denote the sequential time complexity of the problem and the *speedup* obtained by the parallel solution, respectively. If $S_p = p$, then the parallel algorithm is clearly *optimal*. Theoretical work for Parallel Computational Geometry has so far focussed on the case $\frac{n}{p} = O(1)$, also referred to as the *fine grained* case. However, for parallel geometric algorithms to be relevant in practice, such algorithms must be *scalable*, that is, they must be applicable and efficient for a wide range of ratios $\frac{n}{p}$.

Most existing multicomputers (e.g. the Intel Paragon, Intel iPSC/860, and CM-5) consist of a set of p *state-of-the-art* processors (e.g. SPARC proc.), each with considerable local memory, connected to some interconnection network (e.g. mesh, hypercube, fat tree). For these machines, $\frac{n}{p}$ is considerably larger than $O(1)$. Therefore the design of scalable algorithms is listed as one major goal in the 1992 "Grand Challenges" report [4].

Yet, only little theoretical work has been done for designing scalable parallel algorithms for Computational Geometry. Note that, if there exists an optimal fine grained algorithm, then, at least from a theoretical point of view, the problem is trivial. Standard simulation gives an optimal algorithm for any ratio $\frac{n}{p}$. Many multiprocessors provide system software tools, ususally referred to as *virtual processors*, for implementing such simulation.

---

*This presentation reviews recent joint work with A. Fabri (INRIA, Sophia-Antipolis), C. Kenyon (ENS Lyon), and A. Rau-Chaplin (DIMACS).

[†]Research partially supported by the Natural Sciences and Engineering Research Council of Canada.

However, for most interconnection networks used in practice, many problems do not as yet have fine grained algorithms with linear speedup, or such linear speedup parallel algorithms are impossible due to bandwidth or diameter limitations. Such a situation is depicted in Figure 1. It shows the speedup S_p as a function of p, for $1 \le p \le n$. The diagonal, curve "A", represents an algorithm with linear speedup. The vertical line through $p = n$ represents the fine-grained case. Point "a" represents a fine-grained algorithm with linear speedup. As indicated above, linear speedup is impossible to obtain for some networks (e.g. meshes). Let point "b" represent an optimal fine-grained algorithm with less than linear speedup. For $p < n$, curve "B" (the straight line from "b" to the origin) shows the speedups obtained by standard simulation. However, if "b" represents the optimal fine-grained speedup, then the entire curve "B" does not necessarily represent the optimal speedups for all possible ratios $\frac{n}{p}$.

We can show that, for a variety of geometric problems, the actual curve of optimal speedups for all $1 \le p \le n$ is a convex curve similar to curve "C" depicted in Figure 1. That is, for $\frac{n}{p}$ larger than $O(1)$, we present algorithms with speedups considerably faster than what can be obtained through the standard *virtual processor* simulation of fine-grained algorithms. It is common knowledge that many theoretical Parallel Computational Geometry algorithms perform miserably in practice. The above observation seems to be one of the reasons. In contrast, experiments with some of our methods show that our scalable algorithms designed for $\frac{n}{p}$ larger than $O(1)$ seem to have considerable practical relevance.

In a nutshell, the basic idea for our methods is as follows: We try to combine optimal sequential algorithms for a given problem with an efficient global routing and partitioning mechanism. We devise a constant number of partitioning schemes of the global problem (on the entire data set of n data items) into p subproblems of size $O(\frac{n}{p})$. Each processor will solve (sequentially) a constant number of such subproblems, and we use a constant number of global routing operations to permute the subproblems between the processors. Eventually, by combining the $O(1)$ solutions of it's $O(\frac{n}{p})$ size subproblems, each processor determines it's $O(\frac{n}{p})$ size portion of the *global* solution.

The above is necessarily an oversimplification. The actual algorithms will do more than just those permutations. The main challenge lies in devising the above mentioned partitioning schemes. Note that, each processor will solve only a constant number of $O(\frac{n}{p})$ size subproblems, but eventually will have to determine it's part of the entire $O(n)$ size problem. The most complicated part of the algorithm is to ensure that the algorithm will terminate after $O(1)$ global permutation rounds.

Deterministic Methods For Scalable Parallel Computational Geometry

In [2] we presented deterministic algorithms for the following well known geometric problems:

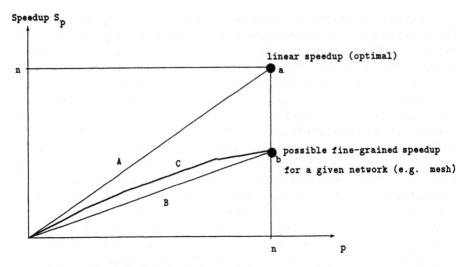

Figure 1: Speedups For Various Values Of p.

(1) Area of the union of rectangles in 2-space.

(2) $3D$-maxima.

(3) All nearest neighbors of a point set in 2-space.

(4) Lower envelope of non-intersecting line segments in 2-space (and with slightly more memory for possibly intersecting line segments).

(5) $2D$-weighted dominance counting.

(6) Multisearch on balanced search trees, segment tree construction, and multiple segment tree search.

We also studied the following applications of (6): the problem of determining for a set of simple polygons in 2-space all directions for which a uni-directional translation ordering exists, and determining for a set of simple polygons a multi-directional translation ordering.

Our scalable parallel algorithms for Problems 1-6 have a running time of

$$O(\frac{T_{sequential}}{p} + T_{sort}(n, p))$$

on a p-processor coarse grained multicomputer with arbitrary interconnection network and local memories of size $O(\frac{n}{p})$ where $\frac{n}{p} \geq p$. $T_{sort}(n, p)$ refers to the time of a global sort operation.

Consider for example the *mesh* architecture. For the fine grained case, $\frac{n}{p} = O(1)$, a time complexity of $O(\sqrt{n})$ is optimal. Hence, simulating the existing results on a coarse grained machine gives $O(\frac{n}{p}\sqrt{n})$ time coarse grained methods. Our methods for the above problems run in time $O(\frac{n}{p}(\log n + \sqrt{p}))$, a considerable improvement over the existing methods. For the *hypercube*, our algorithms are optimal for $n \geq p^{\log p}$, in which case they also yield a considerable improvement over previous methods.

High Probability Methods For Scalable Parallel Computational Geometry

In [3] we present faster and more general *high probability* methods. They can be applied to any p-processor coarse grained multicomputer with arbitrary interconnection network and local memories of size $O(\frac{n}{p})$ where $\frac{n}{p} \geq p^{\alpha}$ and $\alpha > 0$ is a fixed constant. The following problems were studied:

(7) Lower envelope of non-intersecting line segments in $[0, 1]^2$

(8) Visible portion of parallelepipeds in $[0, 1]^d$, $d = O(1)$. For d=3, Problem 8 is the well known visibility problem for parallel rectangles in $[0, 1]^3$.

(9) Convex hull of points in d-space, $d = O(1)$

(10) Maximal elements of points in d-space, $d = O(1)$.

(11) Voronoi diagram of points in $[0, 1]^d$, $d = O(1)$.

(12) All nearest neighbors for points in $[0, 1]^d$, $d = O(1)$.

(13) Largest empty circle for points in $[0, 1]^d$, $d = O(1)$.

(14) Largest empty hyperrectangle for points in $[0, 1]^d$, $d = O(1)$

Our solutions for Problems 7-10 have, with high probability, a time complexity of

$$O(\frac{T_{sequential}}{p} + T_{pSUM}(p) + T_{comp}(n, p))$$

$T_{pSUM}(p)$ and $T_{comp}(n, p)$ denote the parallel time complexity to compute the partial sums of p integers (one stored at each processor) and compress a subset of data, respectively. Our solutions for Problems 11-14 have, with high probability, a time complexity of

$$O(\frac{T_{sequential}}{p} + T_{sort}(n, p))$$

Experimental Results

Our algorithms are simple and easy to implement. The constants in the time complexity analysis are small. Except for a small (fixed) number of communication rounds, all other computation is sequential and consists essentially of solving on each processor a small (fixed) number of subproblems of size n/p. Hence, it allows to use existing sequential code for the respective problem. For the communication rounds, we can use well studied existing code for data compression, partial sum, and sorting.

For coarse grained machines, our methods imply communication through few large messages rather than having many small messages. This is important for machines like the Intel iPSC where each message creates a considerable overhead. Note, however, that our analysis accounts for the length of messages.

We implemented some of our methods on an Intel iPSC/860 and a CM-5, and obtained very good timing results (even without much programming efforts). They indicate that our methods are of considerable practical relevance.

References

[1] S. G. Akl and K. A. Lyons, *Parallel Computational Geometry*, Prentice-Hall, 1993.

[2] F. Dehne, A. Fabri, and A. Rau-Chaplin, "Scalable parallel computational geometry for coarse grained multicomputers," in Proc. *ACM Symposium on Computational Geometry*, 1993, pp. 298-307.

[3] F. Dehne, C. Kenyon, and A. Fabri, "Scalable And Architecture Independent Parallel Geometric Algorithms With High Probability Optimal Time," Technical Report, School of Computer Science, Carleton University, Ottawa, Canada K1S 5B6, 1994.

[4] *Grand Challenges: High Performance Computing and Communications.* The FY 1992 U.S. Research and Development Program. A Report by the Committee on Physical, Mathematical, and Engineering Sciences. Federal Councel for Science, Engineering, and Technology. To Supplement the U.S. President's Fiscal Year 1992 Budget.

Sorting and Selection on Arrays
with Diagonal Connections

Danny Krizanc[1]   Lata Narayanan[2]

Abstract. We examine the problems of sorting and selection on meshes and tori with *diagonal* connections. We are able to achieve dramatic reductions in time required for selection by making use of the diagonal connections. In particular, on an $n \times n$ mesh with diagonal connections, we show how to select the k-th largest of n^2 elements in $0.65n$ steps, and on a torus in $0.59n$ steps. The best known results for the corresponding networks without diagonal connections are $1.15n$ and $1.13n$ respectively. We also give an algorithm for selection that works optimally on a random input on these networks. For the case of sorting, we show the surprising result that there can be no distance-optimal algorithm for sorting on a mesh with diagonal connections, on a model that does not use copies and allows only constant size queues. Combined with results due to Kunde *et al.* [7], our results imply that routing is easier than sorting on the mesh with diagonal connections when copies are disallowed and queues are restricted to size 10. We believe this is the first result of this type in the fixed connection network model. On a torus with diagonal connections, we show that any algorithm that does not use copies must take at least $n - o(n)$ steps regardless of how large the queue size is. We also show non-trivial lower bounds for selection on meshes with diagonal connections.
Keywords: Mesh, torus, diagonal connections, upper bounds, lower bounds, sorting, selection, randomized algorithms, parallel algorithms.

1   Introduction

The mesh-connected processor array has proved itself to be a very viable and attractive architecture for parallel machines. Many new parallel computers are based on the mesh topology including the Maspar MP-1, Intel Touchstone [10], MIT April [1], and the Stanford DASH [9]. Algorithms for solving various problems on the mesh abound in the literature. It has been observed that adding in the toroidal connections can often result in faster algorithms since the diameter

[1] School of Computer Science, Carleton University, Ottawa, Canada K1S5B6. Research was supported in part by the Natural Sciences and Engineering Research Council of Canada, research grant OGP0137991.
[2] Dept. of Computer Science, Concordia University, Montreal, Canada H3G 1M8.

of the torus is half the diameter of the mesh. Another way to reduce the diameter would be to introduce *diagonal* connections. Such a network would still retain many of the properties that make the mesh a desirable topology, such as constant degree (eight), recursive decomposability, scalability etc. Maspar's MP-1 is a torus with diagonal connections. Unfortunately there has been little research involving the design of algorithms that take advantage of the diagonal connections in this machine.

In this paper, we show upper and lower bounds for the problems of sorting and selection on meshes and tori with diagonal connections. We are able to show significantly faster algorithms for selection on arrays by making use of the diagonal connections. In the case of both the mesh and the torus with diagonal connections, we give algorithms that take a little over half the time taken by the best known algorithms on the corresponding networks without the diagonal connections. For sorting, we show some unusual lower bounds. For instance, we show that on a model of the mesh where copying is not allowed, there is no distance-optimal algorithm for sorting that uses constant-size queues. In contrast, on the mesh without diagonal connections, distance-optimal algorithms for sorting are known [3, 13].

Any selection algorithm for the $n \times n$ mesh or torus with diagonal connections must take at least $0.5n - 1$ steps to complete, since this could be the distance from the initial position of the median element to the desired final position. The question of whether there exists a distance-optimal algorithm remains unresolved. For the mesh without diagonal connections, the best known algorithm takes $1.15n$ steps, and the best bound for the torus without diagonal connections is $1.13n$ steps. Clearly, we can use these algorithms on our networks, simply ignoring the presence of diagonal connections. In this paper, we show show how to make use of the diagonal connections to get substantial gains in time. For the mesh with diagonal connections, we show a randomized algorithm that works in $0.65n$ steps with high probability. For the torus with diagonal connections, we show an even faster $0.59n$ step algorithm. In both cases, we cut the time required in the corresponding networks without the diagonal connections almost by half. We also show an algorithm that works optimally on an random input. On a mesh without diagonal connections, Condon and Narayanan [2] showed an optimal bound of $n + o(n)$ steps. We are able to halve that bound by showing a $0.5n + o(n)$ step algorithm on a mesh with diagonal connections.

The obvious distance lower bound for sorting on a mesh with diagonal connections is $n - 1$ steps, and for a torus with diagonal connections is $0.5n - 1$ steps. We are able to show that there can be no distance optimal algorithm for sorting, if elements cannot be replicated, and the queues are restricted to be of constant-size. In contrast, on the mesh without diagonal connections, the only known non-trivial lower bounds are the $3n - o(n)$ bound on a model with queue size 1 [6, 12], and a $2.125n$ bound on a model with queue size 2 [5]. Also sorting can be performed distance-optimally when no diagonal connections are available, albeit using copies of elements [3, 13]. On a torus with diagonal connections, we show that any algorithm for sorting that does not use copies of elements must take at least $n - o(n)$ steps, regardless of the size of the queues. The distance lower bound on this network is only $0.5n$ steps.

Recent work by Kunde *et al.* [7] shows that routing on a mesh with diagonal connections can be done in $1.11n$ steps, and sorting can be done in $1.33n$ steps using queues of size 9. Our lower bound results show that sorting without replication of elements and queue size of 9 must take at least $1.166n$ steps. Thus, our results show that sorting in this setting is provably harder than routing. We believe this is the first result to show a provable separation between the problems of routing and sorting on a fixed connection network. In the same paper, Kunde *et al.* show an algorithm for sorting on a torus with diagonal connections that requires $n + o(n)$ steps. Our lower bound result for sorting on this network shows that this result is optimal (if replication is disallowed).

Finally, we show lower bound results for the problem of c-element selection on meshes with diagonal connections. In this problem, each processor initially has c elements and the objective is to route the selected element to the middle processor in the array. Clearly there is a $0.5n$ step distance lower bound for the problem; we combine this with restrictions imposed by bandwidth constraints to argue a lower bound of $n - 3n/(c + 3)$ steps.

The next section defines the models we use and the problems under consideration. Section 3 contains an algorithm for sorting on a mesh with diagonal connections. Section 4 describes our algorithm for selection on a torus with diagonal connections. Next, we describe our algorithm that works optimally on average. Section 6 contains our lower bounds for sorting on meshes and tori. We conclude with a lower bound for selection on the mesh with diagonal connections.

2 Preliminaries

The $n \times n$ mesh-connected array of processors (or two-dimensional mesh) contains $N = n^2$ processors arranged in a two-dimensional grid without wrap-around edges. More precisely, it corresponds to the graph, $G = (V, E)$, with $V = \{(x, y) \mid x, y \in \langle n \rangle\}$ and $E = E' \cup E_1 \cup E_2 \cup E_3$. The grid edges are specified by $E' = \{((x, y), (x, y + 1)) \mid x \in \langle n \rangle, y \in \langle n - 1 \rangle\} \cup \{((x, y), (x + 1, y)) \mid x \in \langle n - 1 \rangle, y \in \langle n \rangle\}$, where $\langle n \rangle = \{1, \ldots, n\}$. The toroidal connections are specified by $E_1 = \{((n, i), (1, i)) \mid i \in \langle n \rangle\} \cup \{((i, n), (i, 1)) \mid i \in \langle n \rangle\}$. We will refer to a mesh with toroidal connections as a *torus*. The diagonal connections are specified by $E_2 = \{((x, y), (x + 1, y + 1)) \mid x \in \langle n - 1 \rangle, y \in \langle n - 1 \rangle\} \cup \{((x + 1, y), (x, y + 1)) \mid x \in \langle n - 1 \rangle, y \in \langle n - 1 \rangle\}$. The diagonal toroidal edges are specified by $E_3 = \{((n, i), (1, i + 1)) \mid i \in \langle n - 1 \rangle\} \cup \{((1, i), (n, i + 1)) \mid i \in \langle n - 1 \rangle\} \cup \{((n, n), (1, 1)), ((1, n), (n, 1))\}$. When we talk about a torus with diagonal connections, we always mean a mesh with all three additional sets of edges. (However a mesh with diagonal connections only has E_2 added in.) We are interested in meshes and tori with and without diagonal connections. MasPar's MP-1 is a torus with diagonal connections. Notice that putting in either the toroidal or the diagonal edges halves the diameter of the network, and putting in all three sets of edges reduces the diameter of the network to $n/2 - 1$.

The input to both of our problems is a set $X = \{x_1, \ldots, x_N\}$, the elements of which may be linearly ordered. An indexing scheme is a bijection from $\langle N \rangle$ to $\langle n \rangle \times \langle n \rangle$. The sorting problem on the mesh is: Given a set X, stored with one element per processor, and an indexing scheme, I, move the element of rank k in X to the processor labeled $I(k)$. The selection problem is: Given a set X, stored with one element per processor, an integer $1 \leq k \leq N$, and a specified processor (i, j), move the element of rank k in X to the processor labeled (i, j). In what follows we will consider only the case where the specified processor is labeled $(\lceil \frac{n}{2} \rceil, \lceil \frac{n}{2} \rceil)$, referred to as the middle processor below. It is generally straightforward how to modify the algorithm for the case of another designated processor.

2.1 Upper Bound Model

The upper bound model we use is the standard one used to describe algorithms for routing, sorting, and selection [8, 4]. During a single parallel communication step, each processor can send and receive a single packet along each of its incident edges, where a packet consists of at most a single element of X along with $O(\log N)$ bits of header information used for routing and counting purposes. Between communication steps, processors can store packets in their local queues, which are of bounded size. Furthermore, they can perform a constant number of simple operations (e.g., copying, addition, comparison) on the elements and the header information of packets.

Our algorithms for sorting and selection are randomized and therefore have some probability of failure. In this paper, *with high probability* means with probability at least $1 - n^{-\beta}$ for some appropriate constant β. To analyze such probabilities, we make extensive use of the following bounds for the tails of the binomial distribution.

Fact 1 *(Bernstein-Chernoff bounds) Let $S_{N,p}$ be a random variable having binomial distribution with parameters N and p. Then, for any h such that $0 \leq h \leq 1.8Np$,*

$$P(S_{N,p} \geq Np + h) \leq \exp\left(-h^2/3Np\right).$$

For any $h > 0$,

$$P(S_{N,p} \leq Np - h) \leq \exp\left(-h^2/2Np\right).$$

2.2 Lower Bound Model

We define in some detail the lower bound models we consider in this paper. Initially, each processor contains a single element of the set X. The elements cannot be replicated. At the end of the computation it is precisely these N elements which appear in sorted order on the mesh. The processors run in a synchronous manner but they may run different programs. During a single step, each processor may perform arbitrarily complex computations based upon the information it has received upto this step in the algorithm and may communicate with all four of its nearest neighbors. This communication may consist of elements of the set X along with an arbitrary amount of information concerning what the processor has knowledge of upto this step in the algorithm. However, at the end of the step each processor is restricted to contain at most c of the original N elements of X. The case $c = 1$ is the Schnorr-Shamir-Kunde model.

3 Selection on Meshes with Diagonal Connections

In this section, we outline an algorithm to perform selection on the $n \times n$ mesh with diagonal connections. Recall that allowing diagonal connections halves the diameter of the mesh, and hence the lower bound for selection at the middle processor in this case is $0.5n - 1$ steps. We show here that by using the diagonal connections, we are able to reduce the time to select on the mesh almost by half.

We adapt an algorithm of Kaklamanis *et al.* [4] that runs on a mesh without diagonal connections. Briefly, the algorithm is as follows. All packets are routed inside a diamond, which we call the *middle* diamond. At the same time, using sampling techniques, *bracketing elements* are computed by the center processor, and are broadcast to all processors in the middle diamond. With high probability, the number of elements that lie between the bracketing elements is $O(N^{1-\delta})$ for some $\delta > 0$, and the element of desired rank lies between the bracketing elements. All packets that lie between the bracketing elements are routed to a small block near the center. Since there are few of them, standard sorting algorithms can be used to sort them in $o(n)$ time. Once they are sorted and the exact ranks of the bracketing elements are computed, the element of desired rank can be identified and sent to the center. We summarize these steps for the case of selecting the median in Figure 1.

1. All packets are routed inside the middle diamond.
2. The center processor selects bracketing elements $b < b'$ and broadcasts these to all processors in the middle diamond. With high probability, there are $O(N^{1-\delta})$ input elements in the range $[b, b']$ and the median lies in this range.
3. All packets Q with $b \leq Q \leq b'$ are routed to a central block of side $o(n)$, and are sorted.
4. The ranks of b and b' are computed and are broadcast to all processors in the central block.
5. The element in the central block of rank $\lfloor N/2 \rfloor - rank(b)$ is the median. The processor with this element sends it to the center.

Fig. 1. High-level Algorithm Description

We prove the following theorem about meshes with diagonal connections:

Theorem 1 *There exists an algorithm to perform selection on the $n \times n$ mesh* *with diagonal connections that finishes in $0.65n + o(n)$ steps, and uses constant-size queues.*

Proof: As in the algorithm of Figure 1, we route a random sample into the middle block. Each packet chooses a random destination in the middle block. To get to its destination, the packet uses diagonal edges to get to the right row (column), and then uses row (column) edges to get to the destination. Clearly, there are no collisions in the diagonal edges, and the only collisions occur when two packets try to turn into a row (column) edge. Using an analysis analogous to [4], we can show with high probability, the number of packets that turn into any node in the mesh are bounded by a constant a. Thus, the routing takes at most $n/2 + o(n)$ steps. Next, we compute and broadcast the bracketing elements into the center sub-mesh of side $2r$. The diagonal edges enable us to do this in $r + o(n)$ steps.

Meanwhile, we overlap all the packets into the center sub-mesh of side $2r$. This must take at least $n/2 - r$ steps, because it takes a packet at a corner processor that many steps to get to the center processor. The center sub-mesh of side $2r$ has $24r$ edges entering it, and so only $24r$ packets can enter it any given time step. However, there are only $32r^2$ packets that are at a distance $2r$ or less from the boundary of the center sub-mesh of side $2r$, and therefore, only these packe ts can enter it in $2r$ steps. After $2r$ steps, we can conceivably push packets in using the maximum bandwidth, but this requires at least $(n^2 - 32r^2 - 4r^2)/24r$ steps. Thus, overlapping into the center sub-mesh of side $2r$ requires at least $max\{n/2 - r, n^2/(24r) + r/2\}$ steps. This bound can be achieved by ensuring a steady stream of packets after the first $2r$ steps. As in [4], we can do with constant size queues by overlapping into a sub-mesh of side $2r + \epsilon n$ for any $\epsilon > 0$. Those elements that lie in between the bracketing elements now move towards the center processor using an adaptation of the randomized algorithm for routing due to [11]; the only difference is that we use diagonal edges instead of column edges. It is easy to show that this takes $r + \epsilon n + o(n)$ steps. Once again, it is possible to do the overlapping simultaneously with the routing of the sample and the broadcasting. To obtain a good value for r, we equate $n^2/(24r) + r/2 = n/2 + r$ to get $r = n(1/\sqrt{3} - 1/2)$. The total time required by the algorithm is therefore $n/2 + 2r + \epsilon n = n/2 + (2/\sqrt{3} - 1 + \epsilon)n < 0.65n$ steps. $\square$

4 Selection on a Torus with Diagonal Connections

The diameter of the torus implies a $0.5n - 1$ step lower bound for any algorithm for selection on the torus. Owing to the regular interconnection structure of the torus, this bound holds for any location in the torus. Clearly, we can ignore the toroidal connections and use an algorithm for selection on a mesh with diagonal connections described above, thus giving us an upper bound of $0.65n$ steps for selection on the torus.

We show how to make use of the toroidal connections to propagate the random sample throughout the middle diamond. This enables us to eliminate the global broadcast of the bracketing elements, thus reducing the time taken by the algorithm. However, in the algorithm of [4], packets continued to be overlapped during the broadcast, and thus we were able to overlap into a smaller diamond. Since we now have less time for the overlapping, we are able to overlap into a larger middle diamond, but we still make significant gains by omitting the global broadcast.

We describe our algorithm briefly. Consider the mesh to be broken up into square blocks, each of size $n^{1-\delta/2}$. We restrict our attention to the sub-meshes that have a non-empty intersection with the middle diamond of side r, where $r = n/8 + \epsilon n$; there are $O(n^\delta)$ such blocks. Further, let B be the central block. As in [4], we select a random sample of size $S = \alpha N^{5\delta} \ln N$. To do this each processor selects itself with probability S/N. Each element that is at a selected processor picks a random destination in the central block B. We now route a copy of the sample to the *corresponding location in each of the blocks* inside the middle diamond. We sort the sample elements within *every* block in the middle diamond. Since these sorting operations are performed in disjoint sub-meshes that are of the same size as B, they can be done in parallel, and the time taken is $O(n^{1-\delta/2})$. Thus the identity of the bracketing elements is known at some processor within each block. Meanwhile, we overlap all the elements in the mesh into the middle diamond. We broadcast the bracketing elements within each block; this takes $o(n)$ steps. At this point, every element in the middle diamond can know the identity of the bracketing elements, and the global broadcast step in our mesh algorithm can be dispensed with. Now, each element that belongs to the middle bucket routes itself into the middle block; at the same time , we collect information to compute the global ranks of the sample elements. We finish by sorting the elements in B in order to determine the element of rank l in X.

It remains to be proved that a copy of the sample can be routed to every block in the middle diamond simultaneously with the overlapping operation in n steps. The rest of the algorithm is identical to [4], and the same analysis holds for the other steps.

Lemma 1 *In n steps, we can route a copy of the sample into every block inside the middle diamond of side $n/8 + \epsilon n$, using constant sized queues.*

Proof: Since the side of each block is $n^{1-\delta/2}$, we can divide the blocks into $n^{\delta/2}$ groups of blocks such that two blocks are in the same group if the corresponding processors in them belong to the same row of the mesh. The sample packet travels to the same corresponding destination within each block; in particular, for each group of blocks, its destination is in the same row. In other words, there are $n^{\delta/2}$ rows and columns that each packet is destined for. Each packet that is chosen to be a sample packet routes itself around the diagonal that it is in. Recall that this diagonal column is a ring, so there are no delays and any row or column can be reached in $n/2$ steps. When it reaches any of the destination rows or columns , it leaves a copy of itself and moves on. A copy of a sample packet now moves around the row or column (in both directions on the row ring) to reach the right destinations. Using Fact 1 we can show that there is a constant

a such that with high probability, not more than a packets turn into a given row. This implies that no packet can be delayed more than a steps in the second phase, and that the queue size does not exceed a. Since the maximum distance between any two processors is $0.5n - 1$, each sample packet is delivered to its destinations in $0.5n + o(n)$ steps. $\square$

Lemma 2 *The routing of the sample packets can be done simultaneously with the overlapping in $n + o(n)$ steps.*

Proof: Recall that overlapping into a square of side $2r$ can be done in $max\{0.5n - r, n^2/24r + r/2\}$ steps. In $0.5n$ steps, all the packets can be overlapped into a square of radius $0.09n$. We will give the sample packets priority over overlapping packets. In each row and each column, there are at most $o(n)$ sample packets, hence, they can only cause a delay of $o(n)$ for the overlapping packets. $\square$

This proves the following theorem:

Theorem 2 *There exists an algorithm to select the element of rank l at any processor of the $n \times n$ torus with diagonal connections in $0.59n$ steps, using constant sized queues, with high probability.*

5 Average-Case Selection on Meshes with Diagonal Connections

The randomized algorithms of the previous sections selected an element of a given rank in close to optimal time on all inputs, with high probability. In this section, we describe a deterministic algorithm for selection with a running time of $0.5n + o(n)$ steps on a random input, with high probability, on the $n \times n$ mesh or torus with diagonal connections. (By a random input we mean the input is chosen uniformly among all possible permutations. With minor adaptations the algorithm will work in the case where each element appears with equal probability at some processor independently at random.) Since any algorithm that correctly selects an element of specified rank requires at least $0.5n - 1$ steps on every input, the expected running time of the algorithm is optimal.

The key idea of the algorithm is that the elements in any $O(N^\epsilon)$ size block (for an appropriately chosen $\epsilon < 1$) of the network constitute a random sample of the input that can be used to eliminate all but a small fraction of the inputs before any routing is performed. The elements that pass the local test are routed to a $o(n)$-size block near the center where the actual selection is performed. We outline the steps of an algorithm for selecting the median at the center processor of a mesh with diagonal edges in Figure 2. It is straightforward to generalize to selecting an element of rank k. The same algorithm may be used on the torus.

We prove the following theorem concerning the above algorithm:

Theorem 3 *On a random input the above algorithm finds the median on the $n \times n$ mesh with diagonal connections in $0.5n + o(n)$ steps, using constant-size queues, with high probability.*

1. Sort the elements in square blocks of size N^ϵ. Within the ith block let b_i be the element of rank $N^\epsilon/2 - N^\delta$ and c_i the element of rank $N^\epsilon/2 + N^\delta$, for an appropriately chose $\delta < \epsilon$.
2. Route the elements between and inclusive of b_i and c_i in each block i to a central block (including the center processor) of size $O(N^{1-\epsilon+\delta})$.
3. Sort the elements in the central block.
4. Find $b = \max_i b_i$ and $c = \min_i c_i$.
5. Compute the ranks of b and c.
6. If the rank of b is less than $N/2$ and the rank of c is greater than $N/2$ then the median is found in the central block and is routed to the center processor. Otherwise, sort the entire input so that the median resides at the center processor.

Fig. 2. Average-case Algorithm Description

Proof: Clearly every element of the input between b and c is in the central block after step 2. It follows that the algorithm correctly selects the median. We now analyze its running time. The routing in step 2 is a fixed permutation that can be performed in $0.5n + o(n)$ steps using a greedy-like strategy. The sorting in steps 1 and 3 requires $o(n)$ steps using standard algorithms for mesh sorting [12]. Finding b and c (step 4) and computing their ranks (step 6) are both applications of parallel prefix requiring $o(n)$ steps [4]. It can be shown, for appropriate choices of ϵ and δ, each of the (b_i, c_i) pairs form a bracketing pair for the median. It follows that with high probability b and c bracket the median and that step 6 can be completed in $o(n)$ steps. Each of the steps requires only constant size queues. $\square$

6 Lower Bounds for Sorting

We now derive lower bound results for sorting on meshes and tori with diagonal connections. Clearly, the diameter of the mesh with diagonal connections is n steps, and that of the torus with diagonal connections is $n/2$ steps. These provide lower bounds on all the models that we are considering. On the Schnorr-Shamir-Kunde model of the mesh with diagonal connections, we are able to prove the following lower bound analogous to those in [6, 12]:

Theorem 4 *The worst case complexity of any algorithm to sort with respect to row-major order on an $n \times n$ mesh with diagonal connections on the Schnorr-Shamir model of the mesh is $2n - o(n)$ steps.*

Proof: Let T be the lower triangular region of side $\sqrt{n}$. Run the algorithm on input I for $n - \sqrt{n} - 1$ steps, where I is constructed from I_0 by exchanging the values at $(n, 1)$ and $(1, n)$, and adding the value n^2 to all the packets in T, thus making them the largest valued packets in the network. Let V be the value at the processor labeled $(1, 1)$ at this time. Change I to $I(z)$ by negating the z largest values in T. This pushes the rank of V up by z. Since the only difference between the inputs I and $I(z)$ is in T, we can show by induction that in $n - \sqrt{n} - 1$ steps, the processor $(1,1)$ will contain the same packet when the algorithm is run on these two inputs. There exists a value of z ($1 \leq z \leq n$) which moves the sorted location of V to the rightmost column (since the sorting order is row-major). By choosing this value of z to get $I(z)$, we can force the algorithm to take another $n - 1$ steps just to move the packet with value V to the rightmost column. This gives us a lower bound of $2n - o(n)$ steps. $\square$

By restricting the number of elements allowed at any processor at any time, we are able to show the following non-trivial lower bound:

Theorem 5 *The worst case complexity of any algorithm to sort on an $n \times n$ mesh with diagonal connections on a model where copying elements is not allowed, and each processor can hold at most c elements, is $f(c, n)$, where*

(i) $1 < c \leq 9 \implies f(c, n) \geq n + \frac{8cb^2 - 2cnb - n^2}{4cb - 2cn} - o(n)$ *steps, where $b = \frac{n}{2} - \frac{n}{\sqrt{8}}\sqrt{1 - 1/c}$.*

(ii) $9 < c \leq 18 \implies f(c, n) \geq 25n/24 + 3n/4c - o(n)$ *steps.*

(iii) $c > 18 \implies f(c, n) \geq n + n/c - o(n)$.

Proof: Given an algorithm, we will construct an input permutation such that the algorithm requires at least the claimed number of steps on that input. We prove the following three claims which together imply the theorem.

Claim 1 *Any algorithm for sorting on an $n \times n$ mesh with diagonal connections on a model where copying packets is not allowed, and each processor can hold at most c packets, where $c \leq 9$, takes at least $n + \frac{8cb^2 - 2cnb - n^2}{4cb - 2cn} - o(n)$ steps, where $b = \frac{n}{2} - \frac{n}{\sqrt{8}}\sqrt{1 - 1/c}$.*

Proof: Let D be the region defined by $n - \sqrt{2n} \leq i, j \leq n$. Let S_1 be the center (square) sub-mesh of side $2x$ and let S_2 consist of all the processors in the middle $2(x - y)$ columns. Let $S = S_1 \cup S_2$, and S' be all the remaining processors in the mesh (see figure 3). Since each processor can have at most c packets at it, the maximum number of packets that can be in S is $2cn(x - y) + 4cxy$. By choosing $y > \frac{n^2 - 2cnx}{4cx - 2cn}$, we can ensure that there is at least one packet outside of S in any time step.

Change input I_0 to I by exchanging the values at the processors labeled $(n, 1)$ and $(1, n)$. Clearly any algorithm must take $n - 1$ steps on the input I. Let the given algorithm run for $n/2 + x - \sqrt{n}$ steps on this input (y subject to the above constraint will be specified later). There must be a packet in S' at this point in time. Without loss of generality, let it be in the quadrant Q_1 and let V be the value of this packet. Since copies are not allowed, and the values are all distinct, this value does not exist anywhere else in the network. Based on the value of V, we construct a new input $I(z)$ by first adding n^2 to all the values in D and then negating the z highest values in D. Re-run the algorithm on the input $I(z)$. Since the inputs I and $I(z)$ are identical except for the values

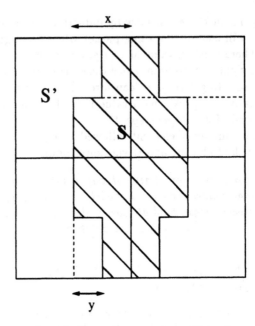

Fig. 3. Sorting on a mesh with diagonal connections: Claim 1.

in D, we can show that in $n/2 + 2x - y - \sqrt{2n}$ step s, the processors contained in $S' \cap Q_1$ contain the same packets when the algorithm is run on the inputs I and $I(z)$. In particular, the packet with value V is in the region $S' \cap Q_1$ at this time, when the algorithm is run on input $I(z)$. Since the indexing function is row-major, there exists a z $(1 \leq z \leq |D| = n)$ such that by negating z of the values in D, we can change the rank of V so as to make it move to the rightmost column. On this input, the algorithm takes at least an additional $n/2 + x - y - 1$ steps simply to move the value V to the rightmost column. This gives us a lower bound of $n + 2x - y - o(n)$ steps. By solving to maximize the value of this lower bound, we obtain the value $x = b$ as specified above. Note that for $c > 9$, the restrictions on y specified above are not satisfied. □

Claim 2 *Any algorithm for sorting on an $n \times n$ mesh with diagonal connections on a model where copying packets is not allowed, and each processor can hold at most c packets, where $c \leq 18$ takes at least $25n/24 + 3n/4c - o(n)$ steps.*

Proof: The proof is similar to that in [5]. Define D to be the lower triangular region of side $2\sqrt{n}$. Define S_1 to be the middle diamond of side x and S_2 to consist of all the processors in the middle $2(x - y)$ columns (see Figure 4). Let $S = S_1 \cup S_2$, and S' be all the remaining processors in the mesh. Since each processor can have at most c packets at it, the maximum number of packets that can be in S is $2cy^2 + 2cn(x - y)$. By choosing $x < \frac{n}{2c} - \frac{y^2}{n} + y$ we can ensure that there is at least one packet outside of S at any step. Change input I_0 to I by exchanging the values at the processors labeled $(n, 1)$ and $(1, n)$. Clearly any algorithm must take $n - 1$ steps on the input I.

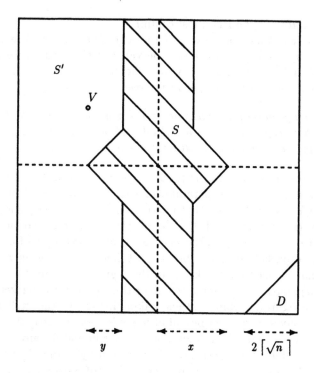

Fig. 4. Sorting on a mesh with diagonal connections: Claim 2

Let the given algorithm run for $n/2 + x/2 - \sqrt{n}$ steps on this input ($1 \leq x < \frac{n}{2c} - \frac{y^2}{n} + y$ will be specified later). ¿From the above argument, there must be a packet in S' at this point. Without loss of generality, let it be in the top left quadrant (Q_1) and let V be that packet. Since copies are not allowed, and the values are all distinct, this value does not exist anywhere else in the network. Based on the value of V, we will construct a new input $I(z)$, by adding the value n^2 to all the elements in D and then negating the z largest inputs in the region D. We now re-run the algorithm on input $I(z)$. Since the only difference in the inputs I and $I(z)$ is in the packets contained in D, we can show by induction that at step t, only the processors that are less than $t + \sqrt{n}$ steps away from the processor (n, n) can contain different packets when the algorithm is given I and $I(z)$ as inputs. Therefore, at the end of $n/2 + x/2 - \sqrt{n}$ steps, the same processors in the region $S' \cap Q_1$ contain the same packets when the algorithm is run on inputs I and $I(z)$. In particular, the packet with value V is in the region $S' \cap Q_1$ after $n/2 + x/2 - \sqrt{n}$ steps after running the algorithm on input $I(z)$. Since the indexing function is row-major, there exists a value of z ($1 \leq z \leq | D | = 2n$) which changes the rank of V such that V has to move to the rightmost column (either in the same row or in the next row). By using this value of z, we can force the algorithm to take an additional $n/2 + x - y - 1$ steps simply to move the packet V across to the rightmost column.

The worst case time complexity of the algorithm is therefore $n + 3x/2 - y - o(n)$, and by solving to maximize the value of this bound, we obtain $x = n/2c + 5n/36$ and $y = n/6$. This gives us the claimed lower bound. Note that for $c > 18$, the conditions on x specified above are violated. □

Claim 3 *Any algorithm for sorting on an $n \times n$ mesh with diagonal connections on a model where copying packets is not allowed, and each processor can hold at most c packets, where $c > 0$ takes at least $n + n/c - o(n)$ steps.*

Proof: The proof is similar to the proof of Claim 1 above. For any given algorithm, we will design an input permutation such that the algorithm will require at least $n + n/c - o(n)$ steps on that input. Define S to be the region consisting of the $2x$ innermost columns. Let S' be all the processors that do not belong to S (see figure 5). Since each processor can hold a maximum of c packets, the total number of packets in S can be at most $2cnx$. By choosing $x < n/2c$, we can ensure that at least one packet is in the region S' at any step. Here, after $n/2 + x - o(n)$ steps, we can force the algorithm to push the packet in $S' \cap Q_1$ to move to the rightmost column, thus making it take another $n/2 + x$ steps to complete. This gives us a lower bound of $n + n/c - o(n)$ steps. □

Finally, we note that Claim 3 gives us a lower bound for any value of c but for some small values of c, better bounds are given by Claims 1 or 2. For $c = 1$, the best bound is given by Theorem 4. For values of c between 2 and 9, Claim 1 gives the best bound, and for values of c between 9 and 18, Claim 2 gives a better bound than Claim 3. □

Thus, there is no distance-optimal algorithm for sorting on a mesh when the processors have constant size queues and copying packets is not allowed. The case $c = 9$ gives us a surprising new result about the relative complexities of sorting and routing. The above theorem shows that sorting on the mesh with diagonal connections must take at least $1.166n$ steps when copies are not allowed. Recent results due to Kunde *et al.* [7] show that routing on the same network can be done using at most 9 packets at every processor in $1.11n$ steps. This implies that routing is easier than sorting when copies are not allowed, and the queue size is restricted to be at most 9.

On a torus with diagonal connections, for any bound on the queue size, any algorithm may need twice as much time as the distance bound, if copying packets is not allowed, as shown below.

Theorem 6 *The worst case complexity of any algorithm to sort on an $n \times n$ torus with diagonal connections with any bound on the queue size, in a model where copying is not permitted is $n - o(n)$ steps.*

Proof: Given an algorithm for sorting on a $n \times n$ torus, we will construct an input such that the algorithm requires at least $n - o(n)$ steps on that input. Change I_0 to I by exchanging the values at the processors $(n/4, 3n/4)$ and $(3n/4, n/4)$.

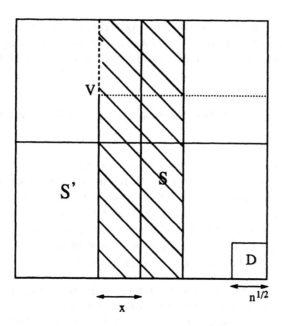

Fig. 5. Sorting on a mesh with diagonal connections: Claim 3.

Clearly the algorithm needs to run at least for $n/2 - 1$ steps on this input. Run the algorithm for $n/2 - \sqrt{n}/2 - 1$ steps on I. At this point, there is at least one processor in the network that has a packet at it. Let P be the processor, and V be the packet at it. Re-label the torus so that P is the middle processor. Let C be the region made up of the triangular corners (according to the new labeling) in all the 4 quadrants, each of side $\sqrt{n}/2$. Add n^2 to each input value at a processor in C. This ensures that every value in C is larger than any value outside C. Based on the value of V, we now negate z of the values in C so that the rank of V is moved up by z; call this input $I(z)$.

Re-run the algorithm for $n/2 - \sqrt{n}/2$ steps on $I(z)$. Since the inputs I and $I(z)$ are identical except for the values in C, and all the packets in C are more than $n - \sqrt{n}/2 - 1$ steps away from the center processor, we can prove (using induction) that the center processor contains the same packets when the algorithm is run on inputs I and $I(z)$. Thus, the packet with value V is at the center processor after $n/2 - \sqrt{n}/2 - 1$ steps of the algorithm on input $I(z)$. Since the indexing function is row-major, there exists a value of z ($1 \leq z \leq n < |C| = 2n$), which changes the rank of V so that V has to move across to the rightmost column. Using this value of z, we can force the algorithm to take an additional $n/2 - 1$ steps to simply move the value V to the rightmost column. Therefore the algorithm takes at least $n - \sqrt{n}/2 - 2$ steps on input $I(z)$. This gives us a lower bound of $n - o(n)$ steps. $\square$

This immediately implies a bound on any algorithm for sorting on the torus on any model when the queue size is bounded, but copies are not allowed. When replicating packets is allowed, the above argument falls through. The best known lower bound in any model where replicating packets is allowed, is due to distance.

Recent work by Kunde *et al.* [7] shows algorithms for $h-h$ sorting on a torus with diagonal connections that take $hn/10$ steps when $h \geq 10$ and $n+o(n)$ steps when $h \leq 10$. Our lower bound result shows that the latter result is optimal.

7 A Lower Bound for Selection on a Mesh with Diagonal Connections

In this section, we will prove a lower bound for selecting the element of rank k out of cn^2 elements, initially placed c elements at every processor of an $n \times n$ mesh with diagonal connections. The selected element is to be moved to the middle processor. Clearly, any algorithm must take $n/2 - 1$ steps on an input where the element of rank k is at the processor labeled $(1,1)$, simply to move the element to the middle processor in a mesh with diagonal connections. The bounds we derive are based on the edge bandwidth.

Theorem 7 *The worst-case complexity for any algorithm for selection on an $n \times n$ mesh with diagonal connections, where initially each packet has c packets at it, on a model of computation where there is no bound on the size of the queues, and replication of packets is allowed, is at least $n - 3n/(c+3) - o(n)$ steps.*

Proof: Define D to be the lower triangular region of side n(the processors (i,j), where $n + 1 \leq i + j \leq 2n$); let D' be the remaining processors. Let I_0 be an input consisting of the values $\{1, 2, \cdots, cn^2\}$, where (roughly) every odd value is in D and every even value is in D'. Now change I_0 to I by moving the value k to the processor labeled $(1,1)$. This ensures that any algorithm must take at least $n/2 - 1$ steps on input I. Define T to be triangular corner C_1 of side x in the upper left quadrant. Each processor has c packets at it initially; there are therefore $cx^2/2$ packets in T initially. There are only $3x$ edges connecting the region T to the rest of the mesh, and so at the end of $cx/6$ steps there is at least one packet left in T.

Run the algorithm on the input I for $(n - x)/2 - 2$ steps. By equating $cx/6 = (n - x)/2$, we get the value $x = 3n/(c + 3)$, and for this value of x, from the above reasoning, there is at least one packet left in T at the end of $(n - x)/2 - 2$ steps. Let V be the value of the packet left in T at this point. It is easy to see that there exists a z such that by changing z of the values in D, the value V is of rank k in the input. Choose this value of z, and let the changed input be $I(z)$. Re-run the algorithm on $I(z)$. Since the distance between D and T is at least $(n - x)/2 - 1$, and the only difference between inputs I and $I(z)$ lies in the packets in D, we can show that the processors in T contain the same packets in $I(z)$ as in I. Therefore, the packet with value V is in T after $(n - x)/2 - 2$ steps of running the algorithm on $I(z)$. But since it has rank k in the input $I(z)$, the algorithm is forced to take an additional $(n - x)/2 - 1$ steps simply to move the it to the middle processor. This gives us a lower bound of $n - x - 3 = n - \frac{3n}{c+3} - 3$ steps. $\square$

The above theorem tells us that the worst-case complexity of any algorithm for k-selection, where each element initially has 6 packets at it, is $2n/3$ steps, and 9 packets at it, is $3n/4$ steps. Notice, however, that we still cannot beat the distance bound for $c \leq 3$. Also, we cannot obtain a bound exceeding n.

References

1. A. Agarwal, B. Lim, D. Kranz, and J. Kubiatowicz. APRIL: A processor architecture for multiprocessing. In *Proceedings of the 17th Annual International Symposium on Computer Architecture*, pages 104–114, 1990.
2. A. Condon and L. Narayanan. Upper and lower bounds for selection on the mesh. Unpublished manuscript, 1993.
3. C. Kaklamanis and D. Krizanc. Optimal sorting on mesh-connected processor arrays. In *Symposium on Parallel Algorithms and Architecture*, pages 50-59, 1992.
4. C. Kaklamanis, D. Krizanc, L. Narayanan, and A. Tsantilas. Randomized sorting and selection on mesh-connected processor arrays. In *Symposium on Parallel Algorithms and Architecture*, pages 17–28, 1991.
5. D. Krizanc, L. Narayanan, and R. Raman. A lower bound for sorting on the mesh. Submitted for publication, 1993.
6. M. Kunde. Lower bounds for sorting on mesh-connected architectures. *Acta Informatica*, 24:121–130, 1987.
7. M. Kunde, R. Niedermeier, and P. Rossmanith. Faster sorting and routing on grids with diagonals. In Symposium on Theoretical Aspects of Computer Science, 1994.
8. F. Leighton, F. Makedon, and I. Tollis. A $2n - 2$ step algorithm for routing in an $n \times n$ array with constant size queues. In *Symposium on Parallel Algorithms and Architecture*, pages 328–335, 1989.
9. D. Lenoski, J. Laudon, K. Gharachorloo, A. Gupta, and J. Hennessy. The directory-based cache coherence protocol for the DASH multiprocessor. In *Proceedings of the 17th Annual International Symposium on Computer Architecture*, pages 148–159, 1990.
10. S. L. Lillevik. Touchstone program overview. In *Proceedings of the 5th Distributed Memory Computing Conference*, Charleston, SC, April 9-12 1990.
11. S. Rajasekaran and T. Tsantilas. Optimal algorithms for routing on the mesh. *Algorithmica*, 8:21–38, 1992.
12. C. Schnorr and A. Shamir. An optimal sorting algorithm for mesh-connected computers. In *Symposium on the Theory of Computation*, pages 255–263, 1986.
13. T. Suel. Nearly optimal deterministic sorting on mesh-connected arrays of processors. In preparation, 1993.

Work-Optimal Thinning Algorithm On SIMD Machines

Ubéda Stéphane

Laboratoire d'Informatique Théorique
Ecole Polytechnique Fédérale de Lausanne
IN-Ecublens, CH-1015 Lausanne, Switzerland

Laboratoire de Traitement du Signal et Instrumentation
CNRS-URA 842, Facultés des Sciences et Techniques
23, rue du Docteur Michelon, 42023 St-Etienne CEDEX 2, France.

Abstract. We proposes a parallel thinning algorithm for binary pictures. Given an $N \times N$ image including an object, our algorithm computes in $O(N^2)$ the skeleton of the object, using a pyramidal decomposition of the picture. With the Exclusive Read Exclusive Write (EREW) Parallel Random Access Machine (PRAM), our algorithm runs in $O(\log N)$ time using $O(\frac{N^2}{\log N})$ processors. Same complexity is obtained using an SIMD hypercube. Both the PRAM and the Hypercube algorithms are work-optimal. We describe the basic operator, the pyramidal algorithm and some experimental results.

1 Introduction

The general problem of pattern recognition lies in efficiently extracting some distinct features from the pattern. The object of this paper is the thinning of an object in a binary image. The thinning transformation is a derivative of the medial axis transformation. Blum has described the medial axis transformation in the continuous plane as a shape descriptor allowing both data reduction and description [1].

It can be defined with the help of the *fire grass concept*, where the object is seen as a meadow. A fire is lit along its contour such as all fire fronts invade the object with the same speed. The medial axis of the object is the set of points reached by more than one fire front at the same time. The thinning transformation is the discrete implementation of the fire grass concept and its result is a *skeleton*, for which giving a simple and rigorous definition is a challenging problem [17].

Extracting a digital skeleton from a binary picture consists in removing, at each iteration, all the contour points except those belonging to the "axis" of objects and proceed until no more change occurs.

Thinning algorithms fall into two classes according to the way contour points are detected. Lets N be the width of the picture. *Area based* algorithms check each pixel of the picture, every iteration (i.e. $O(N^2)$ accesses to a pixel in the image). *Contour based* algorithms instead make, each iteration, a contour tracing of both the exterior and the holes [16] and check the entire contour (i.e. $O(N)$ operations).

The number of iterations a thinning algorithm has to perform in order to compute the skeleton is proportional to the maximum thickness of the object. The thickness of an object increases with the resolution of the digitization. The effective complexities of these two thinning approaches are respectively $O(N^3)$ and $O(N^2)$. However, area based algorithms use parallel operations while contour based algorithms are made of strongly sequential operations.

The goal of this paper is to design a new thinning algorithm preserving the "regularity" of area based algorithms while having no more time complexity than contour based algorithms. We also give the parallel complexity of our new algorithm for SIMD abstract machines.

1.1 PRAM and Hypercube models

From a theoretical point of view, many abstract models for parallel machines exist in the literature, and the PRAM (Parallel Random Access Machine) is by far the most preferred model for describing parallel algorithms. This preference stems from the fact that this model occupies a mid way between physical reality − where VLSI models stand −, and its complete abstraction − the parallel comparison model −, while being able to capture the intrinsic parallelism hidden in most problems [10].

A PRAM is viewed as a collection of processors synchronously working on a single instruction flow that comes from a control unit. Processors have random access to a shared memory, through which interprocessor communication is implemented. Both the size of the central memory (m) and the number of processors (p) are usually defined as functions on the input size of the problem, say n. The time complexity, say $O(f(n))$, of a parallel algorithm is measured in terms of the number $f(n)$ of parallel instructions performed. PRAM algorithms are said to be efficient if $f(n) = \log^k n$, for some constant k. An algorithm is considered work-efficient (respectively, work-optimal) if $p * f(n) = O(t)$, where t is the time complexity of the best known sequential algorithm (respectively, the sequential lower bound) for the problem.

Since processors are allowed to randomly access the shared memory, conflicts can occur while reading from or writing into memory positions. Different protocols exist for ruling memory access, and each one is the basis of a specific PRAM model. Therefore, from the weakest to the strongest, one finds the following.

- Exclusive Read Exclusive Write (EREW): concurrent accesses to the same memory position are disallowed in both read and write modes.
- Concurrent Read Exclusive Write (CREW): processors can read simultaneously from the same memory location, but concurrent access to the same memory position is disallowed in write mode.
- Concurrent Read Concurrent Write (CRCW): concurrent accesses to the same memory position are allowed in both read and write modes. The capability of simultaneously writing in the same memory location creates, again,

conflicts. Different resolving rules lead to different CRCW sub-models, and we could list more than a dozen of them [10].

Throughout this paper we shall use the weakest PRAM model, i.e., an EREW PRAM.

Because of technological constraints, concurrent access to all memory positions, when the number of processors becomes very larges, still seems difficult to be supported practically. Several Distributed Memory Parallel Computers (or *DMPC* for short) were conceived and produced. The interconnection topology characterizes each DMPC. The parallel algorithms for DMPC can not be well evaluated using the PRAM model. In [5] models of SIMD distributed memory parallel computers are presented. For instance, we are interested in the Hypercube Random Access Machine (or *HRAM* for short).

A HRAM is composed of $N = 2^d$ processing elements connected in a Hypercube topology, each having its own memory. As for the PRAM model, the time is discrete and at each time step, all the processors of the HRAM work synchronously. Each processor can access its own memory and memories of its neighbors in the Hypercube. As for the PRAM model, we consider different protocols for ruling memory access: EREW, CREW and CRCW.

Again throughout this paper we shall use the weakest HRAM model, i.e., an EREW HRAM.

1.2 New results

After a short survey, in Section 2, of existing parallel thinning algorithms, Section 3 presents a new parallel thinning operator that works on a 2×2 bloc of pixels. It computes the survival condition of 4 pixels, at iteration k according to the values of the bloc and of its neighbors at iteration $k - 1$. The application of a thinning operator must preserve connectivity of both the object and the background. This property is obtained using the *Euler Number* of the picture and it is discussed in 3.1.

With this operator we design a new parallel thinning algorithm preserving "regularity" while achieving reduction of the complexity. This reduction is obtained with a pyramidal decomposition of the picture and it is presented in 3.2.

In 3.3, resulting skeletons and execution times for both the pyramidal algorithm and a standard neighboring algorithm are presented.

Section 4 analyses the complexity of our parallel thinning algorithm according to two main abstract models of parallel machines. We show in 4.1 that for the PRAM model this algorithm is work-optimal and is running in $O(\log N)$ time. In 4.2 a similar result is given according to the hypercube SIMD model.

We close the paper with some concluding remarks and ways for further research.

2 Parallel thinning algorithms

A thinning algorithm takes as input an $N \times N$ binary image which contains one object (i.e. only one connected component of black pixels), and produces an $N \times N$ image with the skeleton of the original object. It is known that the skeleton produced by a thinning algorithm should satisfy a number of conditions [16] (listed below). Preserving these properties while using parallel processing is a challenging problem. In the following we review, first the characterization of the skeleton and second the design of parallel thinning algorithms.

2.1 Characterization of a skeleton

A digital skeleton is characterized by its topological properties. Topological analysis of such properties can be found in the literature [19]. In this section we simply enumerate those properties.

Be homotopic. It is a basic requirement that any thinning algorithm must preserve connectivity of both the object and the background.

Be significant of elongations. Skeletons are shape descriptors. Therefore, skeleton branches must be meaningful of the elongation of the original object. This condition is well used, although no formal definition for elongation could be found. It is usually expressed through the end points of the object, where an end point is a pixel having a single neighbor in the object.

Have some noise immunity. Since elongations are not formally defined, the condition above can hardly avoid distorsions on the skeleton. Undesirable branches can appear due to some noise in the contour which appears during an iteration of the thinning algorithm. Again, many applications and authors are concerned with immunity to noise.

Be 1-pixel thick. A skeleton is a set of curves. In order to be able to use these curves to their best, they must be as thin as possible.

Be isotropic. A shape descriptor has to be invariant with the position of the object in the image. That means that thinning must be invariant under translation and rotation. Notice that this condition does not hold in the digital plane, since rotations are no longer invariant transformations in the digital plane. Hence, the thinning process must delete pixels symmetrically.

2.2 Parallel area based algorithms

With the development of graphic work-stations, larger and larger pictures are now being processed. The computation time for such pictures is often prohibitive. Parallelism allows to decrease processing time. Many parallel thinning algorithms have been introduced in the literature. Both area based algorithms and contour based algorithms are proposed. This section describes four methods to design parallel area based thinning algorithm. Similar techniques can be used for the contour based algorithm class [11, 16].

First of all a parallel thinning operator must be defined. To be a parallel operation, the survival condition of a pixel at iteration k must be computed only with the values of its neighboring pixels at iteration $k-1$. The survival condition is usually computed with respect to neighbors at distance 1 in the 8 main directions in the picture (the 8-neighborhood). That operator is called 3×3 operator according to the size of the scanning window (or mask). Unfortunately, fully parallel 3×3 thinning operators have difficulties in preserving the connectivity of an image.

Pixels of the set A, B, C in Fig. 1 have congruant 8-neighborhoods. Thus, a parallel 3×3 thinning operator can not distinguish pixel A from pixels B, C. The deletion of non skeleton pixel in 1 by such a parallel thinning operator splits the original object into connected sets.

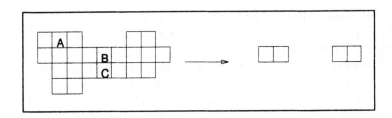

Fig. 1. Possible loss of connectivity with a parallel 3×3 operator.

The first solution is to extend the window dimension. Chin and Wan extend the size of the thinning mask [4]. An algorithm is proposed that uses two subiterations in each iteration with a 4×4 thinning window [18]. Holt presents a thinning algorithm using a 5×5 thinning mask and two subiterations per iteration [9].

A second solution is to serialize the algorithms partially by breaking a given iteration into distinct subiterations, each using different a operator or working on distinct subfields of the original picture.

To define (sub)operators, a common solution is to introduce a directional bias, for example, by favoring north over south and west over east. Distortion is minimized by introducing subiterations that differ only in the directions of the bias. In [3], a systematic approach is used to generate all the compatible couples of thinning suboperators.

The third method is a variation of the Arcelli approach which uses the notion of crossing number. The same operator is applied on two different subfields by partitioning the original image like a chessboard [8]. Some extensions of this idea are presented in [8]. Olszewski in suggests a solution using a 4 domains decomposition of the image [13].

The last parallelization technique is based on a recoding of the image [6]. During a preprocessing stage, each pixel of the picture is labeled according to the shortest distance to the background of the image. The thinning operator maintains this information and avoids loss of connection (without introducing subiterations).

3 Pyramidal thinning algorithm

In this section we describe a pyramidal thinning algorithm. An iterative algorithm is considered to be pyramidal if it reduces the amount of considered data by a constant factor at each iteration. We are interested in a thinning algorithm taking as input a $N \times N$ binary image at a given iteration and having a $\frac{N}{2} \times \frac{N}{2}$ image as a result.

First we must define our thinning operator and then illustrate how it can be used to iteratively reduce the scale of the current picture.

3.1 Basic operator

A thinning operator computes the survival condition of an object pixel in the current image at iteration k. The survival condition is a boolean value which decides whether the considered pixel will be removed from the picture at iteration $k + 1$. Olszewski has proposed a thinning operator based on the *Euler Number* [15]. This notion elegantly solves the homotopy requirement of our operator.

The Euler Number of a picture is equal to the number of object components minus the number of holes (i.e. background connected components which are not connected to the border of the image). This global parameter of the picture can be obtained by counting local configurations. In fact, the Euler Number G of an image is:

$$G = V + F - E$$

where V is the number of object pixels E the number of horizontally or vertically adjacent pairs of object pixels and F the number of complete 2×2 blocs within the object (see Fig. 2). Note that the Euler number define as above considers the object as 4-connected. It is common practice to compute the Euler number of the 4-connected background related to the 8-connected object.

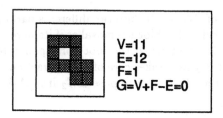

$$V=11$$
$$E=12$$
$$F=1$$
$$G=V+F-E=0$$

Fig. 2. computation of the Euler number.

In this section we present a new thinning operator. This operator takes as input a 2×2 pixel bloc from the image at iteration k, and computes the survival condition of the full block in the image at iteration $k + 1$. In [14] method to design thinning operators of any dimension (i.e. taking a $K \times K$ pixel bloc as input) are proposed. We restrict this idea to an operator having as input a 2×2

blocs. The operator take as input a 2×2 bloc and its neighboring pixels, i.e. the 12 pixels adjacent to a pixel in the bloc. The idea is to compute the variation of the Euler Number if the bloc is removed. In an $N \times N$ image where coordinates take value in $0..N-1$, all considered 2×2 blocs have the top left pixel with both coordinates even. The union of the (disjoint) blocs is the original image. Fig. 5 shows the 5 possible bloc configurations which contain at least one object pixel (a bloc without any object pixel is already removed !).

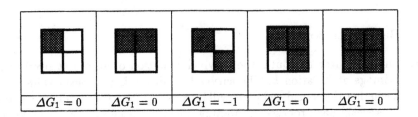

Fig. 3. ΔG_1 for the 5 possible blocs.

As defined for a single pixel size thinning operator, a bloc can be deleted if its removal do not change the Euler Number of the picture. As in [15] we consider the variation ΔG of the Euler Number considering background pixels and 4-connectivity. The Euler Number counting algorithm yields:

$$\Delta G = \Delta V + \Delta F - \Delta E$$

where ΔV (respectively ΔF and ΔE) is the number background pixels (respectively the number of 2×2 blocs of background and the number of horizontally or vertically adjacent pairs of background pixels) pixels added to the image by the removal of the bloc.

The computation of ΔG is split into two differents parts ΔG_1 and ΔG_2. ΔG_1 corresponds to the effect of the bloc's removal as if its were isolated in the image. Notice that $\Delta G_1 = 0$ except when the bloc corresponds to a pair of diagonal object pixels (see Fig. 3). ΔG_2 takes care of the interaction between the removed bloc and its neighborhood. We decompose ΔG_2 into ΔV_2, ΔF_2 and ΔE_2 (see Fig. 4 for some cases).

- $\Delta V_2 = 0$, there is no pixel removed.
- ΔF_2 is equal to the sum of :
    $\Delta F_{2.1}$, the number of horizontal (or vertical) pairs of adjacent background pixels, in the neighborhood excluding corners, and which are 4-adjacent to at least one object pixel in the bloc (there exist at most four such pairs, each along one side of the 2×2 bloc),
    $\Delta F_{2.2}$, the number of corner groups made of 3 background pixels in a right angle adjacent to an object pixel in the corresponding corner of the

bloc (there exist at most four such corner groups, each adjacent to a corner of the 2×2 bloc).

- ΔE_2 is equal to the number of background pixels in the neighborhood which are 4-adjacent to an object pixel of the bloc.

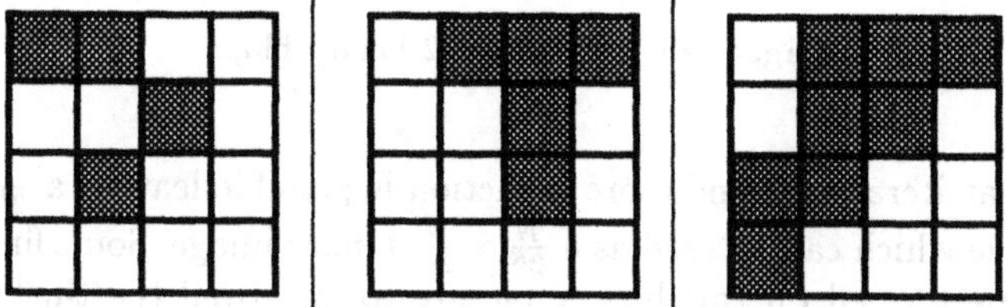

$\Delta G_1 = -1$	$\Delta G_1 = 0$	$\Delta G_1 = 0$
$\Delta F_2 = 4$	$\Delta F_2 = 3$	$\Delta F_2 = 3$
$\Delta E_{2.1} = 3$	$\Delta E_{2.1} = 2$	$\Delta E_{2.1} = 2$
$\Delta E_{2.2} = 2$	$\Delta E_{2.2} = 1$	$\Delta E_{2.2} = 0$
$\Delta G = 0$	$\Delta G = 0$	$\Delta G = -1$

Fig. 4. three examples of ΔG computation.

Our new thinning operator can be considered as a function of $\{0,1\}^{16}$ (the 2×2 bloc and its neighborhood) to $\{0,1\}^4$ (the resulting bloc) or as a function of $\{0,1\}^{16}$ to $\{True, False\}$ (the survival condition of the bloc). Let us show how this operator can be used to design a pyramidal thinning algorithm.

3.2 Pyramidal decomposition

An $N \times N$ binary picture may be considered as a $\frac{N}{2} \times \frac{N}{2}$ array of blocs. Each bloc can take values in $[0, 15]$ where 0 corresponds to the "empty" bloc of 4 background pixels (see Fig. 6). Our bloc thinning operator removes all contour blocs in the object except those belonging to an elongation. All $[1, 14]$ valued blocs are contour blocs. Some of the 15 valued blocs may be also contour blocs.

The initial picture can be interpreted as $\frac{N}{2} \times \frac{N}{2}$ 16-level image. Suppose at the first iteration all contour blocs are removed then the result is a $\frac{N}{2} \times \frac{N}{2}$ binary picture, thus enabling further reduction. We iterate until at least one $[1, 14]$ valued bloc remains in the picture (Fig. 5).

Suppose this occurs at iteration k. We have a $\frac{N}{2^{k+1}} \times \frac{N}{2^{k+1}}$ array I_k of blocs.

Let us call Q the set of blocs in I_k whose values belong to $[1, 14]$. To force a binary image (i.e. to reduce the set Q), any $[1, 14]$ valued bloc in Q may be replaced by a 15 valued bloc as long as topological conditions are preserved. If Q reduces to an empty set the iteration can continue.

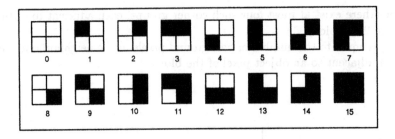

Fig. 5. all possible 2×2 binary blocs.

Suppose at iteration K no more reduction is possible leaving a $\frac{N}{2^{K+1}} \times \frac{N}{2^{K+1}}$ 16-level image which can be read as a $\frac{N}{2^K} \times \frac{N}{2^K}$ binary image. Some final thinning can yet be performed on this binary picture to improved the thickness of the resulting skeleton.

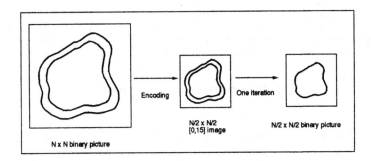

Fig. 6. an iteration of the pyramidal algorithm.

During the K first iteration we reduce each step the width of the current picture by a factor 2. The sequence of size decreasing picture can be viewed as a pyramidal data structure. The amount of processing on the elements of this truncated pyramidal structure has a bound of $\frac{4}{3}N^2$. As a consequence the number K of iterations is lower than $\log N$.

Let N^* be the width of the image at the lowest resolution reached. We want to suggest that N^* is shape dependent but independent of the initial resolution rate (i.e. the initial image width N).

Let us consider a critical situation at a given resolution rate will prevent further scale reduction (see Fig. 7). On Fig. 7 one can see a configuration leading the algorithm to stop. This configuration is characterized by the "strait" present in the continuous picture. The ratio of geometrical dimension of this strait with respect to average thickness (or width) of the object is directly related with N^*. This magnitude, a number without dimension, is a characteristic feature of the object.

Suppose that the initial resolution N produces a situation where the original

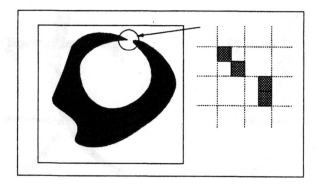

Fig. 7. the "strait", a local configuration prohibiting scale reduction.

strait is represented by 4 non adjacent pixels which belong to two adjacent blocs as in Fig 7. These two blocs cannot be removed since they represent extremities of the object nor can they be replaced by full object pixel blocs. In this case, $N^* = N$. Now suppose that initial resolution where N is twice N^* allows one step of scale reduction before the similar situation is reached. More generally if $N = 2^k N^*$ then k step of scale reduction are possible which shows that $N*$ is independent of the initial resolution. As a result the number of iterations is proportional to $\log N$, where N is the initial resolution rate.

Now let us consider the postprocessing part of the algorithm where thinning is performed at the final scale. This requires a number of operation is a cubic function of N^* and therefore independent of N.

3.3 Experimental results

The goal of this section is to prove that such an approach is efficient and that the obtained skeleton is well formed for subsequent pattern recognition applications. Our pyramidal algorithm and Olszewski's algorithm were implemented on a SIMD (Single Instruction Multiple Data) machine. Namely the MasPar MP-1 composed of $1,024$ processors connected as a $2D$ mesh.

First of all, let us compare the resulting skeleton of both algorithms. Fig. 8 presents both skeletons for a 128×128 Chinese ideogram. One can see that both skeletons are connected and included in the original object. Both skeletons are meaningful of the shape of the original object and both skeletons preserve extremity of elongated parts of the original shape. However skeleton branches are longer in the case of the pyramidal algorithm. Both skeletons are thick. The two skeletons differ in the preservation of symmetries. Olszewski's skeleton is closer to the medial axis of the original object while pyramidal skeletons are made of short broken lines. This is due to the dimensional reduction.

A skeleton is obtained, close to the medial axis of the object at the highest level of reduction. Then, this skeleton is projected onto the original image. A single pixel of the skeleton at highest reduction level became a short line in the

Area based algorithm   Pyramidal algorithm

Fig. 8. Resulting skeletons.

original image. So the resulting skeleton cannot be as close to the medial axis as the Olszewski skeleton. This distortion is less important for larger objects. If strong symmetry preservation is needed the pyramidal algorithm becomes inefficient. But for most pattern analysis post-processing the obtained pyramidal skeletons are meaningful enough. The last condition address noise immunity of the thinning operator. Noise is created because of very special configuration of pixels neighboring. For this reason it is difficult to say that an algorithm is noise immune or not. On our examples set of different sizes and shapes, both algorithms seem to be noise immune. To sum up, we can state that pyramidal algorithms create skeletons as good as the Olszewski algorithm except for a light loss of symmetries for thick objects.

Now let us consider the efficiency of the both algorithm. Our experimental results are compared according to 2 criteria: the number of iterations and the effective execution time.

The number of iterations is a standard efficiency measure of a thinning algorithm. This number is usually strictly proportional to the execution time and is machine-independent and implementation-independent.

One can see on figure 9 that the number of iterations of the pyramidal algorithm is a logarithmic function of the dimension of the input image. This is due to the pyramidal data reduction and it is detailed in the next Section.

Fig. 10 shows execution times of our algorithm compared with the execution time of the Olszewski algorithm.

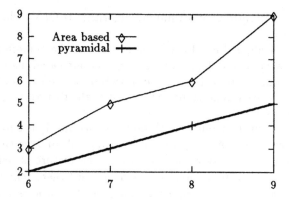

Fig. 9. Number of iterations for each algorithm as a function of log N

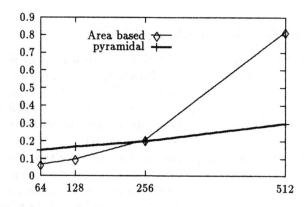

Fig. 10. Execution times for various dimension of a Chinese ideogram

4 Complexity evaluation

There exist two main families of thinning algorithms: area based methods, which apply a thinning operator all over the picture at each iteration and contour based methods, which apply a thinning operator only to the contour at each iteration. The complexity evaluation of thinning algorithms consists in the evaluation of the number of times the thinning operator is applied.

To obtain a skeleton the number of iterations needed can be expressed as a function of the image size. At each iteration, all contour pixels are erased or labeled "skeleton" by application of the thinning operator. The number of such shrinking needed to completely removed or labeled an object is proportional to the thickness of the object. Thickness is a function of the digitalization rate. So the number of iterations of a thinning algorithm is $O(N)$ while N is the width of the image.

Area based algorithms do N^2 applications of the thinning operator per iteration involving a complexity of $O(N^3)$. Contour based algorithms operate on the current contour at each iteration, invading the object until all object pixels have been processed. The complexity of such algorithms is equal to the area of the object i.e. $O(N^2)$.

Now let us consider our new thinning algorithm. It is similar to area based algorithms since the thinning operator is applied all over the picture at each iteration. But at each iteration, the size of the picture is reduced by a factor two. However such a reduction can be done until there remains but one pixel. $O(\log N)$ iterations are necessary to complete the process and the number of applications of operator is $O(N^2)$. This sequential complexity is the same as for the contour based method. However the pyramidal algorithm is made of parallel applications of a local operator. Therefore optimal parallel complexity can be expected.

4.1 On the (EREW) PRAM model

Let us consider the three presented thinning algorithms as if implemented on an Exclusive Read Exclusive Write PRAM machine. Each iteration of an area based thinning algorithm can be done by N^2 processors in a constant time. Thus a parallel area based algorithm can produce a skeleton in $O(N)$ time using N^2 processors. There exists a parallel contour based algorithm which produces a skeleton in $O(N)$ time using $O(N)$ processors [7].

Now let us consider the pyramidal thinning algorithm. Each iteration may be split into two independent parts: application of the operator and reduction of the scale. Both are made of parallel operators. Each pixel may be processed in parallel (considering subiteration rather than global iteration for the operator application). Taking $\frac{N^2}{\log n}$ processors, the first iteration may be achieved in $O(\log N)$; the second one is done in $O(\frac{\log N}{4})$; the ith iteration is completed in $O(\frac{\log N}{4^i})$. Suppose the pyramidal decomposition of the image is made until a single pixel remains, this is done in:

$$O\left(\log N \sum_{i=0}^{\log_4 N} \frac{1}{4^i}\right) = O(\log N)$$

We conclude that a pyramidal thinning algorithm may compute the skeleton of an N size object in $O(\log N)$ with $\frac{N^2}{\log N}$ processors on an (EREW) PRAM.

4.2 On the Hypercube model

Adapting a contour based algorithm on a distributed memory parallel machine is a challenge. The contour thinning algorithm presented can not be directly be used for a distributed architecture [7]. Using a grid of N^2 processors (the 8 neighbors of a pixel can be accessed in a constant time), the parallel contour based algorithm and the parallel area based algorithm become almost identical . They both perform in $O(N)$ time.

As for area based methods pyramidal algorithms are made of local operations in the picture. At first sight, one may think that processors interconnected as a mesh can be sufficient to exhibit maximal parallelism. However, as suggested by the section title a mesh is not sufficient to preserve a logarithmic number of iterations.

To see this, we consider a hypercube network as well as its embedded two dimensional mesh [2]. The number of processors is fixed at $\frac{N^2}{\log N}$. The algorithm runs as described for the PRAM algorithm until there remains a single pixel in each processor. This occurs just before a data reduction stage, at the end of the $(\log_4 \log N)^{\text{th}}$ iteration. After the subsequent reduction stage, only a quarter of the processors are in charge of a pixel. To apply the thinning operator, neighbors of this pixel are needed and these neighbors are in processors at a distance 2 in the mesh. On each further iteration the distance in the mesh increases by a factor 2. Thus preventing the computation to terminate in $O(\log N)$ time.

One solution is to concentrate the image in the left higher quarter of the mesh after at each data reduction stage. This may be done in two steps. First concentration in made in every column of the mesh and then similarly for the rows (see Fig. 11). Let us consider a mesh embedded in a hypercube where columns and rows appear as subcubes [2]. Each step of the concentration operation can be done in $O(\log M)$ if M is the size of a row or a column [12]. Now the left higher quarter of the mesh is also a subcube of the hypercube. The initial size of a row (or a column) is less than N and it decreases by a factor 2 each iteration. So the sum of concentration operations during the algorithm is $O(\log N)$. The algorithm can be performed in $(\log N)$ combining the two concentration steps in each iteration after the $(\log_4 \log N)^{\text{th}}$.

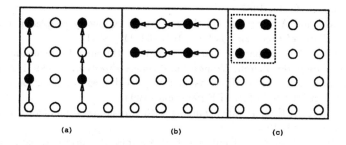

Fig. 11. concentration algorithm in 2 steps.

5 Conclusion

We present a new thinning algorithm for parallel machines whose optimality is based on theoretical considerations and is justified by practical applications.

This algorithm produces skeletons in accordance with standard topological theoretical criteria. However the main progress over previously published algorithms is concerned with the parallel complexity. Like the best contour based algorithms, it has a sequential cost of $O(n^2)$, but it uses a full parallel operator like the best area based algorithms. Area based methods make $O(N)$ iterations while our new algorithm performed only $O(\log N)$ iterations due to its pyramidal structure. We prove that this pyramidal thinning algorithm is work-optimal on a EREW parallel machine as well as on a EREW Hypercube parallel machine.

Most skeletons produced by the new method have branches that seem to be snaked around the exact medial axis of the shape and introduce a loss of accuracy. However the obtained skeleton is meaningful enough for most of pattern recognition applications ; moreover, this apparent drawback can be reduced by an improvement of the projection technique. In fact, it produces a skeleton in multiple levels of resolution, a technique which becomes a common feature of modern pattern recognition applications.

The experimental results show good speed for not too small data ; for large data structures it is faster than the specific Assembler Code Machine subroutine of our target machine.

References

1. H. Blum. *A Transformation For Extracting New Descriptors.* Symp. On Models for perception of speech and visual form, MIT Press, 1964.
2. J. Brandenburg and D. Scott. Minimal mesh embeddings in binary hypercubes. *IEEE Trans. Comp.*, 37(10):1284–1285, 1988.
3. Y.S. Chen and W.H. Hsu. A systematic approach for designing 2-subcycle and pseudo-subcycle parallel thinning algorithms. *Pattern Recognition*, 22:267–282, 1989.
4. R.T. Chin, H.K. Wan, D.L. Stover, and R.D. Iverson. A one-pass thinning algorithm and its parallel implementation. *Computer Vision, Graphics and Image Processing*, 40:30–40, 1987.
5. M. Cosnard and A. Ferreira. On the real power of loosely coupled parallel architectures. *Parallel Processing Letters*, 1(2):103–111, 1991.
6. A. Favre and H.J. Keller. A parallel syntactic thinning by recording of binary pictures. *Computer Vision, Graphics and Image Processing*, 23:99–112, 1983.
7. A. Ferreira and S. Ubéda. Ultra-fast contour tracking with application to thinning. Research Report 87, LITH/EPF Lausanne, 1993.
8. Z. Guo and R.W. Hall. Parallel thinning with two-subiteration algorithms. *Comm. ACM*, 32:359–373, 1989.
9. C.M. Holt, A. Stewart, M. Clint, and R.H. Perrott. An improved parallel thinning algorithm. *Communications of the ACM*, 30:156–160, 1987.
10. R. Karp and V. Ramachadran. A survey of parallel algorithms for shared-memory machines. Technical report ucb/csd 88/408, university of california, Computer Science Division, 1988.

11. P.C.K. Kwok. A thinning algorithm by contour generation. *Comm. ACM*, 31:1314–1324, 1988.
12. D. Nassimi and S.S. Sahni. Data broadcasting in simd computers. *IEEE Trans. on Comp.*, 30(2):101–107, 1981.
13. C. Neusius, J. Olszewski, and D. Scheerer. A flexible thinning algorithm allowing parallel, sequentiel and distributed application. *Pattern Recognition*, 18:47–55, 1992.
14. L. O'Gorman. $k \times k$ thinning. *Computer Vision, Graphics and Image Processing*, 51:195–215, 1990.
15. J. Olszewski. A flexible thinning algorithm allowing parallel, sequentiel and distributed application. *ACM Trans. on Mathematical Sofware*, 1990.
16. T. Pavlidis. A thinning algorithm for discrete binary images. *Computer vision and image processing*, 20:142–157, 1980.
17. C. Ronse. A topological characterization of thinning. *Theoretical Computer Science*, 43:31–41, 1986.
18. S. Suzuki and K. Abe. Binary picture thinning by an iterative parallel two-subcycle operation. *Pattern Recognition*, 20:297–307, 1987.
19. H. Tamura. A comparison of line thinning algorithms from a digital geometry viewpoint. In 4th *International Conference on Pattern Recognition (Kyoto,)*, volume 1, pages 715–719, 1978.

APPENDIX: the pyramidal data structure

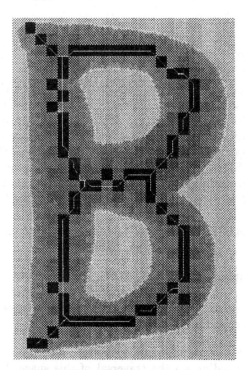

Example of the projection of the resulting skeleton in the original image.

An Efficient Implementation of Parallel A*

Van-Dat Cung[1][2] and Bertrand Le Cun[1][2]

[1] Laboratoire PRiSM, Université de Versailles - St. Quentin en Yvelines. 45, Avenue des États-Unis, F-78000 Versailles, France.
E-mail: van-dat.cung@prism.uvsq.fr, bertrand.lecun@prism.uvsq.fr
[2] INRIA-Rocquencourt, Domaine de Voluceau Bât.17, B.P. 105 F-78153 Le Chesnay Cedex, France. E-mail: Van-Dat.Cung@inria.fr, Bertrand.Le_Cun@inria.fr

Abstract. This paper presents a new parallel implementation of the heuristic state space search A* algorithm. We show the efficiency of a new utilization of data structure the *treap*, instead of traditional priority queues (heaps). This data structure allows operations such as *Insert*, *DeleteMin* and *Search* which are essential in the A* algorithm. Furthermore, we give concurrent algorithm of the treap within a shared memory environment. Results on the 15 puzzle are presented; they have been obtained on two machines, with virtual or not shared memory, the KSR1 and the Sequent Balance 8000.

Keywords : Heuristic search, A*, data structure, binary search tree, priority queue, parallelism, concurrence.

1 Introduction

Search is a technique widely used in Artificial Intelligence (AI) and Operational Research (OR) for solving Discrete Optimization problems [18, 17, 20, 27]. The space of potential solutions of these problems can be specified, but the difficulty is that its cardinality is too large to be enumerated (time exponential in the size of the problem instance). Such problems are Combinatorial Optimization problems (as Traveling Salesman Problem, Quadratric Assignment Problem, ...) or Artificial Intelligence games (as 15 puzzle, ...). Indeed, many of these problems have been classified as NP-complete and search is one of the few available means for solving them.

The space of potential solutions of these problems is generally defined in AI in terms of state space. A state space is defined by :

1. an initial description of the problem called *initial state*,
2. a set of operators that transform one state into another,
3. a termination criterion which is defined by the properties that the solutions or the *set of goal states* must satisfied.

This state space is often huge but not untractable. Thus, parallelism is logically an idea for speeding up the traversal of this space. It could reduce the searching time and therefore increase the size of problems solved. Clearly, if many processors are available, then they can search different parts of the space concurrently. Research in this field is very active [11, 9, 14, 21, 7].

However, a straightforward parallelization of state space search may not be efficient. For many problems, heuristic domain knowledge is available and is gathered during the traversal of a state space. This knowledge can then be used to avoid searching some useless parts of the state space. If the first denerated state are distributed between different processors, this may eventually do more work than a single processor if there is no communication between them . The extra work may definitively reduce the speedup that can be obtained by parallel processing.

In this paper, we propose a new asynchronous parallel implementation of the A* state space algorithm. It differs mainly from the few previous work [11] in the data structures used to implement the OPEN and CLOSED lists of the A* algorithm, respectively a priority queue plus a search tree and a hash table. This implementation is done on shared memory architectures. New massively parallel machines like the KSR1 propose a *virtual* shared memory mechanism, it is interesting to test its effectiveness in the case of A* with respect to other parallel machines with a *classical* shared memory such as the Sequent Balance 8000.

2 Heuristic searches

The implicit representation of the state space is commonly defined as a weighted, directed graph. Each state is represented by a vertex and each transformation from one state to another is represented by a directed edge. Here after, we call without distinction a vertex for a state and vice-versa. The weight of an directed edge is the cost $c(v_i, v_j)$ of generating a new state v_j by applying the corresponding operator on state v_i. Let us denote v_0 the vertex corresponding to the initial state and v_n a vertex corresponding to a goal state. Thus, to find a optimal solution in the state graph is equivalent to find a least cost directed path from the initial state v_0 to the set of goal states.

2.1 The A* algorithm

The A* algorithm [18] is a well-known heuristic search for finding a least cost path between an initial state and goal states of a given state graph. The A* algorithm maintains two lists called OPEN and CLOSED. The OPEN list contains those vertices whose successors have not yet been generated, and the CLOSED list those vertices whose successors have been generated.

Each state in the graph is assigned a cost with an evaluation function f, the cost is denoted by f-value. The traversal strategy of the state graph is a *best-first* strategy according to the f-values. At each iteration of the algorithm, a state with the best f-value in the OPEN list is selected for expansion.

In the A* algorithm, the evaluation function f (cf. Figure 1) is the sum of two other functions g and h: $f = g + h$.

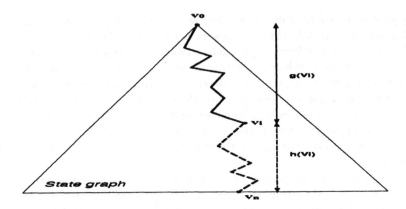

Fig. 1. Evaluation function f.

Definition 1. Let v_0 be the initial state, v_n a goal state and v_i a current state of the search. The functions g and h are defined as follows:

- $g(v_i)$ is the cost of the current path from the initial state v_0 to the current state v_i,
- $h(v_i)$ is a *heuristic* function which estimates the cost of the optimal path from v_i to a goal state v_n; if v_i is a goal state then $h(v_i)$ is equal to 0.

At the beginning of the search, OPEN contains only the initial state v_0 and its corresponding f-value $f(v_0)$. At each iteration, the A* algorithm selects the most promising state according to its f-value. This state becomes the current state v_i. If this state is a goal state or if no more state is available in OPEN, the algorithm would respectively terminate with v_i as solution or proclaim a failure. Otherwise, the current state is expanded by applying the operators which are defined on v_i. Once the state v_i has been expanded, it is removed from OPEN and added to CLOSED. For each successor v_j of v_i, if state v_j has not been generated before, it is added to the OPEN list. If state has been generated before and its new f-value is greater than its current f-value in OPEN or CLOSE then v_j is discarded; otherwise, the previous f-value of v_j is substituted with the new one. If the f-value of state v_j has been updated and v_j is in CLOSED, then transfer the state v_j from CLOSED to OPEN.

This search process continues until a goal state v_n is found or the OPEN list becomes empty.

Several properties of the A* algorithm can be pointed out. They are all founded on two assumptions:

1. the cost $c(v_i, v_j)$ of each edge from state v_i to state v_j is greater than ϵ strictly positive,
2. the heuristic function h is always positive.

Under these assumptions, three main properties [18, 20] hold; the first two concern the termination of the algorithm and the third one the update of states in the CLOSED list. We recall them briefly.

Property 1 : *The A* algorithm would terminate on finding a goal state if there is one (A* is complete).*

Property 2 : *If the heuristic function h is* admissible, *that is,*
if $h(v_i) \leq h^*(v_i)$ for all state v_i where $h^*(v_i)$ is the cost of the optimal path from state v_i to a goal state;
then the A* algorithm is guaranteed to terminate with a least cost path from the initial state to a goal state, if there is one (A* is admissible).

Property 3 : *If the heuristic function h is admissible and monotone, that is,*

$$0 < h(v_i) - h(v_j) \leq c(v_i, v_j), \quad \text{for all successors } v_j \text{ of } v_i,$$

then the values given by the evaluation function f are increasing,

$$f(v_i) \leq f(v_j), \quad \text{for all successors } v_j \text{ of } v_i;$$

The property 3 implies that no state in the CLOSED list will be updated and transfered to the OPEN list for future re-expansion. Under this condition, the delete operation in the CLOSED list becomes useless. However, we should not discard the CLOSED list, because we could re-expand some state already expanded and do redundant work.

On the performance evaluation aspect, the A* algorithm always maintains a search tree during the traversal of a state graph.

Definition 2. For a given search tree of a state space traversal, we call :

- *heuristic branching factor b*, the average number of successors, over all the search tree, that are generated by the application of an operator to a given state;
- *depth d*, the length or the number of the applied operators used to transform the initial state into a goal state of the least cost path.

It has been shown [10] that the average time complexity of A* is $O(b^d)$ and the average space complexity is also $O(b^d)$. These results lead us to think that parallelism may help us to speedup the search and by this way to solve problems of larger sizes.

2.2 The IDA* algorithm

The Iterative-Deepening A* algorithm [10] is essentially, on the opposite of the A* algorithm, a *depth-first* search procedure combined with the technique of iterative deepening. At each iteration, IDA* performs a bounded depth-first search of the state graph by using an admissible monotone evaluation function f.

This function is identical to that used by the A* algorithm. The depth-frist search discards a state when its f-value exceeds a given threshold. For the first iteration, this threshold is the f-value of the initial state. For each new iteration, the threshold used is the minimum of all state f-values that exceeded the previous threshold in the preceeding iteration. The algorithm continues until a goal state is expanded.

The IDA* algorithm is guaranteed to find an optimal solution if the evaluation function f is admissible and monotone. It has also been shown that the IDA* algorithm has a average space complexity $O(d)$ while the average time complexity still is $O(b^d)$. This algorithm seems to be better than A*, the space memory used is linear with respect to the search depth.

But the main drawback of the IDA* algorithm is that many states are re-expanded for thresholds which are less than the cost of the optimal path. Recent results [19] show that in the worst case, the IDA* algorithm generates $N(N + 1)/2$ *surely-expanded* states while the A* algorithm generates only N. A *surely-expanded* state v_i is defined by its f-value $f(v_i)$ which is less than the cost of the optimal path.

3 Parallelization of A*

Several parallelizations have been proposed for the Branch and Bound procedure with *best-first* traversal strategy (see [7] for a large survey). But at our best knowlegde, there are only a few specific parallel implementations of the A* algorithm [11, 5]. Moreover, recent works [25, 22, 9, 23] in this area seem to prefer parallelizing the IDA* algorithm to the A* one.

The reason of this preference is that the A* algorithm seems to be more difficult to parallelize than IDA*, from a point of view of simplicity and overheads [25]. Thresholds in the IDA* algorithm could be searched in parallel [9]. Whereas the only work to do in parallel in A* is the management of OPEN and CLOSED global lists. Furthermore, no suitable data structures for the OPEN and CLOSED lists have been designed in parallel formulations given in the literature.

A simplest parallelization of the A* algorithm is to let all available processors work on one of the current best state in the OPEN list, following an asynchronous concurrent scheme. Each processor gets work from the global OPEN list. This scheme has the advantage that it provides small search overheads, because global information are available for all processors via the OPEN list. This scheme is similar to the one proposed by Kumar and al and is particulary well suited for multiprocessors with shared memory.

However, parallel processing introduces two difficulties. First, the termination criterion of the A* algorithm does not work any more. We can not insure that the first goal state found is the optimal one. The optimal solution could be concurrently computed by another processor. Thus, we propose that termination occurs only after a goal state has been found by one processor, and when no any other processor has a better f-value state to expand.

Secondly, since the global OPEN and CLOSED lists are accessed asynchronously by every available processor, contention for the OPEN list will greatly limit the performances. But this drawback could be reduced if the data structures could be accessed concurrently. Thus we propose to use a *concurrent treap* (see paragraphs 4 and 5) for OPEN and a *hash table* for CLOSED.

We also introduce two improvements in the A* algorithm.

The first one is the *local best state* notion. After expanding a state, instead of adding all the successors in the OPEN list, each processor compares locally the f-value of each newly generated state with the one of the best state in OPEN (i.e. the root of the treap).

If the f-value of the best state in OPEN is better than the one of a newly generated state, this state is added to OPEN. Otherwise, this state is kept by the processor and becomes the local best state for this processor. This notion would save a lot of useless operations (Insert and DeleteMin) and accesses in OPEN, because successors of a state have often better f-values than the one of the best state in OPEN.

The second improvement concerns the tests in the CLOSED list. In the A* algorithm, before adding a new state in OPEN we have to verify if this state exists already in CLOSED. Otherwise, because of the asynchronous concurrent scheme, it may induce contention in the CLOSED list. Since a selected state from OPEN is added to CLOSED after its expansion, we propose to check CLOSED only when a state is selected for expansion. This will reduce the number of accesses to the CLOSED list in parallel.

We implemented this asynchronous concurrent scheme for solving the 15 puzzle [10] which could almost be considered as a benchmark for A*-like algorithms. The game consists of a 4x4 square frame containing 15 square tiles and an empty position. The operator slides any tile adjacent to the empty position into this position. The goal of the game is to rearrange the tiles from an initial configuration (state) into a desired goal configuration. The Figure 2 shows an example of a 3x3 puzzle.

As the previous works, we use the Manhattan distance heuristic for the function h. This distance is defined as follows:

Definition 3. Let A and B two positions of a square frame with coordinates (x_A, y_A) and (x_B, y_B), the Manhattan distance between A and B is

$$d(A, B) = |x_A - x_B| + |y_A - y_B|.$$

However, in case of equal f-values in OPEN, the state with the smallest value of h is selected for expansion [23]. Thus, we have modified a little bit the evaluation function f in order to take in account this tie-breaking rule. Our evaluation function is now $f = C(g + h) + h$, where C is a constant equal to

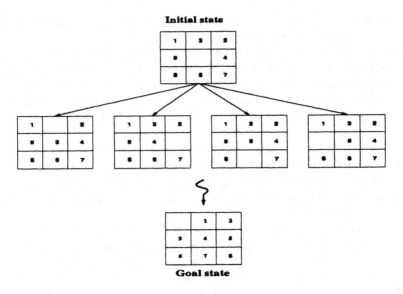

Fig. 2. 3x3 puzzle.

$10^{\lceil \log_{10}(2(n-1)(n^2-1)) \rceil}$, with n the size of a $n \times n$ square frame. For the 15 puzzle, the number n is equal to 4 and the number C is equal to 100.

Let us point out that the evaluation function $g + h$ is monotone, but it is no longer the case for $C(g + h) + h$. Thus, updated states in the CLOSED list could be transfered to the OPEN list for re-expansion.

4 Efficient data structures

In the A* algorithm, the operations we need for the OPEN list are essentially *Insert*, *DeleteMin* and *Search*. Those of the CLOSED list are *Insert*, *Delete* and *Search*. The *Insert* operations consist of adding an element in a list (OPEN or CLOSED), while the *Search* operations identify if an element already exists or not in one of the two lists.

As the A* algorithm uses the *best-first* search strategy, it explores in each iteration the best f-value state in the OPEN list. This is accomplished by the *DeleteMin* operation. The *Delete* operation removes a state from the CLOSED list. It is used when a updated state is transferred from CLOSED to OPEN.

Usually, the data structure used for OPEN is a priority queue (heap) and the one for CLOSED is a hash table. The hash table is relatively easy to implement and the access can be concurrent without any problem. In contrast, concurrent accesses to a priority queue are not simple to achieve.

Also, priority queues [1, 29] may be suitable for the *DeleteMin* operation, they are completely inefficient for the *Search* operation. Other data structures such as AVL-trees [1], splay-trees [6, 30], etc, are efficient for the *Search* operation but not for *DeleteMin*. There were no data structures with both operations. Moreover, the parallelism increases some specific problems as synchronization overheads, because operations have to be done in an exclusive manner.

In the next sections we discuss more precisely this data structure and how implement concurrent accesses to it.

5 Treap Data Structure

Treap was introduced by Aragon and Seidel [2]. The authors use them to implement a new form of binary search tree : Randomized Search Trees. McCreight [16] uses them also to implement a multi-dimensional searching. He called it *Priority Search Tree*.

Let X be a set of n items, a *key* and a *priority* are associated to each item. The keys are drawn from some totally ordered universe, and so are the priorities. The two ordered universes need not to be the same.

A *treap* for X is a rooted binary tree with node set X that is arranged in In-order with respect to the keys and in Heap-order with respect to the priorities [3].

In-order means that for any node x in the tree $y.key \leq x.key$ for all y in the left subtree of x and $x.key \leq y.key$ for all y in the right sutree of x. *Heap-order* means that for any node x with parent z the relation $x.priority \leq z.priority$ holds.

It is easy to see that for any set X such a treap exists. The item with the largest priority is in the root node.

The A* algorithm uses an OPEN list where a *key* (a state of the problem) and a priority (an evaluation of this state) are associated to each item. Thus, we can use a treap to implement the OPEN list of the A* algorithm.

Let T be the treap storing set X. The operations presented in the literature that could be apply on T are *Search*, *Insert* and *Delete*. We add one more operation *DeleteMin* and modify the *Insert* operation to implement the OPEN list of A*. Here are the definitions and the properties of our set of operations :

- *Search(T, key)* finds the item x in T such that $x.key = key$.
- *Insert(T, x)* adds the item x in T but $x.key$ must be unique in T.
  Let y be an item already in T such that $y.key = x.key$,
    • if $y.priority < x.priority$, x is inserted, y is removed,
    • if $x.priority \leq y.priority$, x is not inserted,
- *DeleteMin(T)* selects and removes the item x in T with the highest priority,
- *Delete(T, key)* removes the item x from T such that $x.key = key$,

---

[3] Vuillemin introduced the same data structure in 1980 and called it *Cartesian Tree*. The term *treap* was first used for a different data structure by McCreight, who later abandonned it in favour of the more commonly used *priority search tree* [16].

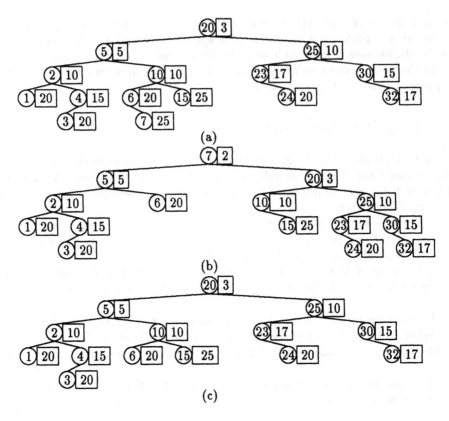

Fig. 3. (a)→(b) Insert(7,2) − (b)→(c) DeleteMin

5.1 Sequential operations

We explain the different sequential operations on the treap.

Given the key of x, an item $x \in X$ can be easily accessed in T by using the usual search tree algorithm.

As several binary search trees [28, 30, 6, 12, 4], the update operations use a basic operation called Rotation (Figure 4).

In the literature, the *Insert* operation is as follows. At first, using the key of x, attach x to T in the appropriate leaf position. At this point, the key of all the node of the modified tree are in In-order. To re-establish Heap-order, simply rotate x as long as its parent have a smaller priority.

To keep the properties of the *Insert* operation as defined above, the *Insert* algorithm can not be used in this form. We design a new algorithm which inserts an item x in T.

Using the key of x, we search the position, with respect to the In-order and the Heap-order. That is, we use the *Search* algorithm but it stops when :

− if an item z is found such that $z.key = x.key$,

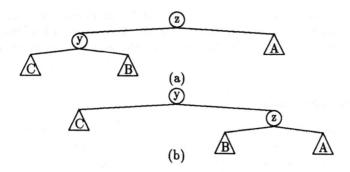

Fig. 4. Rotation used to reorganized the treap.

– if a item y is found such that $y.priority < x.priority$.

If such an item z is found, the algorithm ends, because it is garanteed that $x.prority < z.priority$, thus x must not be inserted in T. However, if such an item y is found, the item x is inserted at this position (between y and the father of y).

Let SBT_x be the subtree rooted in x of T. At this point, the priorities of all the node of the modified tree (SBT_x) are in Heap-order. To re-establish the In-order we use the splay operation [28, 30]. SBT_x is splitted in a Left subtree and a Right subtree. The left subtree (resp: right) contents all the items of SBT_x with the key smaller (resp: bigger) than the key associated with x. Finally left subtree and the right subtree are attached to x. Then, SBT_x and T are in In-order and in Heap-order.

If we find an item y such that $y.key = x.key$, during the splay operation, the item y, is deleted (the y priority will be smaller then x priority).

The *DeleteMin* and *Delete* operations are not very different. The *DeleteMin* operation removes the root of T, and the *Delete* operation removes the root x of a subtree of T such that $x.key = key$. Thus, first we search such a node x and we apply a *DeleteMin* on the subtree rooted in x.

The *DeleteMin* operation is achieved as follows. Let x be the root of the treap T. We rotate x down until it becomes a leaf (where the decision to rotate left or right is dictated by the relative order of the priorities of the children of x), and finally clip away the leaf.

Each node contains one item (a key and a priority), thus, the set occupies $O(n)$ words of storage.

The time complexity of each operation is proportionnal to the depth of the treap T. If the key and the priority associated with an item are in the same order, the structure is a linear list. However, if the priorities are independant, identically distributed continuous random variables, the depth of the treap is $O(logn)$ (the treap is a balanced binary tree). Thus, the expect time to perform one of these operations, is still $O(logn)$ (n number of node in T) [16].

To get an balanced binary treap, in a implementation for the A* algorithm, the problem is *reversed*. The priority order can not be modified. However, we can find an arbitrary bijective function to encode the ordered set of keys into a new set of randomized ordered keys. The priority order and the key order are then different.

5.2 Concurrent operations

In asynchronous parallel A* algorithm implemented on multiprocessors with shared memory, the exclusive use of the entire treap to perform a basic operation serializes the access to the treap. But the speed-up obtained with the parallel algorithm is limited.

Each operation on the treap manipulates the data structure in the same *Top-Down* direction and is made of successive elementary operations. We can use the technique denoted by **Partial Locking** [15, 4, 8, 24] to reduce the contentions. Each processor holds exclusive use of the smallest subset of needed items. Hence, the time delay, during which the access to the structure is blocked, is decreased.

The *tree partial locking protocol* uses the paradigm of *user view serialization* introduced by Lehman and Yao, 1981 [13], Calhoun and Ford, 1984 [3]. Every processor in concurrent implementation sees and modifies the data structure as if it could hold the entire tree excluding access. To hold an exclusive access on each node of the treap, a locking protocol or marking protocol (boolean like) is used.

If each processor respect a well ordering scheme to lock the node, this technique allows us to maintain consistency of the data structure and to avoid deadlock. All the proof and several details can be found in [15, 4].

The Treap implementation on a Sequent Balance use the partial locking with the marked protocol. On the KSR1, the subpage primitives are used to implement the Locking protocol on each node.

6 Empirical results

Here we discuss our experimental results on the KSR1 and the Sequent Balance 8000 in the context of the 15 puzzle problem.

The KSR1 is a massively parallel architecture machine which can scaled up to 1088 processors. Our machine have 32 processors, each of them has 32 megabytes of local memory. The local memories are shared between the processors through the ALLCACHE system which emulates a global shared memory. When a processor needs data which are not in its local memory, it asks the ALLCACHE system to search in and copy the corresponding data from the other processors.

The Sequent Balance 8000 is a classical parallel machine with respect to the KSR1. We have 10 processors and 16 megabytes of global memory. The memory is shared between processors via fast hardware locks. If a processor needs to write

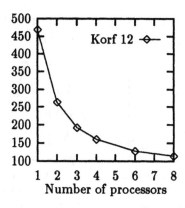

Fig. 5. Time results in second on the Sequent Balance 8000.

data in the memory, it first locks the corresponding cells for exclusive access. Thus, there is no copy of data in the global memory.

We selected four instances of the 15 puzzle presented in [10] according to their difficulties. The total number of states generated varied between approximatively 60 thousand and 1,5 million. The running times are presented in Figures 5 and 6. The figure 7 shows the corresponding speedups.

On the running time curves, we remark that times reduce quickly according to the number of processors used on every instance of the problem. This is verified for both machines. However, we have not been able to compute big instances on the Sequent Balance because of memory lack. Only the smallest instance has been computed.

On the same instance of the problem, we note that the KSR1 is faster than the Sequent Balance, but the speedup obtained by the second one is better (see Figure 7). The fact that the Sequent Balance has a real shared memory explains the latter results. The ratio communication speed over processor speed is greater on the Sequent Balance than on the KSR1.

We also observe a speedup anomaly for the instance Korf78 when 2 and 3 processors are used. This happens because in parallel A*, a goal state could be found earlier with several processors than in sequential [25, 26].

The speedup differences between the instances are essentially due to the sizes of the problem. The instance Korf78 is significantly large with respect to the others.

For the 15 puzzle problem, the expansion of a state and the evaluations of its successors need a few amount of time with respect to the access time on the data structures. That is the reason why speedup curves are not linear to the number of processors used. Applying this parallel A* algorithm on another

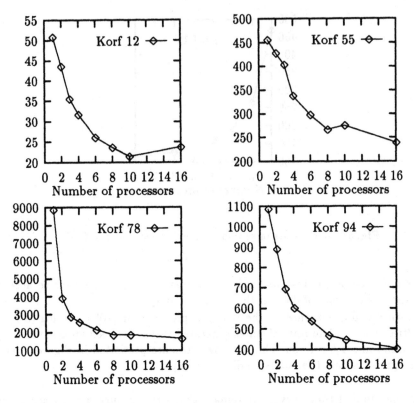

Fig. 6. Time results in second on the KSR1.

problem where the local computation time is greater, better speedups could be obtained.

7 Concluding remarks

We have presented in this paper a specific parallel implementation of the A* algorithm for the 15 puzzle problem on two shared memory machines.

A new utilization of data structure, the treap is proposed with the operations *DeleteMin* and *Search*. This data structure allows us to store a set of items containing a pair of key and priority. We can apply on the same data structure the basic operations of binary search trees and those of priority queues.

To have basic operations of binary search trees and priority queues applied to the same data structure is essential to implement efficiently the A* algorithm.

We have implemented concurrent access for these operations with the technique called Partial Locking. This technique could be easily apply to tree data structures.

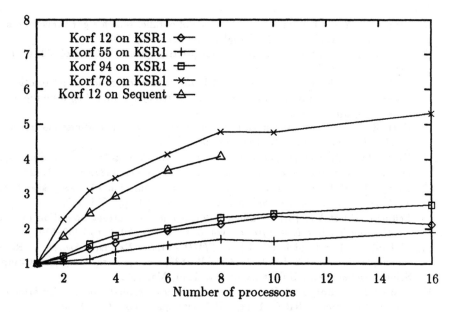

Fig. 7. Speedups.

Results obtained on both machines (KSR1 and Sequent Balance 8000) show that the treap is efficient for a group of ten processors. The speedups presented are as better as those in [25] with the same parallel scheme.

The difference between the speedups curves on both machines lets us conclude that the ratio communication speed over processor speed is greater on the Sequent Balance than on the KSR1. This leads us to implement currently a version of parallel A* algorithm with more data locality for the OPEN and CLOSED lists on the KSR1.

References

1. Aho (A.), Hopcroft (J.) et Ullman (J.). – *The Design and Analysis of Computer Algorithms.* – Addison-Wesley, 1974.
2. Aragon (C.) et R (G. S.). – Randomized search trees. *FOCS 30*, 1989, pp. 540–545.
3. Calhoun (J.) et Ford (R.). – Concurrency control mechanisms and the serializability of concurrent tree algorithms. *In: of the 3rd ACM SIGACT-SIGMOD Symposium on Principles of Database Systems.* – Waterloo Ontario, Avr. 1984. Debut de la theorie sur la serializability.
4. Cun (B. L.), Mans (B.) et Roucairol (C.). – *Opérations concurrentes et files de priorité.* – RR n° 1548, INRIA-Rocquencourt, 1991.
5. Cung (V.-D.) et Roucairol (C.). – *Parcours parallèle de graphes d'états par des algorithmes de la famille A* en Intelligence Artificielle.* – RR n° 1900, INRIA, Avr. 1993. In French.
6. Ellis (C.). – Concurrent search and insertion in avl trees. *IEEE Trans. on Cumputers*, vol. C-29, n° 9, Sept. 1981, pp. 811–817.

7. Grama (A. Y.) et Kumar (V.). – A survey of parallel search algorithms for discrete optimization problems. – Personnal communication, 1993.

8. Jones (D.). – Concurrent operations on priority queues. *ACM*, vol. 32, n° 1, Jan. 1989, pp. 132–137.

9. Kalé (L.) et Saletore (V. A.). – Parallel state-space search for a first solution with consistent linear speedups. *International Journal of Parallel Programming*, vol. 19, n° 4, 1990, pp. 251–293.

10. Korf (R. E.). – Depth-first iterative-deepening : An optimal admissible tree search. *Artificial Intelligence*, no27, 1985, pp. 97–109.

11. Kumar (V.), Ramesh (K.) et Rao (V. N.). – Parallel best-first search of state-space graphs : A summary of results. *The AAAI Conference*, 1987, pp. 122–127.

12. Kung (H.) et Lehman (P.). – Concurrent manipulation of binary search trees. *ACM trans. on Database Systems*, vol. 5, n° 3, 1980, pp. 354–382.

13. Lehman (P.) et Yao (S.). – Efficient locking for concurrent operation on b-tree. *ACM trans. on Database Systems*, vol. 6, n° 4, Déc. 1981, pp. 650–670.

14. Mahanti (A.) et Daniels (C. J.). – *SIMD Parallel Heuristic Search*. – Rapport technique n° UMIACS-TR-91-41, CS-TR-2633, College Park, Maryland, Computer Science Department, University of Maryland, Mai 1991.

15. Mans (B.) et Roucairol (C.). – *Concurrency in priority queues for branch and bound algorithms*. – RR n° 1311, INRIA-Rocquencourt, Oct. 1990.

16. McCreight (E. M.). – Priority search trees. *SIAM J Computing*, vol. 14, n° 2, Mai 1985, pp. 257–276.

17. Nau (D. S.), Kumar (V.) et Kanal (L.). – General branch and bound, and its relation to a* and ao*. *Artificial Intelligence*, vol. 23, 1984, pp. 29–58.

18. Nilsson (N. J.). – *Principles of Artificial Intelligence*. – Tioga Publishing Co., 1980.

19. Patrick (B. G.), Almulla (M.) et Newborn (M. M.). – An upper bound on the time complexity of iterative-deepening-a*. *Annals of Mathematics and Artificial Intelligence*, vol. 5, 1992, pp. 265–278.

20. Pearl (J.). – *Heuristics*. – Addison-Wesley, 1984.

21. Powley (C.), Ferguson (C.) et Korf (R. E.). – Parallel tree search on a simd machine. *In : The Third IEEE Symposium on Parallel and Distributed Processing*. – Déc. 1991.

22. Powley (C.) et Korf (R. E.). – Simd and mimd parallel search. *In : The AAAI symposium on Planning and Search*. – Mars 1989.

23. Powley (C.) et Korf (R. E.). – Single-agent parallel window search. *IEEE Transactions on pattern analysis and machine intelligence*, vol. 13, n° 5, Mai 1991, pp. 466–477.

24. Rao (V.) et Kumar (V.). – Concurrent insertions and deletions in a priority queue. *IEEE proceedings of International Conference on Parallele Processing*, 1988, pp. 207–211.

25. Rao (V. N.), Kumar (V.) et Ramesh (K.). – *Parallel Heuristic Search on Shared Memory Multiprocessors : Preliminary Results*. – Rapport technique n° AI85-45, Artificial Intelligence Laboratory, The University of Texas at Austin, Juin 1987.

26. Roucairol (C.). – *Recherche arborescente en parallèle*. – RR n° M.A.S.I. 90.4, Institut Blaise Pascal - Paris VI, 1990. In French.

27. Roucairol (C.). – Exploration parallèle d'espace de recherche en recherche opérationnelle et intelligence artificielle. *In : Algorithmique parallèle*, éd. par Cosnard (M.), Nivat (M.) et Robert (Y.), pp. 201–211. – Masson, 1992. In French.

28. Sleator (D.) et Tarjan (R.). – Self-adjusting trees. *In: 15th ACM Symposium on theory of computing*, pp. 235–246. – Avr. 1983.
29. Sleator (D.) et Tarjan (R.). – Self-adjusting heaps. *SIAM J. Comput.*, vol. 15, n° 1, Fév. 1986, pp. 52–69.
30. Tarjan (R.) et Sleator (D.). – Self-adjusting binary search trees. *Journal of ACM*, vol. 32, n° 3, 1985, pp. 652–686.

Constant-Time Convexity Problems on Reconfigurable Meshes

V. Bokka, H. Gurla, S. Olariu, J. L. Schwing

Department of Computer Science
Old Dominion University
Norfolk, VA 23529-0162
U.S.A.

Abstract. The purpose of this paper is to demonstrate that the versatility of the reconfigurable mesh can be exploited for the purpose of devising constant time algorithms for a number of important computational geometry tasks relevant to image processing, computer graphics, and computer vision. In all our algorithms we assume that one or two n-vertex (convex) polygons are pretiled, one vertex per processor, onto a reconfigurable mesh of size $\sqrt{n} \times \sqrt{n}$. In this context, we propose constant-time solutions for testing an arbitrary polygon for convexity, solving the point location problem, the supporting lines problem, the stabbing problem, constructing the common tangents for separable convex polygons, deciding whether two convex polygons intersect, and computing the smallest distance between the boundaries of two convex polygons. To the best of our knowledge this is the first time that O(1) time algorithms are proposed for these problems on this architecture for the "dense" case. The proposed algorithms translate immediately into constant-time algorithms that work on binary images.

1 Introduction

The mesh-connected architecture has emerged as one of the most natural choices for solving a large number of computational tasks in image processing, computational geometry, and computer vision. This is due, in part, to its simple interconnection topology and to the fact that many problems feature data that maps easily onto the mesh structure. In addition, meshes are particularly well suited for VLSI implementation. However, due to their large communication diameter, meshes tend to be slow when it comes to handling data transfer operations over long distances.

In an attempt to alleviate this problem, mesh-connected machines have been enhanced by the addition of various bus systems. For example, Aggarwal [1] and Bokhari [5] consider meshes enhanced by the addition of a single global bus. Yet another such system has been adopted by the DAP family of computers [20], and involves enhancing the mesh architecture by the addition of row and column buses. In [22] an abstraction of such a system is referred to as mesh with multiple broadcasting. A common feature of these bus structures is that they are static in nature, which means that the communication patterns among processors cannot be modified during the execution of the algorithm.

Typical computer and robot vision tasks found nowadays in industrial, medical, and military applications involve digitized images featuring millions of pixels. The large amount of data contained in theses images, combined with real-time processing requirements have motivated researchers to consider adding reconfigurable features to high-performance computers. Along this line of thought, a number of bus systems whose configuration can change, under program control, have been proposed in the literature: such a bus system is referred to as *reconfigurable*. Examples include the *bus automaton* [25], the *reconfigurable mesh* [15], and the *polymorphic torus* [11, 13, 14]. The reconfigurable mesh combines two attractive features of massively parallel architectures, namely, constant diameter and a dynamically reconfigurable bus system.

The notion of convexity is fundamental in image processing, computer graphics, pattern recognition, and computational geometry. In image processing, for example, convexity is a simple and important shape descriptor for objects in the image space [2,20,22]. In pattern recognition, convexity appears in clustering, and computing similarities between sets [7]. In computational geometry, convexity is more than often a valuable tool in devising efficient algorithms for a number of seemingly unrelated problems [19].

In this work, we are interested in devising simple constant-time algorithms for performing a number of tasks involving planar (convex) polygons. First, given an arbitrary polygon P, we determine whether P is convex. Next, we address query-related problems including: point location, supporting-lines, and stabbing. Further, given two convex polygons we address the problems of deciding whether they intersect and of computing the common tangents of the two polygons in case they are separable. Finally, we address the problem of computing the minimum distance between the boundaries of the two polygons.

Computational geometry problems have been solved on reconfigurable meshes before [17, 19, 10]. In both these papers however, the input is of size n while the reconfigurable mesh on which the problem is solved is of size $n \times n$. By contrast, in all our algorithms we consider the "dense" case, whereby instances of size n are being processed on reconfigurable meshes of size $\sqrt{n} \times \sqrt{n}$. The problems we address are motivated by, and find applications to, problems in image processing, computer vision, path planning, VLSI design, and computer graphics [2, 4, 6, 8, 12, 21, 23, 30, 32].

The remainder of this work is organized as follows: Section 2 introduces the computational model used throughout the paper; Section 3 presents a number of preliminaries needed in later sections; Section 4 proposes algorithms to solve the problems mentioned; finally, Section 5 summarizes the results and proposes a number of open problems.

2 The Computational Model

The computational model used throughout this work is the *reconfigurable mesh*.[1] A reconfigurable mesh of size $M \times N$ consists of MN identical processors positioned on a rectangular array with M rows and N columns. As usual, it is assumed that every processor knows its own coordinates within the mesh: we let $P(i, j)$ denote the processor placed in row i and column j, with $P(1, 1)$ in the north-west corner

---

[1] When no confusion is possible a reconfigurable mesh will be referred to simply as a mesh.

of the mesh. Every processor $P(i,j)$ is connected to its four neighbors $P(i-1,j)$, $P(i+1,j)$, $P(i,j-1)$, and $P(i,j+1)$, provided they exist. We assume a SIMD model: in each time unit the same instruction is broadcast to all processors, which execute it and wait for the next instruction. Every processor has 4 ports denoted by N, S,

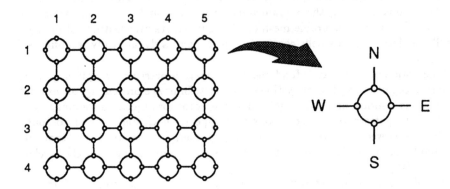

Fig. 1. A reconfigurable mesh of size 4×5

E, and W (see Figure 1). Local connections between these ports can be established, under program control, creating a powerful bus system that changes dynamically to accommodate various computational needs. Our computational model allows at most two connections to be set in each processor at any one time. Furthermore, these two connections must involve disjoint pairs of ports (see Figure 2).

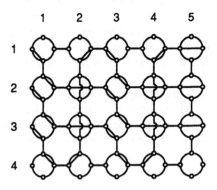

Fig. 2. Examples of allowed connections

At any given time, only one processor can broadcast a value onto a bus. Processors, if instructed to do so, read the bus: if no value is being transmitted on the bus, the read operation has no result. We assume that the processors have a constant number of registers of $O(\log MN)$ bits and a very basic instruction set. Each instruction can consist of setting local connections, performing a simple arithmetic or boolean operation, broadcasting a value on a bus, or reading a value from a specified bus. More sophisticated operations including floor and modulo computation as specifically excluded. As it turns out, these operations can, in fact, be added to the repertoire with very little overhead, thus setting the stage for a very powerful computing environment [16].

It is worth mentioning that at least two VLSI implementations have been performed to demonstrate the feasibility and benefits of the two-dimensional reconfigurable mesh: one is the YUPPIE (Yorktown Ultra-Parallel Polymorphic Image Engine) chip [11, 13] and the other is the GCN (Gated-Connection Network) chip [27, 29]. These two implementations suggested that the broadcast delay, although not constant, is very small. For example, only 16 machine cycles are required to broadcast on a 10^6-processor YUPPIE. The GCN has further shortened the broadcast delay by adopting pre-charged circuits. Recently, it has been shown in [26] that the broadcast delay is even further reduced if the reconfigurable bus system is implemented using fiber optic technology as the underlying global bus system and Electrically Controlled Directional Coupler Switches (ECS) [9] for connecting or disconnecting two fibers. In the light of these experiments, and in accord with other workers [11, 13, 14, 15, 25, 27, 29], we assume that communications along buses take $O(1)$ time.

3 Preliminaries

In this section we discuss several basic results that will be instrumental in developing constant time algorithms for the problems we propose to address.

Specifying an n-vertex polygon P in the plane amounts to enumerating its vertices in some order as $p_1, p_2, \ldots, p_n$ ($n \geq 3$). Here $p_i p_{i+1}$ ($1 \leq i \leq n-1$) and $p_n p_1$ define the edges of P. This representation is also known as the *vertex* representation of P. We note that the vertex representation of a polygon can be easily converted into an *edge* representation: namely, P is represented by a sequence $e_1, e_2, \ldots, e_n$ of edges, with e_i ($1 \leq i \leq n-1$) having p_i and p_{i+1} as its endpoints, and e_n having p_n and p_1 as its endpoints.

A polygon P is termed *simple* if no two of its non-consecutive edges intersect. Recall that Jordan's Curve Theorem [19] guarantees that a simple polygon partitions the plane into two disjoint regions, the *interior* (bounded) and the *exterior* (unbounded) that are separated by the polygon. A simple polygon is *convex* if its interior is a convex set [19]. An arbitrary polygon (not necessarily simple) $P = p_1, p_2, \ldots, p_n$ is said to be in *standard form* if the following conditions are satisfied

- p_1 is the rightmost vertex with least y-coordinate;
- all the vertices are distinct;
- no three consecutive vertices are collinear.

For further reference, we now present the details of a simple routine that finds the supporting line of two upper hulls U and V that do not overlap in the x direction.

We assume that both U and V have size $\sqrt{n}$ and that the points in U and V are sorted by x coordinate; since U and V are non-overlapping, we may assume without loss of generality that all the points in U have smaller x coordinates than those in V. To compute the supporting line we shall use a reconfigurable mesh of size $\sqrt{n} \times \sqrt{n}$ with every processor $P(i,j)$ containing a point of U and a point of V. More precisely, let $u_1, u_2, \ldots, u_{\sqrt{n}}$ and $v_1, v_2, \ldots, v_{\sqrt{n}}$ be the points of U and V in left to right order. Now for every i, j $(1 \leq i, j \leq \sqrt{n})$, processor $P(i,j)$ of the mesh stores u_i and v_j.

For every fixed i $(1 \leq i \leq \sqrt{n})$ every processor $P(i,j)$ in row i checks whether the hull points v_{j-1} and v_{j+1} (provided they exist) lie below the line determined by u_i and v_j. Clearly, in every row of the mesh, either one, or at most two adjacent processors detects this condition. By using suitably constructed horizontal buses, the leftmost processor $P(i,j)$ detecting the condition above sends v_j to $P(i,i)$. In turn, $P(i,i)$ broadcasts the ordered pair (u_i,v_j) vertically, to all the processors in column i of the mesh: recall that these processors store the points of the hull U. Now every processor in column i checks whether the hull point in U it stores lies below the line determined by u_i and v_j. It is easy to see that $\overline{u_i v_j}$ is a supporting line if and only if every processor in column j detects this condition. Clearly, there exists exactly one such supporting line for the hulls U and V. Note that the entire computation is performed in O(1) time. Therefore, we can state the following intermediate result.

Theorem 3.1. Let U and V be two non-overlapping upper hulls of size at most $\sqrt{n}$ stored in the first row and the first column, respectively, of a reconfigurable mesh of size $\sqrt{n} \times \sqrt{n}$. The supporting line of the two hulls can be computed in O(1) time. □

Consider the following problem. We are given two polygons P and Q of $\sqrt{n}$ vertices each stored one vertex per processor in the first row and column, respectively of a reconfigurable mesh of size $\sqrt{n} \times \sqrt{n}$. We are interested in detecting whether P and Q intersect. As a preprocessing step, we convert P and Q to edge form. Trivially, this task can be done in O(1) time. After having mandated every processor to connect its ports N and S, we replicate the first row containing the edges of P throughout the mesh. Next, we establish horizontal buses in every row of the mesh. Every processor in the first column broadcasts the edge of Q it stores horizontally. Every processor that detects that the edge of Q received on the horizontal bus, intersects the edge of P it contains, sets up a local flag. It is now an easy task to inform processor $P(1,1)$ that an intersection was detected.

Finally, if no intersection is detected, then we need to test one vertex from P and one vertex from Q for inclusion in the opposite polygon. As we shall see in Theorem 4.2 this task can be performed in O(1) time. Consequently, we have proved the following result.

Theorem 3.2. The task of testing whether two polygons P and Q of $\sqrt{n}$ vertices each stored one vertex per processor in the first row and column, respectively of a reconfigurable mesh of size $\sqrt{n} \times \sqrt{n}$ intersect can be performed in O(1) time. □

4 The Algorithms

The purpose of this section is to provide constant time algorithms for the problems stated below.

CONVEXITY: given an n-vertex polygon decide whether it is convex.

INCLUSION: given an n-vertex convex polygon P and a query point q, does q belong to the interior of P?

SUPPORTING-LINES: given an n-vertex convex polygon P and a query point q, in case q lies outside P, compute the supporting lines from q to P.

STABBING: given an n-vertex convex polygon P and a query line λ, does λ intersect P?

INTERSECTION: given two n-vertex convex polygons P and Q decide whether they intersect.

COMMON-TANGENT: given two separable n-vertex convex polygons P and Q determine their common tangents.

MIN-DISTANCE: given two n-vertex non-intersecting convex polygons P and Q, find the smallest Euclidian distance between them.

4.1 Problems Involving One Polygon

Consider a reconfigurable mesh $\mathcal{M}$ of size $\sqrt{n} \times \sqrt{n}$. The input is a polygon $P = p_1, p_2, \ldots, p_n$, in standard form, stored in row major order in the mesh $\mathcal{M}$, one vertex per processor. To be more specific, let $r(i)$ and $c(i)$ stand for $\lceil \frac{i}{\sqrt{n}} \rceil$ and $(i-1) \bmod \sqrt{n} + 1$, respectively. In this notation, for every i ($1 \leq i \leq n$) the point p_i of P is stored by processor $P(r(i), c(i))$.

Our solution to the CONVEXITY problem relies on the following proposition which is an adaptation of a result of Shamos ([28], page 28). In the following discussion, by edge angles we mean the angles determined by the edges of the polygon with the positive direction of the x axis.

Proposition 4.1. A polygon P, in standard form, is convex if and only if its edge angles are monotonic. $\square$

To check for the condition specified in Proposition 4.1, every processor $P(r(i), c(i))$ ($1 \leq i \leq n-1$) computes the edge angle, α_i, determined by the vector $p_i p_{i+1}$ and $p_n p_1$ with the positive direction of the x-axis. By Proposition 4.1, P is a convex polygon if and only if the sequence $\alpha_1, \alpha_2, \ldots, \alpha_n$ is strictly monotonic.

To determine if the sequence $\alpha_1, \alpha_2, \ldots, \alpha_n$ is strictly monotonic, every processor $P(r(i), c(i))$ ($1 \leq i \leq n$) checks whether α_i is local maximum or local minimum. Every processor is mandated to link its ports N and S, thus creating vertical buses in every column of the mesh. Every processor that detects a local maximum or minimum in the previous computation removes its NS connection. Next, every processor that has removed its connection broadcasts a "*" northbound. It is easy to see that some processor in the first row of $\mathcal{M}$ receives a "*" if and only if the polygon is not convex. In another $O(1)$ time processor $P(1, 1)$ can determine if any processor in the first row has received a "*". To summarize our findings, we state the following result.

Theorem 4.2. Consider an arbitrary n-vertex polygon P stored in row-major order in a reconfigurable mesh of size $\sqrt{n} \times \sqrt{n}$. The task of testing P for convexity can be solved in $O(1)$ time. $\square$

Next, we present a constant time algorithm to solve the INCLUSION problem. As before, consider an n-vertex convex polygon $P = p_1, p_2, \ldots, p_n$, stored in row-major order in a reconfigurable mesh of size $\sqrt{n} \times \sqrt{n}$, and let q be an arbitrary query point in the same plane as P. Begin by broadcasting the coordinates of q to all the processors in the mesh. Upon receiving this information, every processor $P(r(i), c(i))$

determines whether q lies in the positive or negative halfplane determined by the infinite line collinear with the edge $p_i p_{i+1}$. It is easy to see that q lies in the interior of P if and only if it lies in the positive halfplane of every edge. To check whether this is true, the processors are next mandated to connect their ports N and S. Every processor that detects that q lies in the negative halfplane determined by the corresponding edge, broadcasts a signal northbound. It is now an easy data movement to inform $P(1, 1)$ that the signal was received by one of the processors in the first row of the mesh. The entire computation takes constant time. Thus, we have proved the following result.

Theorem 4.3. Let P be an n-vertex convex polygon stored in row-major order in a reconfigurable mesh of size $\sqrt{n} \times \sqrt{n}$. Given an arbitrary query point q in the same plane as P, the task of testing whether q lies in the interior of P can be solved in $O(1)$ time. $\square$

Our solution to the SUPPORTING-LINES problem proceeds along similar lines. Given an n vertex polygon P and a query point q as above, we first determine whether q lies outside of P. By Theorem 4.3 this can be done in $O(1)$ time. Next, we broadcast the coordinates of q to all the processors in the mesh. Every processor $P(r(i), c(i))$ checks whether both p_{i-1} and p_{i+1} lie to the same side of the line determined by q and p_i. Clearly, two processors detect this condition. A simple data movement operation informs processor $P(1, 1)$, in $O(1)$ time, of the identity of these supporting lines. Consequently, we have the following result.

Theorem 4.4. Let P be an n-vertex convex polygon stored in row-major order in a reconfigurable mesh of size $\sqrt{n} \times \sqrt{n}$. Given an arbitrary query point q in the same plane as P, the task of determining the supporting lines to P from q, if any, can be solved in $O(1)$ time. $\square$

We are now in a position to provide a solution to the STABBING problem. We are given an n-vertex convex polygon P and a query line λ. The problem is to detect whether the query line intersects the polygon. Our solution uses the solutions to INCLUSION and SUPPORTING-LINES problems developed above. Specifically, after having selected an arbitrary point q on λ, we proceed to test whether q is interior to P. If so, λ must intersect P. In case q lie outside of P we find the supporting lines d_1 and d_2 to P from q. Finally, we test whether λ lies in the wedge determined by d_1 and d_2. By virtue of Theorems 4.3 and 4.4, the whole algorithm runs in constant time. Therefore, we have proved the following result.

Theorem 4.5. Let P be an n-vertex convex polygon stored in row-major order in a reconfigurable mesh of size $\sqrt{n} \times \sqrt{n}$. Given an arbitrary query line λ in the same plane as P, the task of testing whether q intersects the interior of P can be solved in $O(1)$ time. $\square$

For further reference, we now observe that the intersection (if any) of line λ and polygon P can be determined within the same time complexity.

Corollary 4.5.1. Let P be an n-vertex convex polygon stored in row-major order in a reconfigurable mesh of size $\sqrt{n} \times \sqrt{n}$. The line segment corresponding to the intersection of P with a query line λ in the same plane as P, can be computed in $O(1)$ time. $\square$

4.2 Computations Involving Two Polygons

We now develop a constant time solution for the COMMON-TANGENT problem. Consider separable convex polygons $P = p_1, p_2, \ldots, p_n$ and $Q = q_1, q_2, \ldots, q_n$, with P and Q stored in row-major order in a reconfigurable mesh $\mathcal{M}$ of size $\sqrt{n} \times \sqrt{n}$. We only show how the upper tangent is computed, the lower tangent and the "interior" tangents are computed similarly.

In the sequel we consider upper hulls only. For convenience we shall continue to refer to them as P and Q. A *sample* of P is simply a subset of points in P enumerated in the same order as those in P.

To simplify the notation we shall assume without loss of generality that the upper hull of P is $(p_1, p_2, \ldots, p_k)$. Consider an arbitrary sample $A = (p_1 = a_0, a_1, \ldots, a_s = p_k)$ of P. It is easy to see that the sample A partitions P into s pockets $A_1, A_2, \ldots, A_s$, such that A_i involves the points in P lying between a_{i-1} and a_i (we assume that a_{i-1} belongs to A_i and that a_i does not). Similarly, consider a sample $B = (q_1 = b_0, b_1, \ldots, b_t = q_l)$ of Q. As noted, these two samples determine pockets $A_1, A_2, \ldots, A_s$ and $B_1, B_2, \ldots, B_t$ in P and Q, respectively. Let the supporting line of A and B be achieved by a_i and b_j, and let the supporting line of P and Q be achieved by p_u and q_v. The following technical result has been established in [3].

Proposition 4.6. (Lemma 1 in [3]) At least one of the following statements is true:
(a) $p_u \in A_i$;
(b) $p_u \in A_{i+1}$;
(c) $q_v \in B_j$;
(d) $q_v \in B_{j+1}$. $\square$

The samples A and B will be chosen to contain all the points in the last column of the mesh. Specifically, our assumption about the layout of the two polygons guarantees that the samples are points in the two polygons whose subscripts are a multiple of $\sqrt{n}$. By Theorem 3.1, computing a supporting line for A and B takes $O(1)$ time. As before, let the supporting line of A and B be achieved by a_i and b_j. Note that condition (a) in Proposition 4.6 holds only if p_u lies to the right of a_{i-1} and to the left of a_i. To check (a), the supporting lines l and l' from a_{i-1} and a_i to processor holding a_i detects in constant time whether the left neighbor of a_i in P lies above l'. Similarly, the processor holding a_{i-1} checks whether the right neighbor of a_{i-1} in P lies above l. It is easy to confirm that p_u belongs to A_i only both these conditions hold. The other conditions are checked similarly. Suppose, without loss of generality, that (a) holds. Our next task is to compute a supporting line for A_i and Q. The main point to note is that in order to apply Theorem 3.1, the points in pocket A_i have to be moved to the first row of the mesh. It is easy to confirm that this task takes $O(1)$ time. Therefore, Theorem 3.1 and Proposition 4.3 combined imply the following result.

Theorem 4.7. Computing the common tangent of two separable n-vertex convex polygons stored in row-major order in a reconfigurable mesh of size $\sqrt{n} \times \sqrt{n}$ takes $O(1)$ time. $\square$

Next, we are interested in devising a simple constant-time algorithm to decide whether two n-vertex convex polygons intersect. To this end, let the convex polygons $P = p_1, p_2, \ldots, p_n$ and $Q = q_1, q_2, \ldots, q_n$, with P and Q stored in row-major order in a reconfigurable mesh of size $\sqrt{n} \times \sqrt{n}$. Consider arbitrary samples $A = (p_{\sqrt{n}} =$

$a_0, a_1, \ldots, a_k = p_n)$ and $B = (q_{\sqrt{n}} = b_0, b_1, \ldots, b_t = q_n)$ of P and Q, respectively obtained by retaining the points in the last column only. Our INTERSECTION algorithm proceeds to determine whether A and B intersect. Notice that A and B are convex polygons. Now Theorem 3.2 guarantees that the previous task can be computed in $O(1)$ time. In case A and B intersect, the answer to INTERSECTION is "yes".

Therefore, from now on we shall assume that A and B do not intersect, and so A and B are separable. Now the algorithm detailed in Theorem 4.7 can be used to compute a separating line δ of A and B, which is nothing but an interior common tangent of the two polygons. Let s_1 and s_2 be the line segments corresponding to the intersection of δ with P and Q, respectively. It is easy to see that P and Q intersect if and only if s_1 and s_2 overlap. By Corollary 4.5.1 both s_1 and s_2 can be computed in $O(1)$ time. Consequently, we have the following result.

Theorem 4.8. The task of deciding whether two n-vertex convex polygons, stored in row-major order in a reconfigurable mesh of size $\sqrt{n} \times \sqrt{n}$, intersect can be solved in $O(1)$ time. $\square$

Consider separable convex polygons $P = p_1, p_2, \ldots, p_n$ and $Q = q_1, q_2, \ldots, q_n$, with P and Q stored in row-major order in a reconfigurable mesh of size $\sqrt{n} \times \sqrt{n}$. We propose to show that in this setup, the task of computing the smallest distance between P and Q can be solved in $O(1)$ time.

To make the exposition easier to follow, we assume without loss of generality that the two polygons are separable in the x direction, with P to the left of Q. Let T_u and T_l be the upper and lower common tangent of P and Q. Let T_u touch P and Q at p_1 and q_1 and let T_l touch P and Q at p_r and q_s. Let C_P and C_Q be the mutually *visible* chains in P and Q respectively. In other words, C_Q involves vertices $q_1, q_2, \ldots, q_s$, while the chain C_P involves the vertices $p_1, p_2, \ldots, p_r$. Simple geometric considerations confirm that to compute the minimum distance between P and Q we only need examine the distance from vertices in C_P to C_Q. Now a result in [7] guarantees that the distance function of vertices in these chains is unimodal. Specifically, for every vertex u in C_P, the distance to any point v (not necessarily a vertex) of C_Q first decreases and then increases, as v moves from q_1 to q_s, with the minimum achieved by either a vertex of Q or by the perpendicular projection of u on the boundary of Q. An identical property holds for points in C_Q and their distance function to C_P. Once the minimum distance is computed for every point in the chains C_P and C_Q, the minimum distance between the two polygons can be determined in $O(1)$ time by exploiting the convexity of the two polygons.

We now present the details of our algorithm. To begin, the upper and lower tangents T_u and T_l are computed in $O(1)$ time by using the algorithm detailed in Theorem 3.1. Once T_u and T_l have been computed, identifying the chains C_P and C_Q is achieved by a simple broadcasting and marking operation.

Consider arbitrary samples $A = (p_1 = a_0, a_1, \ldots, a_k = p_r)$ and $B = (q_1 = b_0, b_1, \ldots, b_t = q_s)$ of C_P and C_Q, respectively. It is easy to see that the sample A partitions P into k pockets $A_1, A_2, \ldots, A_k$, such that A_i involves the points in P lying between a_{i-1} and a_i (we assume that a_{i-1} belongs to A_i and that a_i does not). Similarly, sample B of Q determines the pockets $B_1, B_2, \ldots, B_t$.

We assume that the points of sample A are stored in the first column of $\mathcal{M}$ while the points in B are stored in the first row of $\mathcal{M}$. We now show how the minimum distance from a vertex a_i in A to B is computed in one row of $\mathcal{M}$. Of course, the same computation is performed, in parallel, in all rows in which $P(i,1)$ stores a vertex in A. Begin by setting up vertical buses and replicate the contents of the first row containing the points in B throughout the mesh. Processor $P(i,1)$ broadcasts a_i horizontally to the whole row i. Every processor in row i $(1 \leq i \leq n)$, storing a vertex b_j in B computes the distance $d(a_i, b_j)$. Notice that a unique processor $P(i,k)$ detects that $d(a_i, b_k) < d(a_i, b_{k-1})$ and that $d(a_i, b_k) \leq d(a_i, b_{k+1})$. In addition, this processor computes the intersection points of each of the edges $b_{k-1}b_k$ and $b_k b_{k+1}$ with the perpendiculars from a_i to these two edges. If one of these points is interior to one of the edges $b_{k-1}b_k$ or $b_k b_{k+1}$, then $P(i,k)$ reports the corresponding perpendicular distance back to $P(i,1)$. Otherwise, $P(i,k)$ reports $d(a_i, b_k)$.

Now every processor in the first column of the mesh that contains a vertex in A compares the minimum distance achieved by its own vertex with the minimum distances achieved by the vertices stored by its two neighbors. Convexity guarantees that exactly one of them will detect the minimum distance.

The unimodality of the distance [7] guarantees that to compute the minimum distance between C_P and C_Q we can restrict out attention to one of the pockets adjacent to a_i and b_j. The previous steps are then repeated four times, thus obtaining the minimum distance between C_P and C_Q. To summarize our findings we state the following result.

Theorem 4.9. The MIN DISTANCE problem involving two separable n-vertex polygons stored in row major order in a reconfigurable mesh of size $\sqrt{n} \times \sqrt{n}$ can be solved in $O(1)$ time. $\square$

5 Conclusions and Open Problems

In an attempt to reduce their communication diameter, mesh-connected computers have recently been augmented by the addition of various types of bus systems. One of the most interesting such systems, referred to as reconfigurable mesh involves augmenting the basic mesh-connected computer by the addition of a dynamically reconfigurable bus system.

In this paper, we demonstrated that the versatility of the reconfigurable mesh can be exploited for the purpose of devising constant time algorithms for a number of important computational geometry tasks relevant to image processing, computer graphics, and computer vision. In all our algorithms we assumed that one or two n-vertex (convex) polygons are pretiled, one vertex per processor, onto a reconfigurable mesh of size $\sqrt{n} \times \sqrt{n}$. In this context, we propose constant-time solutions for testing an arbitrary polygon for convexity, solving the point location problem, the supporting lines problems, the stabbing problem, constructing the common tangents for separable convex polygons, deciding whether two convex polygons intersect, and computing the smallest distance between the boundaries of two convex polygons. To the best of our knowledge this is the first time that $O(1)$ time algorithms are proposed for these problems on this architecture in the "dense" case.

Recently, Olariu et al. [18] have proposed an efficient algorithm to compute the sum of all the entries of a binary matrix of size $\sqrt{n} \times \sqrt{n}$ in $O(\log \log n)$ time. It is easy to see that the same algorithm applies, within the same time complexity, to geometric problems in which the computation results in a binary result being stored at each processor. For example, the point location problem in a simple polygon falls into this category. It would be interesting to see whether the point location problem for a simple polygon can be solved faster.

Yet another intriguing problem involves detecting whether a polygon lies within another one. Even for convex polygons, we cannot solve the problem in O(1). Is it the case the the approach we used to detect intersection works for containment? Next, we don't know how to determine the largest inscribed triangle in a given convex polygon. There exists a well-known O(n) time sequential algorithm for this latter problem, but it does not seem to be parallelizable to run in O(1) time. Next, it would be nice to solve the symmetric problems of computing the smallest-area enclosing triangle as well as the largest enscribed circle and rectangle.

Acknowledgement: This work was supported by NASA under grant NCC1-99. The authors wish to thank Jingyuan Zhang for his comments.

References

1. A. Aggarwal, Optimal bounds for finding maximum on array of processors with k global buses, *IEEE Trans. on Computers*, C-35, 1986, 62-64.

2. S. G. Akl and K. A. Lyons, *Parallel Computational Geometry*, Prentice-Hall, Englewood Cliffs, New Jersey, 1989.

3. M. J. Atallah and M. T. Goodrich, Parallel algorithms for some functions of two convex polygons, *Algorithmica* 3, (1988) 535–548.

4. D. H. Ballard and C. M. Brown, *Computer Vision*, Prentice-Hall, Englewood Cliffs, New Jersey, 1982.

5. S. H. Bokhari, Finding maximum on an array processor with a global bus, *IEEE Transaction on Computers* vol. C-33, no. 2, Feb. 1984. 133–139.

6. R. Cahn, R. Poulsen, and G. Toussaint, Segmentation of Cervical Cell Images, *Journal of Histochemistry and Cytochemistry*, 25, (1977), 681–688.

7. F. Chin and C. A. Wang, Optimal algorithms for the minimum distance between two separated convex polygons, University of Alberta, Tech. Report, January 1983.

8. R. O. Duda and P. E. Hart, *Pattern Classification and Scene Analysis*, Wiley and Sons, New York, 1973.

9. D. G. Feitelson, *Optical Computing*, MIT Press, 1988.

10. J. Jang and V. Prasanna, Parallel geometric problems on the reconfigurable mesh, *Proc. of the International Conference of Parallel Processing*, St. Charles, Illinois, 1992, vol. III, 127–129.

11. H. Li and M. Maresca, Polymorphic-torus network, *IEEE Transactions on Computers*, vol. C-38, no. 9, (1989) 1345–1351.

12. T. Lozano-Perez, Spatial Planning: A Configurational Space Approach, *IEEE Trans. on Computers*, C-32, 1983, 108-119.

13. M. Maresca and H. Li, Connection autonomy and SIMD computers: a VLSI implementation, *Journal of Parallel and Distributed Computing*, 7, (1989) 302–320.

14. M. Maresca, H. Li, and P. Baglietto, *Proc. International Conference on Parallel Processing*, St. Charles, Illinois, 1993, vol. I, 282–289.

15. R. Miller, V. K. P. Kumar, D. Reisis, and Q. F. Stout, Parallel Computations on Reconfigurable Meshes, *IEEE Trans. on Computers*, in press.

16. S. Olariu, J. L. Schwing, and J. Zhang, Fundamental Data Movement for Reconfigurable Meshes, *Proc. of the International Phoenix Conf. on Computers and Communications*, Scottsdale, Arizona, April 1992, 472–480.

17. S. Olariu, J. L. Schwing, and J. Zhang, Time-Optimal Convex Hull Algorithms on Enhanced Meshes, *BIT*, 33 (1993) 396–410.

18. S. Olariu, J. L. Schwing, and J. Zhang, Fast Computer Vision Algorithms on Reconfigurable Meshes, *Image and Vision Computing Journal*, 10 (1992), 610–616.

19. S. Olariu, J. L. Schwing, and J. Zhang, Constant Time Computational Geometry on Reconfigurable Meshes, *SPIE Conference on Vision Geometry*, Boston, November 1992, SPIE Vol. 1832, 111–121.

20. D. Parkinson, D. J. Hunt, and K. S. MacQueen, The AMT DAP 500, 33rd *IEEE Comp. Soc. International Conf.*, 1988, 196–199.

21. T. Pavlidis, *Computer Graphics*, Computer Science Press, Potomac, MD, 1978.

22. V. K. Prasanna and C. S. Raghavendra, Array processor with multiple broadcasting, *Journal of Parallel and Distributed Computing*, vol 2, 1987, 173-190.

23. B. Preas and M. Lorenzetti, Eds., *Physical design and automation of VLSI systems*, Benjamin/Cummings, Menlo Park, 1988.

24. A. Rosenfeld and A. Kak, *Digital Picture Processing*, Academic Press, New York, 1982.

25. J. Rothstein, Bus automata, brains, and mental models, *IEEE Trans. on Systems Man, and Cybernetics* 18, (4), 1988, 522–531.

26. A. Schuster and Y. Ben-Asher, Algorithms and optic implementation for reconfigurable networks, *Proceedings of the 5th Jerusalem Conference on Information Technology*, October 1990.

27. D. B. Shu, L. W. Chow, and J. G. Nash, A content addressable, bit serial associate processor, *Proceedings of the IEEE Workshop on VLSI Signal Processing*, Monterey CA, November 1988.

28. M. I. Shamos, Computational Geometry, Doctoral Dissertation, Yale University, 1979.

29. D. B. Shu and J. G. Nash, The gated interconnection network for dynamic programming, S. K. Tewsburg *et al.* (Eds.), Concurrent Computations, Plenum Publishing, 1988.

30. G. T. Toussaint, Movable Separability of Sets, in G.T. Toussaint ed., *Computational Geometry*, Elsevier Science Publishers, North-Holland, Amsterdam, 1985.

31. G. T. Toussaint Ed., *Computational Geometry*, Elsevier Science Publishers, North-Holland, Amsterdam, 1985.

32. D. Vernon, *Machine vision, automated visual inspection and robot vision*, Prentice-Hall, Englewood Cliffs, New Jersey, 1991.

Deepness Analysis :
Bringing Optimal Fronts to Triangular Finite Element Method

Jérôme Galtier*

Université de Versailles Saint-Quentin, France
Laboratoire PRISM

Abstract. A scheme is presented for analizing finite-element triangulations. The method takes a random triangulated planar graph and gives a multifrontal way to solve the corresponding physical problem, so that the maximum bandwidth of each front is guaranteed to be optimal. The interesting characteristic of this scheme is that it introduces large-grained parallelism dictated by the domain structure.

A way to extend these results to unsymmetric systems is given. Some experimental results are also presented.

1 Introduction

Solving sparse matrices arising from finite element method has always been a problem of major interest for its numerous industrial applications. Two main categories of solvers have appeared so far; direct methods perform Gaussian elimination while indirect ones use some iterative schemes to converge to the final solution.

As a major way of applying direct methods, frontal schemes have known many reverses in the past, mainly because of their confrontation with indirect methods such as the conjugate gradient. In the 70's, they took the lead thanks to their officialization and their extension to non-elliptic problems (see [12, 11]). With the remarkable results of George on regular finite element meshes, people may have had the feeling that the limit was found in terms of direct methods. Moreover George's result cannot be improved by an order of magnitude (see [9, 10, 15]).

To complete these apparently closed theoretical achievements, direct methods were finally found to be practically unefficient in front of conjugate gradient methods with the introduction of preconditioning ([7, 13]). However, frontal or multi-frontal methods remain to solve non-elliptic problems ([5]), and are still quite useful in practice since finding a good preconditionner for a given matrix is not obvious.

In this paper, we shall introduce a particular analysis of triangular finite element methods. It is particularly suited to multifrontal solutions and leads

---

* This work was done while the author was at the McGill University, ACAPS Laboratory, under the supervision of Laurie Hendren

to a special kind of fronts which are said to be optimal. Of course, we shall discuss a way to do a partial pivot while keeping fundamental properties - as an alternative to Hood's method.

1.1 Main Objectives

We should like to show that planar triangulated domains have some important properties that may very quickly give us an idea about the kind of structure we are considering. Another interesting thing to look at is to see if "mathematical" division could lead to an interesting renumbering of the mesh.

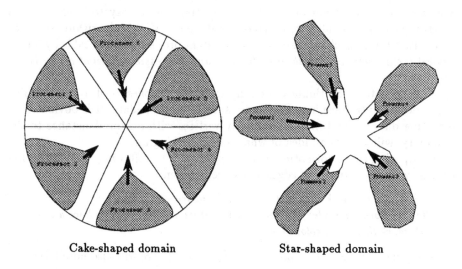

Cake-shaped domain   Star-shaped domain

Fig. 1. Some interesting cases to exploit

In particular, we would like to automatically detect some main properties which are to be exploited to draw some multifrontal schemes - or in a simpler way to evaluate them.

1.2 Subject Concerned

Here we do not take the whole subject of fronts into consideration, but we restrict ourselves to *triangulated planar graphs*. Even if this restriction is quite important, we stress the fact that triangles are a very popular structure in Numerical Analysis for their simple characteristics. In particular, bijections from a random triangle to the standard one are naturally linear, so that errors are less important and computation is faster.

More practically, a graph - or mesh - will be a pair $G = (X, E)$, where X is the set of *vertices* or *nodes*, and $E \subseteq \{\{x, y\} \in X^2, x \neq y\}$ is the set of edges. Two nodes x, y so that $\{x, y\} \in E$ are said to be *adjacent*. The graph is bi-dimensional or planar in the sense that it may be drawn on a map. We are concerned here with simply-connected triangulated domains, that is to say that the mesh is issued from the triangulation T of a simply connected domain Ω of $\mathbb{R}^2$. Then we distinguish a particular subset B of X, including vertices of the boundary (Ω simply connected iff B connected). For a given triangle T, we note T_v the set of vertices of T, and T_e the set of edges. Finally, for $V \subset X$, we note $< V >$ the graph induced by V in G.

2 Deepness Analysis

We saw precedently that "cigar-shaped" domains were easy to analyze. Lead on straight from this idea we remarked that even if King's algorithm could not detect it, "star-shaped" domains had some latent properties exploitable by multi-frontal techniques. *So the thinner the domain, the easier elimination by bands or fronts.*

How to detect the thinness - or inversely the deepness - of a graph ? The basic idea we introduce here lies on distance to the boundary. It is obvious that detecting distance and more particularly vertices with maximal distance to the boundary, will give us an idea of what we are looking for.

Computing this information is not expensive. The following basic algorithm (linear with the number of edges) has been chosen :

- *Initialization : mark each vertex on the boundary with colour 1.*
- *Step n : For each vertex of colour n, check if its adjacent vertices have a colour. If not, mark them with colour n + 1.*

We have then $\forall x \in X \quad colour(x) = \delta(x, B) + 1$. We will also note later

$$\forall T \in \mathcal{T} \qquad colour(T) = \sum_{v \in T_v} colour(v)$$

2.1 Optimal Fronts

The notion of fronts Irons introduced was very general. This was a "line of computation" on the domain which was allowed to shut on itself to become circular and then involved no more vertices on the boundary. However, we shall stick to the idea of a front which starts and ends on vertices of the boundary.

Definition 1. An optimal front is a couple of series of vertices
$((u_i)_{1 \leq i \leq n}, (v_i)_{1 \leq i \leq n})$ with same length n verifying :

(i) $\forall i, 1 \leq i \leq n$ u_i and v_i have same colour i

(ii) $\forall i, 1 \leq i \leq n - 1$ $\begin{cases} u_i \text{ and } u_{i+1} \text{ are adjacent} \\ v_i \text{ and } v_{i+1} \text{ are adjacent} \end{cases}$

(iii) Either $u_n = v_n$ or u_n and v_n are adjacent

(iv) If $n > 1$, then $u_n \neq v_n$ or $u_{n-1} \neq v_{n-1}$

Properties (ii) and (iii) guaranty front cohesion. Property (i) means optimality, which is closely linked to the actual representation. Finally, property (iv) avoids empty fronts.

The following property is straightforward.

Proposition 2. *A front which touches the boundary, contains u_n and v_n, and holds two distinct arms has at least as many vertices as an optimal front.*

We can see that considering two ordered arms ($(u_i, v_i) \neq (v_i, u_i)$) allows us to define an orientation. This way, we can call one arm "right" and the other "left", and "inside" or "outside" the front becomes meaningful regardless of the assembly state - provided the fact the arms do not "cross" each other (ie don't have common nodes). In this context we do not need notions like macro- or generalized element anymore, for it is replaced in terms of optimal fronts.

But is it possible to restrict computational fronts to optimal fronts ? In other terms, can we move these fronts step by step along the graph ?

We shall see that in case it is possible, nodes of a same colour cover some lines named lines of strong connectivity.

2.2 Strong Connectivity

Definition 3. We call strong connected component of a given colour i, each connected component in the graph $G_i = (X_i, V_i)$ with :

$$X_i = \{x \in X; colour(x) = i\}$$
$$V_i = \{v \in V; \exists T \in \mathcal{T}; colour(T) \geq 3i \text{ and } v \in T_v\}$$

The set V_i emphasizes the "elimination" of all triangles having a vertex of colour $< i$, which lets some "islands" appear. Hence the following notion :

Definition 4. We call *local boundary* of one strong connected component C_i of a given colour i the set B_i of edges such that

$$B_i =< C_i > \cap \{ e \in E; \exists T_1, T_2 \in \mathcal{T}; e \in T_1 \cap T_2$$
$$\text{and } colour(T_1) = 3i - 1 \text{ and } colour(T_2) \geq 3i \}$$

The local boundary B_i is in fact the set of edges which forms the boundary of the remaining domain after the "elimination" of triangles having a vertex of colour $< i$. An edge belonging to one B_i will later be called a *strong edge*.

Lemma 5. *Each vertex involves an even number of strong edges.*

Proposition 6. *If Ω is simply connected, the path between two vertices of a same strong component may be drawn on strong edges (ie on the local boundary).*

These statements are proved in [8]

We saw that edges with a given connected component draw level lines on the mesh. That is why it is named strong connectivity.

Among a same colour, some strong components are closer than others. They form what we call weak components.

2.3 Weak Connectivity

Definition 7. We call weak connected component of a given colour i each connected component in the graph $< X_i >$ induced by the vertices of colour i.

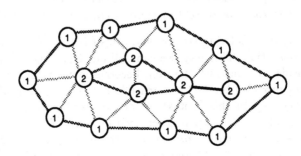

Fig. 2. Strong (grey) and weak (black) edges

The following property follows :

Proposition 8. *Weak connected components are over-sets of strong connected ones.*

In the next paragraphs we shall call *weak edge* each edge linking two different strong connected components. Another interesting property concerns the general structure of typically weak edges :

Proposition 9. *If Ω is simply connected, the graph that weak edges draw on strong components is acyclic.*

2.4 Super-strong connectivity

The structure of strong components is sometimes more complicated than single cycles. The notion of super-strong connectivity is meant to make one understand this more efficiently. The interest of super strong connectivity is concentrated on vertices with more than two strong edges. These introduce some additional cycles among a single strong component. We have the following statements :

Definition 10. A multiple vertex is a vertex with at least four strong edges.

Proposition 11. *The extra loops introduced by multiple vertices have globally a structure of acyclic graph - provided Ω is simply connected.*

Definition 12. We call super-strong connected component of a strong component, the set of all triangles included in one of its "loops".

For further details, see [8]. We now have a better grasp of the inheritance among vertices with successive colours.

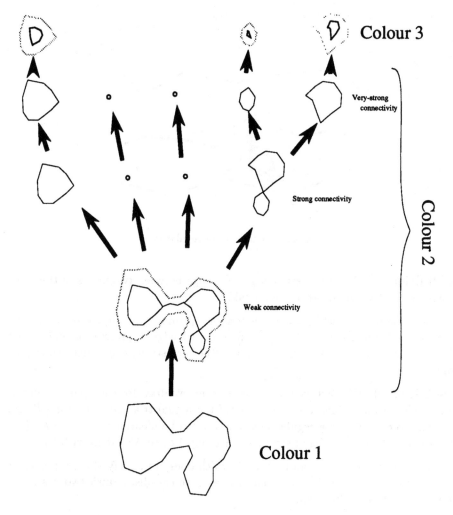

Fig. 3. Connectivity hierarchy

3   Focus On Irregularity - Triangulation Spine

How to define what a regular triangulation is ? According to previous notions, we would like to say that it is a triangulation which can be described only by optimal fronts. Of course, description means step-by-step improvement, where a step involves a few number of elements, minimal if possible.

What is this minimal number ? If we modify the front one vertex after the other, we shall involve two elements each time - except in the case of vertices on the boundary. So we shall try to keep this minimal case.

Structures leading to irregularity are not so common. The previous chapter showed some of them. The following definition will be useful.

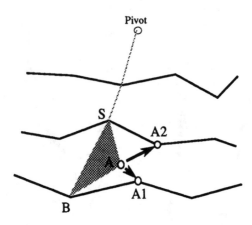

Fig. 4. Definition of regularity

Definition 13. We call internal edge each edge between two nodes of the same colour which is neither strong nor weak.

The general case which occurs before an optimal front improvement can be represented as figure 4 shows. In this figure, dark edges represent current strong edges ("current" means here belonging to the very-strong component where the pivot is).

Definition 14. For each virtual optimal front, we draw for each pair of successive nodes B and S, the last vertex A (of the actual triangle containing B and S). This vertex A is in a regular position if it can be identified either to A1 (the vertex next to B in terms of strong connectivity) or to A2 (idem with S).

In other words, if we can find an A which does not satisfy this property, we have found some irregularity. In such a case, we can distinguish two cases (we suppose that B's colour is n).

I A's colour is n, and since all dark edges are strong edges, A and B are adjacent to an "internal" edge.

II A's colour is $n + 1$ and then we distinguish three other cases :

   α A and S are adjacent by a "strong" edge, ie S is a multiple vertex

   β A and S are adjacent by a "weak" edge

   γ A and S are adjacent by an internal edge

This last case needs further explanation. Turning around S, we see that if one of the vertices preceding or following A has the colour n, then A forms a strong edge with S (absurd !). So A belongs to a very-strong connected component linked to S. This means that S is a multiple vertex.

As a consequence, we can check irregularities by locating the following singularities :

     a – Internal edges

     b – Multiple vertices

     c – Weak edges

However, detecting irregularities independently does not really lead to some exploitable data. We also have to show how these irregularities are connected, and eventually find a means to get round them. In the next section we shall see how these structures can naturally draw some "spines" on the graph.

3.1 Union of Natural Irregularities : Spectacular Examples

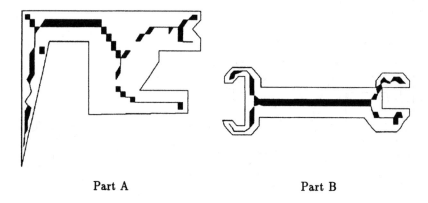

Part A   Part B

Fig. 5. Natural spine on some meshes ...

The real problem that "irregularity" checking raised at once was mainly concentrated on the internal edges' structure. Indeed, multiple nodes are quite rare and finally quite simple : they define a kind of crossroads between very strong components. And even weak edges are nothing more than connexions between strong components. On the contrary, internal edges can draw some very complex structures inside each very-strong component. And more precisely, they define the subdivision between very-strong components of colour n and weak components of the colour $n + 1$. The problem is that an internal edge never points to what it connects - or disconnects.

So one idea consists in locating not only internal edges, but triangles of which they form an edge. In the following pictures, we checked these structures :

a Triangles having an "internal" edge
b Multiple nodes
c Weak edges
d Triangles forming a complete very-strong component

We add the last category since these triangles are local summits and consequently have an obvious interest for optimal front's way. Indeed, more mathematically, we may consider that the pivot also is a case of irregularity.

Let us take a few examples and see what is going on. In figure 5, part A, irregularities define a long line along the structure. On the right we can see a branch dividing itself into two parts. On the left, we note a square, which is one unexpected irregularity. In part B, irregularities are naturally connected one to another. Perhaps there is an unexpected branch down on the right. To conclude, we can say that irregularities form a very interesting set to study, in particular with the aim of describing globally the structure.

3.2 Building The Machine-Tree - Related Properties

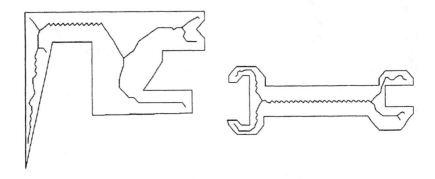

Fig. 6. Machine-tree for previous examples

Having an idea of what a triangulation's spine is is one thing. But making a computer able to understand this is not always very easy. Some apparently simple or very rare problems can take on a great importance on irregular graphs.

As the reader may have already noticed, the natural spine is not always naturally connected. So we have to insure this connexion while keeping fundamental properties. Another task consists in connecting intuitive structures correctly. So we propose a set of rules having those properties. For this, we will use three kinds of connexions :

(i) Between two triangles having a common edge
(ii) Between two adjacent nodes
(iii) One triangle may be connected to one of its vertices

Then we will apply the following rules (the colour of one triangle is the sum of the colours of its three vertices) :

1 *Two triangles sharing an internal edge are connected (connexion i).*
2 *The two ends of a weak edge are connected (connexion ii).*

3 *Each triangle of the colour $3i + 1$ touching an internal edge is connected to its vertex having the colour $i + 1$ (connexion iii).*

4 *The following vertices are named "departure vertices" :*
  - *Multiple vertices*
  - *Vertices which are ends of a weak edge*
  - *Vertices with the colour $i + 1$ belonging to a triangle of colour $3i + 1$ which touches an internal edge*

5 *From each departure vertex, we draw a string of vertices having (strictly) increasing colour. (For a multiple vertex, we do so for each very-strong component that it is linked to, as far it is possible : if the multiple vertex has the colour n, it does not imply that we shall find in each very-strong component one adjacent node of its with the colour $n + 1$.) The process ends as soon as we join a vertex belonging to another string, or one summit of an irregular triangle. Successive vertices of the string are connected one to another (connexion ii), and the final vertex will be connected to one and only one irregular triangle it touches (connexion iii), if no string fusion has occurred.*

We notice that many of those choices are not unique. We could imagine some schemes where those artificial connexions were a little bit more guided. However, for the time being we consider this approach as a minor improvement.

Theorem 15. *If Ω is a simply-connected domain, the machine-tree has a structure of acyclic graph.*

The demonstration should be available on [8].

Apart from those irregularities, the graph is very regular, so that optimal fronts can "move" freely. So a natural idea is to choose this irregularity spine to guide fronts in a parallel computation : irregularities are taken in consideration on the top of the front, while the two arms are guarantied to be optimal.

Of course, if the graph is very regular, the irregularity spine will have lots of useless branches, and many generalized elements will be created and then merged into others. George discovered that excessive use of nested dissections introduces a great latency in computations. We shall see in the next chapter that for non-elliptic problems, this may cause an heavy additional cost, as expected.

Theorem 16. *Under the following restrictions,*

  a *Elliptic problem*
  b *Simply-connected domain Ω*
  c *Uniformly regular triangulation*

the convergence of this scheme is in $O(n^2)$, where n is the number of triangles.

In fact, this result reflects more the hard irregular case, since the convergence is dominated by front fusions. We also note that using the algorithm described in the following paragraph and additionnal properties due to uniform regularity, we will be able to remove the statement "a".

4   Local Pivot and Dual Elimination

It is not always possible to insure that a matrix built from finite element is symmetric positive. This is why introducing numerical pivot is an interesting challenge. Of course, Hood [11] described a way to introduce pivots on local matrices especially adapted to the general concept of the frontal method. However, this is important to keep some control on the frontal matrix size and the previous article assumes that this size may increase arbitrarily (The solution given to excessive size of matrix is to change the maximal size parameter and rerun the program).

So choosing for a given column a pivot row which may not be diagonally associated brings in a difficult alternative : eliminating with an incomplete row or assembly some unwanted elements.

Another significant problem - to which we have no complete solution yet - is about the double elimination of rows. Imagine that the computation is such that two fronts are to fusion. On the one side variable i has been eliminated with "frontal" row k. But on the other side, row k has already been selected to eliminate another variable j. Of course, we have to avoid that for algebraic consistency.

Why wonder about those typically fusion-associated problems ? Normally the assymptotical rate of fusion time should tend to zero, but it is however important for realistic domains. We have to carry on these problems which are doubtless the bottleneck of *Deepness Analysis*.

4.1   A Frontal Approach : Single Front With Postponed Elimination

Principle We present here an alternative method to Hood's one which uses only partial pivots but (minimizes and) warranties the maximum size of the frontal matrix. This algorithm consists in considering elimination like a data-consumption : we give in entrance successive coefficients resulting from assembly, and variable elimination commands - in the same order as the elliptic case. The processor in charge of elimination has to handle these commands and manage its frontal matrix.

So, how to perform the pivot ? In fact, while starting the computation, our processor knows not only the column (ie the variable number) but it may also compute the row which it should use for elimination (maximum absolute value). So it is able to do the pivot choice, but generally not to achieve the elimination.

A good way to avoid the problem consists in introducing *symbolic* or *virtual* elimination : instead of forcing elimination completion, the main data to achieve it in the future is stored. For this, we introduce a *list of incomplete eliminations*. Whenever an unachievable elimination occurs, a new element is added to the list. Typically we are going to store pivot column.

Meanwhile, it is obvious that with such a list, a frontal matrix is never up-to-date. Some operations will be needed to recover an accurate form at some points. Moreover, some additional data will be necessary : if elimination on uncomplete row j has been decided, we will have to store somewhere incomplete coefficients of the row. One idea consists in storing them at the end of the frontal matrix. This additional data is called *array for row completion*.

To conclude, we can put all these structures in a unique array called *extended frontal matrix*. Data will be here stored in different areas, depending on how we use it. We remember that each row corresponds to a variable number. Let us classify those variables according to their current state; this way we obtain six categories (See figure 7).

In the elliptic case, only two categories remain : A (set of active variables) and F (eliminated ones). Introducing this partial pivot, we obtain an $(A + B + C) - by - (A + D)$ frontal matrix, an $(A+B+C)-by-(B+E)$ list of incomplete eliminations and a $D - by - (A + D + B + E)$ array for row completion. So the extended frontal matrix M will be an $(A + B + C + D) - by - (A + D + B + E)$ array.

What kind of control do we have over M's size ? We shall show that, during the computation, the sizes of previous sets verify

$$\#D = \#B + \#C = \#B + \#E$$

So M is a square array, and if n is the number of "elliptic" variables,
$$\#M \leq 4n^2$$

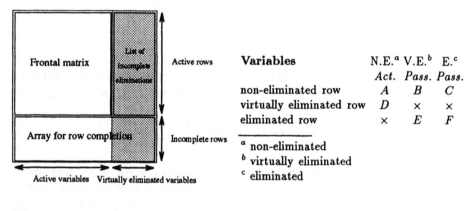

Variables	N.E.[a]	V.E.[b]	E.[c]
	Act.	*Pass.*	*Pass.*
non-eliminated row	A	B	C
virtually eliminated row	D	$\times$	$\times$
eliminated row	$\times$	E	F

[a] non-eliminated
[b] virtually eliminated
[c] eliminated

▓ Complete coefficients area

Fig. 7. Extended frontal matrix

Practical algorithm More precisely, we shall note $x : A \to B$ the passage of variable x from category A to category B. We will also note $pivot(x)$ the index of the row (and the corresponding variable) chosen to eliminate x fictively.

$$pivot : (B + E) \mapsto D$$

So let us eliminate variable x.

(1) Complete column x (note that we first store results in an array m).

$$\text{Initialization} \quad m_y := M_{y,x} \qquad\qquad \forall y \in (A+B+C+D)$$

$$\text{Completion} \quad m_y := M_{y,x} - \frac{M_{y,z} \times M_{pivot(z),y}}{M_{pivot(z),z}}, \quad \begin{cases} \forall y \in (A+B+C) \\ \forall z \in (B+E) \end{cases}$$

(2) Do the pivot choice : we note the row $y \in (A+B+C)$.

(3) Proceed to elimination. Two cases are possible.

- *$y \in A$ ie y active variable. Then we make virtual elimination; $x : A \to B$ or $x : D \to E$, $y : A \to D$ with $pivot(x) = y$. And we have to update :*

$$M_{.,x} := m$$

- *$y \in (B+C)$ ie y passive variable. Then elimination can be completed. We have the following transfers among categories :*

$$x : A \to C \text{ or } x : D \to F$$
$$y : B \to E \text{ or } y : C \to F$$

We have also to proceed to concrete elimination

$$M_{\alpha,\beta} = M_{\alpha,\beta} - \frac{M_{\alpha,x} \times M_{y,\beta}}{M_{y,x}} \quad \begin{cases} \forall \alpha \in (A+B+C+D) \\ \forall \beta \in (A+D) \end{cases}$$

and there we shall use the fact that $m_z = M_{z,x}$.

(0) Perform (now ready) postponed eliminations

- *Add newly assembled coefficients to frontal matrix and array of row completion.*
- *If $x \in D$, proceed to effective elimination of variable y such that $pivot(y) = x$.*

$$M_{\alpha,\beta} = M_{\alpha,\beta} - \frac{M_{\alpha,y} \times M_{x,\beta}}{M_{x,y}} \quad \begin{cases} \forall \alpha \in (A+B+C) \\ \forall \beta \in (A+D) \end{cases}$$

This algorithm shows that in fact managing local pivot introduces many backward calculations. However, while computing time at most four times longer, an average behaviour should lead us to a cost of $1.5 \times 1.5 = 2.25$ times the elliptic one, which seems acceptable.

Another interesting comment concerns variable managing. We noticed that embarrassing coefficients of variables were stored in columns B and C of the matrix. We have to note that these rows are either pivot remembering or dependence coefficients. So it is perfectly possible that variable x starts being virtually eliminated, then that y is completely eliminated before x's problem is solved. Indeed, coefficients of the array of row completion have a very different meaning to those of the frontal matrix. Moreover backward substitution will be slightly modified. It will include a progressive "building" of variables during the time separating virtual elimination from effective one ...

Meanwhile, the problem of double elimination is still not solved. The next paragraph gives a means to avoid big losses in the case of frontal fusions.

4.2 Dual Elimination To Prevent From Double Elimination

One may argue that we could put flags so that one variable would be assigned to a unique front. Not only this method will increase the front's size but it will also introduce global communications. Let us focus on a way to avoid this.

Partial Gauss pivot is warranted to be effective by determinant analysis - which specify that at each stage, a non-zero coefficient will be found in the current column.

We shall here be slightly thinner : one is not obliged to use columns for pivoting. A dual alternative consists in using rows for that purpose. The interest we have there, is that if we choose which row will be used for an elimination, we avoid the possible case in which the front of the other side uses frontal rows.

Of course, inside one front, the way to advance must be uniform : we cannot handle both incomplete rows and incomplete columns. That is why a similar front must have a unique elimination mode.

Total fusion These ideas are particularly well-fitted to the bifrontal scheme; each of the two fronts may use a different elimination mode and final elimination (ie front fusion) will be done without any problem of double elimination.

How to extend this result to multifrontal schemes ?

Partial fusion - Bicoloring problem Partial fusion is more difficult to handle. The problem is that they make a mixed front appear which is then totally available to move. We have to include some post-frontal variables, or even use a third front of appropriate mode to escape from this.

So each spine node should have an even number of parallel fronts. We may then notice that such a way of working will increase the total number of variables to fusion - simply because each 3-front node has to be replaced by a 4-front node. However, elimination is then much easier and some more parallelism is introduced.

Transforming each odd node in to an even one, choosing for each front one of the dual elimination modes belongs to a well-known problem : bicoloring.

5 Experimental results

It is interesting to see what kind of improvement we get, regardless of the machine (or architecture) used. So here we will focus on the efficiency parameter (see [19, 18]), and compare for some examples our efficiency to the King's algorithm one ([14]). To be more honnest, all vertices have been tested as pseudo-peripheral nodes, so that finding a "good" pseudo-diameter should not been taken into account in the use of this scheme. Of course, we give the best performance among all these various choices.

In order to to give a better idea on how parallelizable is the new algorithm, we also give which part of the efficiency parameter is concerned with the front fusion process.

Example (#nodes)	King alg.	*Deepness Analysis*	Improvement	Fusion Share	%
1 (295)	67538	20789	69.2%	6789	32.7%
2 (171)	16393	5190	68.3%	738	14.2%
3 (1015)	622922	222180	64.3%	69442	31.3%
4 (400)	82537	42789	48.2%	4466	10.4%
5 (553)	133160	91289	31.4%	9717	10.6%
6 (253)	21786	9528	56.3%	564	5.9%

Fig. 8. King algorithm vs Deepness analysis

In figure 8, matrices 1 and 3 are purely artificial triangulated planar graphs. Matrice 2 comes from a real-life problem raised by Hydro-Quebec : it is a modelisation of the "Saint-Laurent", used, for instance, to study the propagation of pollution. Triangulations 4, 5 and 6 are issued from a Delaunay algorithm and are typically regular.

We notice that not only *Deepness Analysis* improves the efficiency parameters but also tends to reduce front fusion for regular cases. The sequential part is not a simple mono-processor phase : the computation is shared by only two processors so that meanwhile some other tasks may be performed. However, the parallelism directly introduced by *Deepness Analysis* leads definitively to coarse-grained parallelism.

Conclusion

Today, it is difficult to see how helpful *deepness analysis* is for common-life meshes. Even if theoretically, it costs $O(n^2)$, we saw that its latency constant is completely comparable with normal frontal methods and using George's considerations on regular meshes, we can draw an interesting scheme for each triangulation with $O(n^{1.5})$ convergence rate.

As we said at the beginning, using triangles is probably more an advantage than an inconvenient of our method. However, it could be interesting to see if such schemes have an interest for other kinds of domain structures like quadrangulations. Of course, we could transform the quadrangulation into a triangulation and introduce some additional degrees of liberty to give new elements a "variable-independent" form.

So the main challenge is to export these ideas to the third dimension. In particular, the main notions on connectivity remain but many hard problems may appear; if a very-strong component of the colour n is simply connected, that does not imply that its generated very-strong components of the colour $n + 1$ will be simply connected, and we know the importance of simple-connectivity in the demonstrations given here.

Acknoledgements

I would like to thank Laurie Hendren for her most helpful comments. I am also grateful towards the ACAPS lab., and more particularly towards P.Panangaden, C.Verbrunge, and R.Duquette. The refere's comments were very useful.

References

1. J.J. Dongarra, I.S. Duff, D.C. Sorensen and H.A. Van der Vorst. Solving Linear Systems on Vector and Shared Memory Computers. 1991.

2. I.S. Duff. Parallel implementation of multifrontal schemes. *Parallel Computing*, 3:193-204, 1986.

3. I.S. Duff, A.S. Erisman and J.K. Reid. Direct Methods for Sparse Matrices.. *Oxford Science Publications*, 1986.

4. I.S. Duff and J.K. Reid. The multifrontal solution of indefinite sparse symmetric linear systems. *ACM Trans. Math. Softw.*, 9:302-325, 1983.

5. I.S. Duff and J.K. Reid. The multifrontal solution of unsymmetric sets of linear equations. *SIAM J. Sci. Stat. Comput.*, 5:633-641, 1984.

6. I.S. Duff and J.A. Scott. MA42 - A new frontal code for solving sparse unsymmetric systems. *Rutherford Appleton Laboratory*, Report RAL-93-064, September 93.

7. D.J. Evans. The use of preconditioning in iterative methods for solving linear equations with symmetric positive definite matrices. *J. Inst. Maths. Applics.*, 4:295-314, 1967.

8. J. Galtier Deepness analysis : bringinig optimal fronts to triangular finite element analysis. *ACAPS memo 73*, ACAPS Lab., McGill University, Montreal, Canada.

9. A. George. Nested dissection of a regular finite element mesh. *SIAM J. Numer. Anal. Vol.10 No2*, April 1973

10. A.J. Hoffman, M.S. Martin and D.J. Rose. Complexity bounds for regular finite difference and finite element grids. *SIAM J. Numer. Anal.*, Vol.10 No2, April 1973

11. P. Hood. Frontal Solution Program for Unsymmetric Matrices. *Int. J. num. Meth. Engng.*, 10:379-399, 1976.

12. B. Irons A Frontal Solution Program for Finite Element Analysis. *Int. J. num. Meth. Engng.*, 2:5-32, 1970.

13. D.A.H. Jacobs. Preconditioned conjugate gradient methods for solving systems of algebraic equations. Note No. RD/L/N193/80, CERL, Leatherhead, 1980.

14. P. King. Automatic reordering scheme for simultaneous equations derived from network systems. *Int. J. num. Meth. Engng.*, 2:523-533, 1970.

15. R.J. Lipton, D.J. Rose and R.E. Tarjan. Generalized nested dissection. *SIAM J. Numer. Anal.*, Vol.16, No 2:346-358, 1979.

16. J.K. Reid. Frontal methods for solving finite-element systems of linear equations. *Sparse Matrices and Their Uses*, I.S. Duff(editor), Academic Press Inc. (London) LTD, 1981.

17. D.J. Rose. Triangulated graphs and the elimination process. *J. of Math. Anal. and Appl.*, 32:597-609, 1970.

18. S.W. Sloan and M.F. Randolph. Automatic Element Reordering for Finite Element Analysis with frontal solution schemes. *Int. J. num. Meth. Engng.*, 19:1153-1181, 1983.

19. W.F. Tinney and J.W. Walker. Direct solutions of sparse network equations by optimally ordered triangular factorization. Proc. of the IEEE, Vol.55, No11, November 1967.

20. M. Yannakakis. Computing the minimum fill-in is NP-complete. *SIAM J. Alg. Disc. Meth.*, Vol.2, No1, March 1981.

Communications in Bus Networks[*]

André Hily and Dominique Sotteau

LRI, UA 410 CNRS, bât 490, Université de Paris-Sud, 91405 Orsay, France

Extended Abstract

1 Introduction

Bus interconnection networks generalise the notion of point to point interconnection networks. In a bus network the processors share a communication medium called a bus. In a very natural way they are modeled by hypergraphs in which the vertices represent the processors and the hyperedges represent the buses that group the processors which share the same communication link. Communications in point to point interconnection networks have been much studied (see [5] for a survey). Bus networks have also been considered and the hypergraph model has been used to study several implementation of the networks (see for example [3], [4], [12], [13]). The results, however, are sparse and in very different contexts and not much seems to have been done in terms of a systematic theoretical study of models and modes of communication in bus networks in general.

We have chosen to present here some of the possible modes of communications and, for each of them, to study broadcasting and gossiping, two communications processes essential to many algorithms. We first give some results on bounds for broadcasting and gossiping in general bus networks in the various models (see [2], [10], [11]).. We then concentrate on mesh-bus networks. mesh-bus networks have already been extensively studied and shown to be efficient for many algorithms, in particular routing and sorting (see [9] for references). The results presented here are not necessarily all new but most of the old ones only deal with 2-dimensional mesh-bus networks (see for example [6], [7] where more references can be found).

Let us first give some definitions and notation (see also [1]). A hypergraph H is defined by its set V of vertices $\{x_1, x_2, \cdots, x_N\}$ and its set of edges $E = \{E_1, E_2, \cdots, E_m\}$ where each E_i is a non empty subset of V, the union of the E_i's being V (so that there are no isolated vertices). The degree of a vertex is the

[*] The work was supported partially by NSERC of Canada while the second author was visiting McGill University and by PRC C3 of France.

number of edges it belongs to. A path between two vertices x and y is defined by an alternating sequence of vertices and edges, $x = x_1, E_1, x_2, E_2, \cdots E_l, x_{l+1} = y$, such that each E_i contains vertices x_i and x_{i+1}. The distance between two vertices x and y is the length of a shortest path between x and y, i.e. the number of edges in the path. The diameter of a hypergraph is the maximum of the distances between any two vertices. We will talk without distinction of a bus network or of the hypergraph which represents it. Thus we can talk about the degree of a node (processor) or of a vertex, the distance between two nodes, the diameter of a bus network, etc. All our networks are assumed to be connected i.e. there exists a path between any two nodes.

A mesh-bus network of dimension d, denoted $M(n_1, n_2, \cdots, n_d)$, is represented by the following hypergraph. The vertex set is $\{(x_1, x_2, \cdots, x_d) \mid x_i \in \{0, 1, \cdots, n_i - 1\}\}$. An edge contains the vertices which agree on all coordinates but one. Clearly the diameter of a d-dimensional mesh-bus network is equal to d. The number of its vertices is $n_1 \cdot n_2 \cdots n_d$. For any i, $1 \le i \le d$, the mesh-bus network has $n_1 \cdot n_2 \cdots n_{i-1} \cdot n_{i+1} \cdots n_d$ buses of size n_i, which are called buses in dimension i, and are determined by a choice of all coordinates but the ith. For the sake of simplicity we will assume in the following that all the n_i's are equal to n, so that the number of vertices of $M(n, n, \cdots, n)$ is n^d and the number of buses is $d \cdot n^{d-1}$, each of size n, with n^{d-1} buses in each dimension (all results extend easily to the general case).

Broadcasting refers to the process of sending a message from one node of the networks to all the other nodes. In gossiping, initially every node has its own message (that we will here assume to be of the same - unit - length) and at the end of the process every node should know the message of all the other nodes.

We will consider here communications where messages are sent in *store-and-forward* mode, which means that any node can only send the content of a message after it has received all the bits. Also the network is assumed to function in a synchronous way, so that all receptions or emissions of messages are done simultaneously at each step, and take one time unit. At each step any node can either send a message on a bus to which it is connected or receive a message on this bus, but not both at the same time on the same bus. This hypothesis will be automatically satisfied here since we also assume that any bus has width 1, which means that it can carry at most one message in each step. For a similar analysis of communications with buses of width k, $k > 1$, see [8].

We will study our communications with each of the following two hypothesis:

- 1-*port hypothesis* where each node can only use one of the buses to which it is connected at each step (either to send or receive on this bus)
- Δ-*port hypothesis* where each node can either send or receive on every bus to which it is connected. However, as we said above, a node can never receive a message on one bus and *at the same step* send this message on another bus.

We will also distinguish between the two following possibilities when forwarding messages:

- *unit length message*: any node can only send (forward) a message of unit length at each time unit on all the buses he can use (the same message),
- *arbitrary length message*: any node can concatenate any number of messages he has received (with also its own message) and send the whole new message in only one time unit on all buses he can use. In other words any bus can carry a message of arbitrary length in each step.

This last model was studied in [7] and the intermediate model where each message can carry up to l of the unit length messages was studied in [6] but only under the Δ-port hypothesis and for 2-dimensional meshes.

This is an extended abstract of our talk. Not all proofs are given. The notation are taken from [5]. For any bus network H, $b_1(H)$ (resp. $g_1(H)$) stands for the minimum time (i.e. minimum number of steps) necessary to broadcast (resp. gossip) in H under the 1-port hypothesis. The index 1 is replaced by $*$ under the Δ-port hypothesis. To avoid abusive indices we use the same notation for the minimum broadcast or gossip time with unit or arbitrary length messages, but we make it clear by the context.

2 Broadcasting

2.1 General results on broadcasting in a bus network

In the broadcasting model, only one message is travelling among the various processors, so the we don't need to take care of the length of the messages that might be transmitted along the information process. The results on the minimum time required to send the message from its originator to all the nodes of the network only depends on the choice of hypothesis 1-port or Δ-port.

The minimum time required to send a message from any given node x of a bus network H to all the other nodes under the Δ-port hypothesis is equal to the excentricity of x, that is the maximum distance from x to any other node of H. Thus we have the obvious following result.

Theorem 1. *The broadcast time of any network H under the Δ-port hypothesis is equal to its diameter D*

$$b_*(H) = D.$$

The diameter is necessarily also a lower bound on the broadcast time under the 1-port hypothesis. More precisely we can state the following.

Theorem 2. *The broadcast time of any bus network H on N nodes with diameter D and bus size r under the 1-port hypothesis satisfies the following:*

$$\max\{D, \lceil \log_r N \rceil\} \le b_1(H) \le 2 \cdot \lfloor \frac{N}{r} \rfloor$$

Proof. The number of nodes which have the message is at most multiplied by r at any step, since any node which has the message can inform at most $(r-1)$ new nodes contained in a bus to which it belongs.

To get the upper bound, we give an algorithm in which at least r new nodes are informed in two consecutive steps. Let us introduce some notation.

For any $k \geq 1$, let I_k be the set of nodes which have the message after step k, V_k be the set of nodes not in I_k but at distance 1 of some node in I_k and L_k be the set of all the other nodes (not in I_k or V_k).

Phase 1. At step 1, the originator sends its message along one of the bus to which it is connected. Clearly $r - 1$ new nodes are informed in step 1.

Phase 2. While L_k is not empty, choose a bus E containing a node of L_k and a node y of V_k. Such a bus exists since the network is connected, and it contains no node of I_k by definition of L_k. Let us inform node y at step $k + 1$ and then the other nodes of E at time $k + 2$. Then in two steps at least r new nodes get the message. If V_k is also empty at the end of this second phase, the algorithm stops. Otherwise, the algorithm goes on with the following phase.

Phase 3. Now L_k is empty. Repeat the following as long as V_k is not empty. Let us define a sequence $E_1, E_2, \cdots, E_l$, $1 \leq l \leq i_1$, of buses together with a sequence $x_1, x_2, \cdots, x_l$ of distinct vertices of I_k respectively in each of these buses as follows.

Let E_1 be a bus containing as few nodes of I_k as possible, say i_1. Choose for x_1 any vertex of E_1 which is in I_k. Let $E_1, E_2, \cdots, E_j$ and $x_1, x_2, \cdots, x_j$ be defined for some j , $j \geq 1$.

If there is no vertex of V_k left outside of these buses, or if $j = i_1$, then we stop and we take $l = j$.

Otherwise let E_{j+1} be a bus containing a vertex of V_k outside of the previous buses, and containing as few vertices of I_k as possible, say i_{j+1}. Necessarily $l \leq i_1 \leq i_2 \cdots \leq i_j \leq i_{j+1}$, and therefore since $j < i_1$ it is possible to choose a vertex x_{j+1} of E_{j+1} in I_k which is different from $x_1, x_2, \cdots, x_j$. The construction of the sequence stops either because $l = i_1$ or because there is no vertex left in V_k.

For any k, the sequence being constructed, at step $k + 1$ each vertex x_i sends the message along the bus E_i. If a vertex of V_k is on several buses E_j, it will receive the message only from one of them to avoid collision. Either the construction of the sequence was stopped because all the vertices of V_k were covered by buses of the sequence, and this step $k + 1$ is the last one of the algorithm. Or at step $k + 1$ at least $r - 1$ new vertices receive the message: the $r - i_1$ vertices of V_k which are in E_1 and one vertex of each bus E_j, $2 \leq j \leq i_1$.

Phase 3 stops during a step k where all vertices of V_k get the message, and therefore all the vertices of the network are informed.

Phase 1 takes one step to inform $r - 1$ new vertices. During phase 2, at least r new vertices get the message every two steps. During phase 3, at least $r - 1$ vertices get the message at each step except maybe the last step where every vertex which didn't have the message yet gets it. Clearly phase 2 is the slowest. Therefore the number of steps other than the first one is at most equal to $2 \cdot \lfloor \frac{N-1-r-1}{r} \rfloor$ if N is a multiple of r (in that case the algorithm may end after Phase 2). If N is not a multiple of r the number of steps other than the first one and the last one is at most equal to $2 \cdot \lfloor \frac{N-r-1}{r} \rfloor$. $\square$

2.2 Broadcasting in the d-dimensional mesh-bus network

Broadcasting in the mesh-bus network can be done in d steps under the 1- or Δ-port hypothesis (i.e. to a number of steps equal to the diameter). Indeed, from any node $(a_1, a_2, \cdots, a_d)$ the algorithm proceeds as follows. At step 1, node $(a_1, a_2, \cdots, a_d)$ sends the message to nodes $\{(x_1, a_2, \cdots, a_d) \mid 1 \leq x_1 \leq n\}$ which are all contained in the same bus which is a bus in dimension 1. For any j, $2 \leq j \leq D$, at step j, the n^{j-1} nodes $\{(x_1, \ldots, x_{j-1}, a_j, \ldots, a_d) \mid 1 \leq x_i \leq n\}$ send the message they have received to nodes $\{(x_1, \ldots, x_j, a_{j+1}, \ldots, a_d) \mid 1 \leq x_i \leq n\}$ using disjoint buses along dimension j.

3 General results on gossiping

We assume that our bus networks have N nodes, m buses of size at most r, are regular of degree Δ (i.e. every node is on Δ buses) and have diameter D.

3.1 Unit length message

Theorem 3. *The gossiping time of any bus network H of degree Δ, with N nodes and m buses of size at most r satisfies the following inequalities,*
under the 1-port hypothesis : $\frac{(N-1)r}{r-1} \leq g_1(H) \leq 2N \cdot \lfloor \cdot \frac{N}{r} \rfloor$,
under the Δ-port hypothesis : $\max\{\lceil \frac{(N-1)r}{(r-1)\Delta} \rceil, \lceil \frac{(N-1)N}{(r-1)m} \rceil\} \leq g_*(H) \leq N \cdot D$.

Proof.
Lower bounds. To get to all the vertices of the network, the message from any vertex has to be sent on at least $\frac{N-1}{r-1}$ buses since at most $r-1$ new vertices receive the message when it is sent on one bus. So altogether, since N messages have to reach all the vertices of the network, $N \cdot \frac{N-1}{r-1}$ emissions have to be made, since only one message of unit length can be sent on a bus. Since the network has at most m buses which can be used in parallel, the number of steps needed, irrespective of the 1- or Δ-port hypothesis, is at least equal to $\frac{(N-1)N}{(r-1)m}$. Also since the network has N vertices, at least one vertex has to realise at least $\frac{N-1}{r-1}$ emissions in addition to the reception of the $N-1$ messages other than its own. Since at each step every vertex can either send or receive on only one bus under the 1-port hypothesis, and on at most Δ buses under the Δ-port hypothesis the gossiping needs at least $\frac{N-1}{r-1} + N - 1$ steps under the 1-port hypothesis, and at least $\frac{(N-1)r}{(r-1)\Delta}$ steps under the Δ-port hypothesis.
The *upper bounds* are the ones obtained by assuming that every vertex broadcasts its message to all the other ones successively, using the results of Theorem 1 and Theorem 2. $\square$

3.2 Arbitrary length message

Theorem 4. *The gossiping time of any bus network of degree Δ, with N nodes and m buses of size r satisfies the following inequalities, under the 1-port hypothesis :*

$$\max\{\lceil \log_2 N \rceil, \left\lceil \frac{N}{m} \right\rceil\} \leq g_1(H) \leq \min\{(N-1),(r-1)D\Delta\} + 2\lfloor \frac{N}{r} \rfloor$$

under the Δ-port hypothesis :

$$\max\{\lceil \log_\Delta N \rceil, \left\lceil \frac{N}{\Delta} \right\rceil\} \leq g_*(H) \leq \min\{N + D - 1, r \cdot D\}$$

Proof:

Lower bounds. Let $m_{i,j}$ denote the number of elementary messages at node i after j steps, and $M_j = max \{m_{i,j} \mid i \in \{1,\ldots,N\}\}$. Then under the 1-port hypothesis $M_j \leq 2M_{j-1}$, and thus $M_j \leq 2^j$. Under the Δ-port hypothesis $M_j \leq \Delta M_{j-1}$ and thus $M_j \leq \Delta^j$. This leads obviously to the respective lower bounds of $log_2 N$ and $\log_\Delta N$ since the number of messages at any node must be N when the gossiping is completed. The other lower bound, under the Δ-port hypothesis, is the one we already had for the more constrained model of unit length message. Any vertex has to send its message at least once. Since the number of buses is m, at most m vertices can send their message in one step. This gives a lower bound of $\lceil \frac{N}{m} \rceil$ which gives something better only for the 1-port hypothesis.

Upper bounds. To get the upper bound, the algorithm consists in gathering all messages in one of the nodes, say x_0, and then broadcast all messages from x_0 to all the other nodes. Under either hypothesis (1-port or Δ-port), the gathering takes at most $N - 1$ steps. Indeed nodes at distance D from x_0 send their message sequentially to nodes at distance $D - 1$. Then nodes at distance $D - 1$ send their information (messages they have received together with their own message) sequentially to nodes at distance $D - 2$ and so on. But also, under the Δ-port hypothesis, the gathering takes at most $(r - 1)D$ steps, since in $r - 1$ steps all nodes at distance D from x_0 can send their message to some nodes at distance $D - 1$ and so on. And under the 1-port hypothesis, the gathering takes at most $(r-1)\Delta D\}$ steps, since $(r-1)\Delta$ is the maximum number of steps needed for all nodes at distance l from x_0 to send their message to some node at distance $l - 1$. The result then follows by adding the upper bound on the broadcast time from Theorems 1 and 2. □

4 Gossiping in the mesh-bus network

All the previous results obviously hold for the mesh-bus network with $D = d$, $\Delta = d$, $r = n$, $N = n^d$ and $m = d \cdot n^{d-1}$. We will improve some of the general results in case of a mesh-bus network. The lower bounds on the communication times in a d-dimensional mesh-bus network are generally the same for any dimension d But very often we have found better algorithms (often optimal) in the case of 2-dimensional meshes than in the general case. For a 2-dimensional mesh we will naturally talk about horizontal or vertical buses for the buses of the two dimensions.

4.1 Unit length message

Δ-port hypothesis.

Theorem 5. *For any integer $d \geq 2$, the minimum time to gossip in a d-dimensional mesh-bus network $M(n, \cdots, n)$ with message of unit length under the Δ-port hypothesis is bounded as follows.*

$$\frac{n^d + n^{d-1} + \cdots + n^2 + n}{d} \leq g_*(M(n, \cdots, n)) \leq n^{d-2}\left(\frac{n^2 + n}{2} + (n \bmod 2)\right)$$

Proof.
The *lower bound* is the one obtained from Theorem 3 with $N = n^d$, $r = n$ and $\Delta = d$.
Upper bound. Let $\theta(n, d)$ denote the number of steps of the algorithm given below to gossip in the d-dimensional mesh $M(n, \cdots, n)$.

1. If $d > 2$, a_d being fixed, every d-dimensional sub-mesh defined by $x_d = a_d$ follows independently the algorithm recursively.
2. Meanwhile, a gossiping is realised in n steps independently on each of the n^{d-1} buses in dimension d, which contain, for any given $a_1, \cdots, a_{d-1}$, the vertices $\{(a_1, \ldots, a_{d-1}, x_d) \mid 1 \leq x_d \leq n\}$. This is possible in parallel because of the Δ-port hypothesis and because $\theta(n, d-1) \geq n$ (this will be justify later). After the $\theta(n, d-1)$ steps, each $(d-1)$-dimensional sub-mesh has realised the gossiping of the messages of its own vertices, and moreover it contains globaly all the messages of the vertices of the d-dimensional mesh, with $n-1$ messages from the vertices of other sub-meshes in each of its vertices.
3. Now, in each of these $d-1$-dimensional sub-meshes, $n-1$ gossiping are realised sequentially, with one of the $n-1$ new messages from the other sub-meshes on each vertex for each new gossiping.
4. The induction stops when $d = 2$ in which case the algorithm used in the sub-mesh of dimension 2 is the one given in Theorem 6.

Thus we have the following recurrence equation.
For $d > 2$, $\theta(n, d) = n \cdot \theta(n, d-1)$, with $\theta(n, 2) = \frac{n(n+1)}{2} + (n \bmod 2)$.
Solving this recurrence equation gives the expression of $\theta(n, d)$, which is more than n for $d > 2$ (which justifies the simultaneity of the first two phases) and thus we have the upper bound. $\qquad\square$

In the case of a 2-dimensional mesh we have the following optimal result.

Theorem 6. *The minimum time to gossip in a 2-dimensional mesh-bus network* $M(n, n)$ *with message of unit length under the Δ-port hypothesis is*

$$g_*(M(n, n)) = \frac{n(n + 1)}{2} + (n \bmod 2).$$

Proof.

Lower bound. Theorem 5 gives a lower bound of $n(n + 1)/2$ when $d = 2$. By looking at the proof given for Theorem 3 to obtain the lower bound, it is easy to see that, if the gossiping can be done in exactly $n(n+1)/2$ steps, then necessarily the $2n$ buses are used at each step, and each message is sent $n + 1$ times exactly, once in one dimension, the other n times in the other dimension. Let a (resp. b) be the number of messages which are first sent on a horizontal (resp. vertical) bus and then sent on n vertical (resp. horizontal) buses. During the gossiping $n(n+1)n/2$ horizontal buses are used and as many vertical buses. Thus we must have

$$a + nb = n^2(n + 1)/2$$
$$b + na = n^2(n + 1)/2$$

Adding the two equalities gives $(n + 1)(a + b) = n^2(n + 1)/2$ and therefore $a + b = n^2$. Substracting them gives $(a - b) + n(b - a) = 0$. Therefore we must have $a = b = n^2/2$ which is impossible if n is odd. So we have the lower bound.

Upper bound. The idea is to give an algorithm to gossip which uses the $2n$ buses at each step if n is even as follows. Partition the n^2 vertices of the 2-dimensional mesh into $\lfloor n/2 \rfloor$ classes C_i, $1 \le i \le \lfloor n/2 \rfloor$, of $2n$ vertices each: $C_i = H_i \cup V_i$ with
$H_i = \{(x_1, x_2) \mid x_2 = (x_1 + 2i - 3) \bmod n + 1\}$,
$V_i = \{(x_1, x_2) \mid x_2 = (x_1 + 2i - 2) \bmod n + 1\}$
plus, if n is odd, $n = 2q + 1$, one class C_{q+1} of n elements: $C_{q+1} = H_{q+1} \cup V_{q+1}$ with
$H_{q+1} = \{(x_1, x_2) \mid x_1 \text{ odd and } x_2 = (x_1 - 2) \bmod n + 1\}$,
$V_{q+1} = \{(x_1, x_2) \mid x_1 \text{ even and } x_2 = (x_1 - 2) \bmod n + 1\}$.
This partition satisfies $H_i \cap V_i = \emptyset$. Also for each i, all the vertices of H_i are in distinct horizontal buses, and all the vertices of V_i are in different vertical buses. At step 1 of each phase i each of the vertices of H_i sends its message along its horizontal bus, and each of the vertices of V_i send its message along its vertical bus. This can be done in parallel because of the Δ-port hypothesis. During the next steps of phase i, a gossiping of the messages of the vertices of H_i is realised on each of the n vertical buses and in parallel a gossiping of the messages of the vertices of V_i on each of the n horizontal buses of the mesh. This second part of phase i takes n steps for $1 \le i \le \lfloor n/2 \rfloor$ since each H_i and each V_i has cardinality n. It takes $q + 1$ steps for $i = q + 1$ if n is odd since $|H_{q+1}| = q + 1$ and $|V_{q+1}| = q$. After Phase i, all the messages of the vertices of C_i have reached all the vertices of the mesh. These phases i, $1 \le i \le \lceil \frac{n}{2} \rceil$ are independent and are executed sequentially.

If n is even the gossiping is completed after $\frac{n}{2}$ phases of $n + 1$ steps each.
If n is odd the gossiping is completed after $\frac{n-1}{2}$ phases of $n + 1$ steps each and one phase of $\frac{n-1}{2} + 2$ steps. $\qquad\square$

1-port hypothesis. Under the 1-port hypothesis it is not difficult to get the following optimal result, valid for any dimension $d \geq 2$.

Theorem 7. *For any integer $d \geq 2$, the minimum time to gossip in a d-dimensional mesh-bus network $M(n, \cdots, n)$ with message of unit length under the 1-port hypothesis is*

$$g_1(M(n, \cdots, n)) = n^d + n^{d-1} + \cdots + n^2 + n.$$

Proof.
The *lower bound* is the one of Theorem 3 applied with $N = n^d$ and $r = n$.
Upper bound. The following algorithm completes the gossiping in $n^d + n^{d-1} + \cdots + n^2 + n$ steps.
For a fixed a_1, each of the n^{d-1} vertices $(a_1, x_2, \ldots, x_d)$ sends its message in dimension 1 to all vertices of its bus $\{(x_1, x_2, \ldots, x_d) \mid x_1 \neq a_1\}$, in one step. This is done for each possible value of a_1 between 1 and n. Therefore, after n steps each sub-mesh of dimension $n-1$ determined by the value of a_1 contains globaly the messages of all the vertices of the d-dimensional mesh, each vertex containing n unit length messages. We now repeat this step recursively, but n times because the messages sent can only be of unit length. Therefore if $\theta_d(n)$ is the number of steps required by the algorithm to complete the gossiping in the d-dimensional mesh-bus, then $\theta_d(n) = n + n \cdot \theta_{d-1}(n)$ and by induction on d, since $\theta_1(n) = n$, we get the expected result. $\square$

4.2 Arbitrary length message

Δ-port hypothesis. The following result was obtained recently by Fujita and Yamashira [7].

Theorem 8. *The minimum time necessary to gossip in a 2-dimensional mesh-bus network $M(n, n)$ with message of arbitrary length under the Δ-port hypothesis is bounded as follows*

$$\left\lfloor \frac{n}{2} \right\rfloor + \lceil \log_2 n \rceil - 1 \leq g_*(M(n, n)) \leq \left\lfloor \frac{n}{2} \right\rfloor + \lceil \log_2 n \rceil + 1$$

The proof of the lower bound in the next theorem is a generalisation of the one for dimension 2. However the algorithm given for the 2-dimensional mesh doesn't seem to generalise easily to d-dimensional meshes with $d > 2$.

Theorem 9. *The minimum time necessary to gossip in a d-dimensional mesh-bus network $M(n, \cdots, n)$ with message of arbitrary length under the Δ-port hypothesis is bounded as follows*

$$\left\lfloor \frac{n}{d} \right\rfloor - 1 + \lceil (d-1) \log_d n + \log_d(d-1) \rceil \leq g_*(M(n, \cdots, n)) \leq n + 2d - 2$$

1-port hypothesis. The proof of Theorem 8 from [7] can also be easily adapted to the 1-port hypothesis for 2-dimensional meshes, however the result is not so good, as can be seen in the following statement.

Theorem 10. *The minimum time necessary to gossip in a 2-dimensional mesh-bus network $M(n, n)$ with message of arbitrary length under the 1-port hypothesis is bounded as follows*

$$\left\lfloor \frac{n}{2} \right\rfloor + \lceil \log_2 n \rceil \leq g_1(M(n, n)) \leq \left\lfloor \frac{n}{2} \right\rfloor + 2\lceil \log_2 n \rceil + 1$$

Theorem 11. *If $n \geq d$, the minimum time necessary to gossip in a d-dimensional mesh-bus network $M(n, \cdots, n)$ with message of arbitrary length under the 1-port hypothesis is bounded as follows*

$$\left\lfloor \frac{n}{d} \right\rfloor + \lceil (d-1) \log_2 n \rceil - 1 \leq g_1(M(n, \cdots, n)) \leq d \left\lceil \frac{n}{d} \right\rceil + d^2 - d$$

The lower bound can be given a bit more precisely depending on the rest of the division of N by d, but we omit it in this extended abstract.

References

1. C. Berge. Graphes et Hypergraphes. *Dunod* (1973).
2. J.-C. Bermond. Hypergraph Gossip Problem. *Proceedings of the International Conference on Combinatorial Problems and Graph Theory, Orsay* pages 31-34 (1976).
3. F. Ergincan and J.-C. Bermond. Bus Interconnection Networks *submitted to Networks.*
4. A. Ferreira, A. Goldman vel Lejbman and S.W. Song. Bus based parallel computers and new communication patterns *Technical Report LIP 93-18* (1993)
5. P. Fraigniaud and E. Lazard. Methods and Problems of Communications in Usual Networks. *to appear in Discrete Applied Mathematics* (1994)
6. S. Fujita. Gossiping in mesh-bus computers by packets with bounded length. IPS Japan SIGAL 36-6:41-48 (1993)
7. S. Fujita and M. Yamashira. Optimal Gossiping in Mesh-Bus Computers. *to appear in Parallel Processing Letters* (1994).
8. A. Hily. Algorithmes d'échange d'information dans les réseaux à bus. Rapport de DEA, LRI Université de Paris-Sud (1993)
9. K. Iwama and E. Miyano. Routing problems on the mesh of buses. *Submitted to Journal of Algorithms, also Proceedings of the 3rd International Symposium on Algorithms and Computation, Lecture Notes in Computer Science 650* (1992).
10. R. Labahn. Information Flow on Hypergraphs. *Discrete Mathematics* 113:71-97 (1993)
11. D. Richards and A.L. Liestman. Generalizations of Broadcasting and Gossiping. *Networks* 18:125-138 (1988)
12. I. Scherson. Orthogonal graphs for a class of interconnection networks. *IEEE Transactions on Parallel and Distributed Systems*, Vol2, pages 3-19 (1991)
13. T. Szymanski. Graph-theoretic models for photonic networks. *Proceedings of the New Frontiers, A workshop on future directions of massively parallel processing* McLean, VA, pages 85-96 (1992)

Fault-Tolerant Linear Broadcasting

Krzysztof Diks[1]* and Andrzej Pelc[2]**

[1] Instytut Informatyki, Uniwersytet Warszawski, Banacha 2, 02-097 Warszawa,
Poland
[2] Département d'Informatique, Université du Québec à Hull, Hull, Québec J8X 3X7,
Canada

Abstract. In linear broadcasting, packets originally stored in one node,
called the source, have to visit all other nodes of the network. Every
packet has a predetermined route indicating in which order it visits the
nodes. A faulty link or node of the network destroys all packets passing
through it. A linear broadcasting scheme consisting of packets' routes is
f-fault-tolerant if every fault-free node is visited by at least one packet
for any configuration of at most f link or node failures. We estimate the
minimum number of packets for which there exists an f-fault-tolerant linear broadcasting scheme in complete networks, and we construct schemes
using few packets. Variations of this problem when faults can occur only
in links or only in nodes are also considered.

keywords: broadcasting, fault-tolerance, packet, route.

1 Introduction

Broadcasting consists in transmitting information from one node of a communication network, called the source, to all other nodes. Traditional broadcasting
in packet-switching networks is performed by replicating obtained messages by
every node of the network and sending copies to its neighbors [14]. In high bandwidth networks, however, this method may be inefficient: in order to fully utilize
very fast communication lines, a processor should not waste time processing
messages passing through it but not destined to it. Network designs in this case
tend to liberate processors from communication tasks. In the PARIS network,
for example, a special hardware is added to general purpose processors. This
switching subsystem (SS) can only switch a packet from one link to another and
send it to the general purpose processor which is, itself, liberated from the task
of propagating information (cf. [7]). Thus, although SS does not support traditional broadcasting algorithms, it supports linear broadcasting in which packets
travel from the source via predetermined routes [3, 6]. The sequence of nodes
forming the route is carried along by the packet. At every node of the network

---

* Research partly supported by grant from KBN. This work was done during the
author's stay at the Université du Québec à Hull, supported by NSERC International
Fellowship.
** Research supported in part by NSERC grant OGP 0008136.

the switching subsystem removes the name of the currently visited node from the packet's route and directs the packet to the following node in the route. A linear broadcasting scheme consists of a set of routes, one for each packet, ensuring that every node is visited by at least one packet.

While it is obvious that in a fault-free connected network one packet is sufficient to visit all nodes, the situation becomes more complex when components of the network (links and/or nodes) may be faulty and a faulty component destroys all packets passing through it. A linear broadcasting scheme is called f-fault-tolerant if every fault-free node is visited by at least one packet for any configuration of at most f link or node failures. The study of such schemes is a natural extension of research concerning traditional fault-tolerant broadcasting (cf. [1, 2, 4, 5, 10, 12, 13]). It has been also suggested by Greenberg [11] as a report dispersal problem.

The aim of this paper is to estimate the minimum number of packets for which there exists an f-fault-tolerant scheme for complete networks and to construct schemes using few packets. We consider link and node failures as well as scenarios where faults are restricted only to links or only to nodes. In particular, it follows from our results that the minimum number of packets sufficient to perform 2-fault-tolerant linear broadcasting in n-node networks is $\Theta(\log \log n)$, while it is 2 for 1-fault-tolerant linear broadcasting.

The paper is organized as follows. In section 2 we fix terminology and notation. Section 3 is devoted to the study of f-fault-tolerant linear broadcasting schemes in the general case: when both links and nodes are subject to faults. In section 4 we show how situation changes when faults are restricted only to links or only to nodes. Finally, section 5 contains conclusions and open problems.

2 Terminology and Notation

We consider $(n + 1)$-node complete networks consisting of a fault-free node s, called the source, originally holding all packets, and of nodes $1, \ldots, n$ to be visited (informed). Every packet is equipped with a route which is a sequence of nodes $1, \ldots, n$ containing each of these nodes at least once. This latter condition does not impose any additional restriction: every sequence of nodes which does not satisfy it can be extended by adding all missing nodes at the end. For sequences α and β, $\alpha \bullet \beta$ denotes their concatenation.

A linear broadcasting scheme using k packets is a set $\{\pi_1, \ldots, \pi_k\}$ of routes. Such a scheme is called f-fault-tolerant, for a positive integer f, if for every set A of at most f components of the network (links or nodes), other than s, and every node $v \notin A \cup \{s\}$ there exists an index $i \leq k$ such that no component from A precedes (the first occurence of) v in the sequence (path) $(s) \bullet \pi_i$. It is easy to see that this definition corresponds to the informal meaning of f-fault-tolerance indicated in the introduction.

For positive integers f, n, we denote by $P_f(n)$ the minimum number of packets for which there exists an f-fault-tolerant linear broadcasting scheme in an $(n + 1)$-node complete network. Most of our results are formulated as estimates of

$P_f(n)$, where upper bounds are always constructive: we present schemes using a particular number of packets.

We use $\log n$ to denote the logarithm with base 2. For any set X, $\mid X \mid$ denotes the size of X. For a sequence $\alpha = (a_1, \ldots, a_k)$, α^R denotes the sequence $(a_k, \ldots, a_1)$. If α is a sequence of distinct integers and β is any sequence of integers, we write $\alpha \leq \beta$ if α is a subsequence of β', where β' is the sequence of first occurences of all terms of β. For example, if β is the route of a packet, $\alpha \leq \beta$ means that the order of first visits of the packet to nodes which are listed in α is as they appear in α.

3 Link and Node Failures

In this section we consider the general case when both links and nodes are subject to faults. First note that the number n of nodes to visit must exceed the possible number of faults, for otherwise faults of all links incident to the source preclude any f-fault-tolerant linear broadcasting. Thus throughout this section we assume $n > f > 0$. Also, it is obvious that $P_f(n) \geq P_{f'}(n)$ for $n > f > f'$. For one possible fault the minimum number of packets is easy to determine.

Theorem 1. $P_1(n) = 2$, for any $n > 1$.

Proof. If the source sends only 1 packet, a single fault of the link on which it leaves the source precludes visiting any node. Thus $P_1(n) \geq 2$. In order to show $P_1(n) \leq 2$ it suffices to observe that routes $(1, \ldots, n)$ and $(n, \ldots, 1)$ form a 1-fault-tolerant linear broadcasting scheme. $\qquad\Box$

The situation changes dramatically for $f > 1$. In this case we show that more than $\log \log n$ packets are necessary to perform f-fault-tolerant linear broadcasting in $(n+1)$-node complete networks. Later on we will also prove that for $f = 2$, $O(\log \log n)$ packets suffice.

Theorem 2. $P_f(n) \geq \log \log n + 1$, for $n > f > 1$.

Proof. We will use the following result of Erdös and Szekeres (cf. [9]):

Lemma 3. *In every sequence of n distinct numbers there exists a monotonic subsequence of length at least $\sqrt{n}$.* $\qquad\Box$

For $n \leq 4$, $\log \log n \leq 1$ and the theorem follows from Theorem 1. Thus assume $n > 4$. Let k be an integer smaller than $\log \log n + 1$ and suppose that there exists an f-fault-tolerant scheme $\{\pi_1, \ldots, \pi_k\}$ using k packets. Without loss of generality we may assume that $(1, \ldots, n) \leq \pi_1$.

Lemma 4. *For every $i = 1, \ldots, k$ there exists an increasing sequence of nodes $J_i = (j_1^i, \ldots, j_{r(i)}^i)$ of length $r(i) \geq n^{\frac{1}{2^{i-1}}}$ such that, for all $t \leq i$, either $J_i \leq \pi_t$ or $J_i^R \leq \pi_t$.*

Proof. Induction on i. For $i = 1$ take $J_1 = (1, \ldots, n)$. Suppose that the lemma is true for $i - 1$. Let $J_{i-1} = (j_1^{i-1}, \ldots, j_{r(i-1)}^{i-1})$ be the increasing sequence given by the inductive hypothesis. Consider the permutation π of J_{i-1} such that $\pi \leq \pi_i$. By Lemma 3 it contains a monotonic subsequence σ of length at least $\sqrt{r(i-1)}$. Let $r(i)$ be the length of σ and let $J_i = (j_1^i, \ldots, j_{r(i)}^i)$ be the sequence of its terms in increasing order. Thus $J_i = \sigma$ if σ is increasing and $J_i = \sigma^R$ if σ is decreasing. In any case $J_i \leq J_{i-1}$. By the inductive assumption $J_i \leq \pi_t$ or $J_i^R \leq \pi_t$, for all $t \leq i - 1$, and by construction $J_i \leq \pi_i$ or $J_i^R \leq \pi_i$. Moreover,

$$r(i) \geq \sqrt{r(i-1)} \geq \sqrt{n^{\frac{1}{2^{i-2}}}} \geq n^{\frac{1}{2^{i-1}}},$$

which proves the lemma by induction. $\quad\square$

In order to finish the proof consider the sequence J_k. For every $i \leq k$, $J_k \leq \pi_i$ or $J_k^R \leq \pi_i$ and

$$r(k) \geq n^{\frac{1}{2^{k-1}}} > n^{\frac{1}{2^{\log\log n}}} = 2.$$

Thus $r(k) \geq 3$ and faulty nodes j_1^k and $j_{r(k)}^k$ preclude visiting node j_2^k by any packet, contradicting f-fault-tolerance of the scheme for $f \geq 2$. $\quad\square$

Our next result establishes a lower bound on $P_f(n)$ depending on the number of faults.

Theorem 5. $P_f(n) \geq f(\lceil \frac{f}{n-f} \rceil + 1)$, *for $n > f > 0$.*

Proof. Let $n = f + r$, for $r \geq 1$. Assume that there exists an f-fault-tolerant scheme $\{\pi_1, \ldots, \pi_k\}$ using $k = f(\lceil \frac{f}{r} \rceil + 1) - 1$ packets. Let p_i be the number of packets whose routes start with node i. Without loss of generality we may assume $p_1 \geq p_2 \geq \ldots \geq p_n$. We will show that $\sum_{i=f+1}^n p_i < f$. Suppose otherwise. Thus $p_{f+1} \geq \lceil \frac{f}{r} \rceil$. Hence the total number of packets is at least

$$f\left\lceil \frac{f}{r} \right\rceil + f = f\left(\left\lceil \frac{f}{r} \right\rceil + 1\right) > k,$$

contradiction.

Hence there are less than f packets whose routes start with one of nodes $f + 1, \ldots, n$. Call them key packets. Call a node $v \in \{1, \ldots, f\}$ black if it is the first node from $\{1, \ldots, f\}$ on the route of some key packet and call it white otherwise. There are less than f black nodes, hence there is at least one white node. Suppose that f faults are distributed as follows: all links from the source to white nodes are faulty and all black nodes are faulty. In this case white nodes will not be visited which contradicts f-fault-tolerance of the scheme $\{\pi_1, \ldots, \pi_k\}$. $\quad\square$

We next turn attention to upper bounds on $P_f(n)$.

Theorem 6. $P_f(n) \leq (f+1)(n-1)$, for $n > f > 0$.

Proof. Consider the set of routes $\{\pi_{ij} : i = 1, \ldots, f+1; j \in \{1, \ldots, n\} \setminus \{i\}\}$ defined as follows: $\pi_{ij} = (i, j) \bullet \alpha_{ij}$, where α_{ij} is any permutation of $\{1, \ldots, n\}$. Every node j is joined with s by at least $f+1$ node-disjoint initial segments of routes. Hence the scheme consisting of routes π_{ij} is f-fault-tolerant. $\square$

Theorems 5 and 6 imply immediately the following result in case when the number f of possible faults is only by one less than the number of nodes to be visited (that is largest possible for an f-fault-tolerant scheme to exist).

Corollary 7. $P_f(f+1) = f(f+1)$, for $f > 0$. $\square$

We are able to give much better upper bounds on $P_f(n)$ for small number of faults. To this end we show how to construct an f-fault-tolerant scheme consisting of $T_f(n)$ routes, where $T_f(n)$ is a recursive function to be defined later. In what follows we informally describe our method. The construction is recursive. We partition the set of nodes $\{1, \ldots, n\}$ into $\sqrt{n}$ pairwise disjoint sets $S_1, \ldots, S_{\sqrt{n}}$ of sizes $\sqrt{n}$. (In this informal description we assume $\sqrt{n}$ to be an integer.) Next, we find recursively r-fault-tolerant schemes $\Pi_1^r, \ldots, \Pi_{\sqrt{n}}^r, \Gamma^r$, for the sets $S_1, \ldots, S_{\sqrt{n}}$ and the set $\{S_1, \ldots, S_{\sqrt{n}}\}$, respectively, for every $r = 1, \ldots, f$. Every such scheme consists of $T_r(\sqrt{n})$ routes. We call the schemes Π_i^r local and the scheme Γ^r global. We combine the local and global schemes to get an f-fault-tolerant scheme Π for the whole network. In order to describe how to combine the schemes it is better to look at the problem from the point of view of a fixed set S_i. If S_i is not affected by faults (i.e. there is no faulty component in the network on $\{s\} \cup S_i$) then clearly at most f of the remaining sets are affected by faults. Observe that there is a route in the global scheme Γ^f in which no affected set S_j precedes the set S_i. On the other hand, if S_i is affected by f faults then there are no faults in the remaining sets and for every fault-free node v in S_i there is a route in Π_i^f such that v is not preceded by any faulty component in this route. The routes in Π handling both above cases are obtained by replacing occurences of sets S_j in the routes in Γ^f by different routes from Π_j^f, for $j = 1, \ldots, \sqrt{n}$. This contributes $T_f(\sqrt{n})$ routes to the scheme Π.

Now assume that S_i is affected by r faults, $1 \leq r \leq f-1$. Then there are at most $f - r$ affected sets outside S_i. To handle this case we combine the global scheme Γ^{f-r} and the local schemes Π_j^r, for $j = 1, \ldots, \sqrt{n}$. For every route γ in Γ^{f-r} we construct $T_r(\sqrt{n})$ routes for the whole network replacing the occurence of every set S_j in γ by every route in Π_j^r, for $j = 1, \ldots, \sqrt{n}$. Since there is a route in Γ^{f-r} in which no affected set S_j precedes S_i and for every fault-free node v in S_i there is a route in Π_i^r in which no faulty component precedes v, it guarantees that in the final scheme Π there is a route in which no faulty component precedes v, in the case of r faults affecting S_i. In this case we add

$T_r(\sqrt{n})T_{f-r}(\sqrt{n})$ new routes to Π. Hence the total number of routes in Π is

$$T_f(n) = T_f(\sqrt{n}) + \sum_{r=1}^{f-1} T_r(\sqrt{n})T_{f-r}(\sqrt{n}).$$

Now we define the function $T_f(n)$ precisely. For $f = 1$ let $T_f(n) = 2$, for all $n > 1$. For $n > f > 1$:

$$T_f(n) = \begin{cases} max\{P_f(m) : f < m < (f+1)(f+2)\} & n < (f+1)(f+2) \\ T_f(\lceil\sqrt{n}\rceil) + \sum_{i=1}^{f-1} T_i(\lceil\sqrt{n}\rceil)T_{f-i}(\lceil\sqrt{n}\rceil) & n \geq (f+1)(f+2). \end{cases}$$

Following the ideas described above it is possible to obtain the following technical lemma. Its formal proof will appear in a complete version of the paper.

Lemma 8. $P_f(n) \leq T_f(n)$, for $n > f \geq 1$. □

The next theorem shows that for two faults the lower bound from Theorem 2 gives an exact order of magnitude for $P_2(n)$.

Theorem 9. $P_2(n) \leq 4\lceil\log\log n\rceil + 2$, for every $n > 2$.

Proof. Using a case by case analysis it is possible to show that $max_{2<n<12} P_2(n) = 6$. Since $(2+1)(2+2)=12$, by definition of $T_f(n)$ we get:

$$T_2(n) = \begin{cases} 6 & n < 12 \\ T_2(\lceil\sqrt{n}\rceil) + 4 & n \geq 12. \end{cases}$$

Using this recursive formula it is easy to show that $T_2(n) \leq 4\lceil\log\log n\rceil + 2$. This concludes the proof in view of Lemma 8. □

We finally consider consequences of Lemma 8 concerning explicit upper bounds on $P_f(n)$ for arbitrary f. Let c_n, for $n \geq 1$, be an integer defined as follows:

$$c_1 = c_2 = 1, c_n = c_1 c_{n-1} + c_2 c_{n-2} + \cdots + c_{n-1} c_1, \text{ for } n > 2.$$

The number c_n is the $(n-1)$-st Catalan number (cf. [8]).

Theorem 10. $P_f(n) \leq \sqrt{6}^f c_f \lceil\log\log n\rceil^{f-1}$, for $n > f > 1$.

Proof. In view of Lemma 8 it is enough to prove $T_f(n) \leq \sqrt{6}^f c_f \lceil\log\log n\rceil^{f-1}$. We will prove this inequality by induction on f and n. It is clearly true for $f = 2$ and all $n > f$. We first show that it is true for all $f > 2$ and $f < n < (f+1)(f+2)$. Case 1. $f = 3, 3 < n < 20$.
It is easy to compute that $\sqrt{6}^3 c_3 \lceil\log\log n\rceil^2 > 20$. On the other hand we will show that $P_3(n) \leq 20$. For $n = 4, 5$ this follows from Theorem 6. For $5 < n < 20$

we construct the following scheme. Let $m = \lfloor \frac{n}{2} \rfloor$ and consider sets of nodes $S_1 = \{1, \ldots, m\}$ and $S_2 = \{m+1, \ldots, n\}$. Let $p = max(6, \lceil \frac{n}{2} \rceil)$. Using a case by case analysis it is possible to show that $P_2(\lceil \frac{n}{2} \rceil) \leq 6$. Let $\{\pi'_1, \ldots, \pi'_6\}$ and $\{\pi''_1, \ldots, \pi''_6\}$ be 2-fault-tolerant schemes for complete networks on $\{s\} \cup S_1$ and $\{s\} \cup S_2$, respectively. Now define routes $\pi^*_1, \ldots, \pi^*_p$ and $\pi^{**}_1, \ldots, \pi^{**}_p$ for the complete network on $\{s, 1, \ldots, n\}$ in the following way.

$$\pi^*_i = \pi'_i \bullet \alpha_i, \text{ for } i = 1, \ldots, 6$$

$$\pi^*_i = \alpha_i \bullet (1, \ldots, m), \text{ for } i = 7, \ldots, p$$

$$\pi^{**}_i = \pi''_i \bullet \beta_i, \text{ for } i = 1, \ldots, 6$$

$$\pi^{**}_i = \beta_i \bullet (m+1, \ldots, n), \text{ for } i = 7, \ldots, p,$$

where α_i is any permutation of S_2 starting with $min(m + i, n)$ and β_i is any permutation of S_1 starting with $min(i, m)$. Informally, 6 packets visit each S_1 and S_2 according to a 2-fault-tolerant scheme and then these packets, together with $p - 6$ other packets, visit each a different node of the other set S_i, following further an arbitrary permutation.

Our scheme for the network on $\{s, 1, \ldots, n\}$ is $\Pi = \{\pi^*_1, \ldots, \pi^*_p\} \cup \{\pi^{**}_1, \ldots, \pi^{**}_p\}$. By definition of p, $\mid \Pi \mid \leq 20$. It suffices to show that Π is 3-fault-tolerant. If both S_1 and S_2 contain at most 2 faults, all nodes are visited by Π in view of 2-fault-tolerance of $\{\pi'_1, \ldots, \pi'_6\}$ and $\{\pi''_1, \ldots, \pi''_6\}$. Suppose that 3 faults are in the network on $S_1 \cup \{s\}$ and 0 faults in the network on $S_2 \cup \{s\}$ (the symmetric is analogous). In this case all nodes in S_2 are visited. Moreover, no faulty component precedes the segment β_i in all $(s) \bullet \pi^{**}_i$ and since all nodes from S_1 begin some segment β_i, all nodes from S_1 are visited as well.

Case 2. $f > 3, f < n < (f+1)(f+2)$.
In view of Theorem 6

$$P_f(n) \leq (f+1)(n-1) \leq (f+1)((f+1)(f+2) - 2) = f^3 + 4f^2 + 3f < 4f^3.$$

On the other hand

$$\sqrt{6}^f c_f \lceil \log \log n \rceil^{f-1} \geq \sqrt{6}^f \cdot 5 \cdot 2^{f-1} > \frac{5}{2} 4^f > 4f^3.$$

Hence our inequality is proved for all $f > 2$ and $f < n < (f+1)(f+2)$. Suppose now that $f > 2$, $n \geq (f+1)(f+2)$ and assume by induction that the inequality holds for every pair (f', n') such that $f' < f$ and $f' < n'$ or $f' = f$ and $f' < n' < n$. We have

$$T_f(n) = T_f(\lceil \sqrt{n} \rceil) + \sum_{i=1}^{f-1} T_i(\lceil \sqrt{n} \rceil) T_{f-i}(\lceil \sqrt{n} \rceil) \leq$$

$$T_f(\lceil \sqrt{n} \rceil) + \sum_{i=1}^{f-1} \sqrt{6}^i c_i \lceil \log \log n \rceil^{i-1} \sqrt{6}^{f-i} c_{f-i} \lceil \log \log n \rceil^{f-i-1} =$$

$$T_f(\lceil\sqrt{n}\,\rceil) + \sqrt{6}^f \lceil\log\log n\rceil^{f-2} \sum_{i=1}^{f-1} c_i c_{f-i} = T_f(\lceil\sqrt{n}\,\rceil) + \sqrt{6}^f c_f \lceil\log\log n\rceil^{f-2}.$$

The above inequality holds as well if n is replaced by $m < n$. Applying it $\lceil\log\log n\rceil$ times (for n, $\lceil\sqrt{n}\,\rceil$, $\left\lceil\sqrt{\lceil\sqrt{n}\,\rceil}\,\right\rceil$, etc.) gives

$$T_f(n) \le \sqrt{6}^f c_f \lceil\log\log n\rceil^{f-1},$$

which proves our corollary by induction. $\qquad\square$

Since

$$c_n = \frac{1}{n}\binom{2n-2}{n-1} \in \frac{4^{n-1}}{\sqrt{\pi}(n-1)^{\frac{3}{2}}}\left(1 + O\left(\frac{1}{n-1}\right)\right)$$

(cf. [8]), we obtain

Corollary 11. $P_f(n) \in O((4\sqrt{6}\log\log n)^f).$ $\qquad\square$

4 Failures of One Type.

In this section we consider linear broadcasting under a restricted fault scenario: when only links or only nodes are subject to failures. It will be seen that for faulty nodes and fault-free links our problem is not much easier than in the general case (in many situations the minimum number of packets in an f-fault-tolerant scheme can be reduced by at most half). For faulty links and fault-free nodes, however, the situation changes dramatically : the minimum number of packets does not depend any more on the number of nodes to be visited, but only on the number of possible faults.

4.1 Faulty Nodes, Fault-Free Links

First note that n packets are always sufficient to visit n nodes: it suffices to send each packet to a different node. The lower bound from Theorem 2 still holds: in its proof we have not used the possibility of link failures. All upper bounds clearly hold as well in the restricted case. The following result shows the connection between minimum size of an f-fault-tolerant scheme in the restricted and the general case. Let $N_f(n)$ be the minimum size of an f-fault-tolerant linear broadcasting scheme for a complete $(n+1)$-node network with fault-free links.

Theorem 12. If $N_f(n) \le \lfloor\frac{n}{2}\rfloor$ then $P_f(n) \le 2N_f(n)$.

Proof. Consider routes of $N_f(n)$ packets which visit all fault-free nodes among $1,\ldots,n$, with any configuration of f faulty nodes. Without loss of generality we may assume that each route starts with one of nodes $1,\ldots,\lfloor\frac{n}{2}\rfloor$.

CLAIM *For any configuration of at most f node or link faults, every fault-free node $v \in \{\lfloor\frac{n}{2}\rfloor + 1,\ldots,n\}$ will be visited by at least one of these $N_f(n)$ packets.*

In order to prove the claim suppose that for some configuration F of f link or node faults, a fault-free node $v \in \{\lfloor \frac{n}{2} \rfloor + 1, \ldots, n\}$ will not be visited. If link $e = v - s$ is faulty, it is easy to see that v will not be visited for the fault configuration $F \setminus \{e\}$ as well. Hence we may assume that link $v - s$ is fault-free. Now change the fault configuration replacing each fault of link $x - y$ by a fault of one of its ends, say x, different from v and from s. The new configuration does not have more faults and faults are now restricted only to nodes. Notice that v will not be visited under this new fault configuration either, because if it was preceded by a link $x - y$ from F in a route π it will now be preceded in this route by node x. This contradicts the choice of the $N_f(n)$ packet routes and hence the claim is proved.

Now, in order to guarantee that all nodes $1, \ldots, \lfloor \frac{n}{2} \rfloor$ are visited as well, relabel all nodes so that node i gets label $n - i + 1$ and use additional $N_f(n)$ packets to visit nodes as before but according to new labels. □

4.2 Faulty Links, Fault-Free Nodes

Let $L_f(n)$ be the minimum size of an f-fault-tolerant linear broadcasting scheme for a complete $(n + 1)$-node network with fault-free nodes.

Theorem 13. $L_f(n) = f + 1$, for $n > f > 0$.

Proof. In order to show $L_f(n) \geq f + 1$, suppose that only f packets are used. They can leave the source on $\leq f$ links. Faults of these links preclude visiting any node.

In order to show $L_f(n) \leq f+1$ we construct the following routes $\pi_1, \ldots, \pi_{f+1}$. For $i = 1, \ldots, f + 1$, let

$$\pi_i' = (i, f + 2, i, f + 3, i, f + 4, \ldots, i, n - 1, i, n)$$

and

$$\pi_i'' = (i, i + 1, i, i + 2, i, \ldots, f, i, f + 1, i, 1, i, 2, i, \ldots, i - 2, i, i - 1, i).$$

Then define $\pi_i = \pi_i' \bullet \pi_i''$. That is, the packet starting at node i first visits consecutive nodes $f + 2, \ldots, n$, always returning to i after each visit, and then visits consecutive nodes $i+1, \ldots, f+1$, and $1, 2, \ldots, i-1$, also always returning to i after each visit. We will prove that each node is visited with any configuration of at most f link faults.

The argument is simple for nodes $v \in \{f + 2, \ldots, n\}$. It is enough to observe that packets using routes π_i, π_j for $i \neq j$ reach v by link-disjoint paths: indeed, route π_i' consists of links incident to i and not incident to j; likewise, route π_j' consists of links incident to j but not incident to i.

Now consider sequences π_i'' for $i = 1, \ldots, f + 1$. Links from these sequences do not appear in $(s) \bullet \pi_i'$, for $i = 1, \ldots, f + 1$. Let r be the total number of faulty links in sequences $(s) \bullet \pi_i', i = 1, \ldots, f + 1$. For at least $f + 1 - r$ routes

π_i no faulty link precedes the segment π_i''. We will show that for every node $v \in \{1, \ldots, f+1\}$, v is not preceded by a faulty link in at least one of these routes. Consider two cases.

a) Every faulty link $i - j$, for $i, j \leq f+1$, is the first faulty link in at most one of those $f + 1 - r$ routes.

There are at most $f - r$ such links hence at least one of the $f + 1 - r$ routes does not contain any of them.

b) There is a faulty link $i - j$, for $i < j \leq f + 1$, which is the first faulty link in at least 2 of those $f + 1 - r$ routes.

The link $i - j$ can be the first faulty link only in routes π_i and π_j. Since nodes appear in cyclic order $i+1, i+2, \ldots, j-1$ in route π_i and in cyclic order $j+1, j+2, \ldots, i-1$ in route π_j, it follows that none of the nodes $i+1, \ldots, j-1$ is preceded by a faulty link in route π_i and none of the nodes $j+1, j+2, \ldots, i-1$ is preceded by a faulty link in route π_j. Hence, for every node $v \in \{1, \ldots, f+1\}$, v is not preceded by a faulty link either in π_i or in π_j. □

5 Conclusion

We showed several estimates of the size $P_f(n)$ of a minimum f-fault-tolerant linear broadcasting scheme for complete $(n+1)$-node networks. Our main results concern the general case when both links and nodes are subject to faults. These results can be summarized in the following table (c_f is the $(f - 1)$-st Catalan number defined in section 3).

Table 1. Bounds on $P_f(n)$

parameters	lower bounds	upper bounds
$n > f = 1$	2	2
$f \geq 1, n = f + 1$	$f(f+1)$	$f(f+1)$
$n > f = 2$	$\log \log n + 1$	$4 \lceil \log \log n \rceil + 2$
$n > f \geq 2$	$\log \log n + 1$ $f(\lceil \frac{f}{n-f} \rceil + 1)$	$\sqrt{6}^f c_f \lceil \log \log n \rceil^{f-1}$ $(f+1)(n-1)$

For faults restricted to nodes we proved that the minimum size p of an f-fault-tolerant scheme for a complete $(n+1)$-node network is at least one half the size of such a scheme for general faults, if $p \leq \lfloor \frac{n}{2} \rfloor$. This should be contrasted with the scenario of faults restricted to links, where an f-fault-tolerant scheme of size $f + 1$ always exists, regardless of the size of the network (if $n > f$).

Apart from special cases our bounds are not tight and improving them is a natural problem left by this research. It also seems interesting to investigate optimal linear broadcasting schemes for important non-complete networks, e.g. hypercubes, meshes etc.

References

1. K.A. Berman & M. Hawrylycz, Telephone problems with failures, SIAM J. Alg. Disc. Meth. 7 (1986), 13-17.
2. D. Bienstock, Broadcasting with random faults, Disc. Appl. Math. 20 (1988), 1-7.
3. S. Bitan & S. Zaks, Optimal linear broadcast, J. of Algorithms 14 (1993), 288-315.
4. B.S. Chlebus, K. Diks & A. Pelc, Sparse networks supporting efficient reliable broadcasting, Proc. ICALP'93, LNCS 700, 388-397.
5. B.S. Chlebus, K. Diks & A. Pelc, Optimal broadcasting in faulty hypercubes, Digest of Papers, FTCS'21 (1991), 266-273.
6. C.T. Chou & I.S. Gopal, Linear broadcast routing, J. of Algorithms 10 (1989), 490-517.
7. I. Cidon & I.S. Gopal, PARIS: An approach to private integrated networks, Intern. J. Analog Digital Cable Systems 1 (1988), 77-85.
8. T. Cormen, C. Leisserson, & R.L. Rivest, Introduction to algorithms, The MIT Press, 1990.
9. P. Erdös & Szekers, A combinatorial problem in geometry, Compositio Mathematica 2 (1935), 463-470.
10. L. Gargano, Tighter bounds on fault-tolerant broadcasting and gossiping, Networks 22 (1992), 469-486.
11. D. Greenberg, Report dispersal, in: Open problems for the International Workshop on Networks and Information Dissemination, Bowen Island, B.C. (1992).
12. R.W. Haddad, S. Roy & A.A. Schafer, On gossiping with faulty telephone lines, SIAM J. Alg. Disc. Meth. 8 (1987), 439-445.
13. S.M. Hedetniemi, S.T. Hedetniemi & A.L. Liestman, A survey of gossiping and broadcasting in communication networks, Networks 18 (1988), 319-349.
14. A. Segal, Distributed network protocols, IEEE Trans. Inf. Theory IT-29 (1983), 23-35.

The Minimum Broadcast Time Problem

Klaus Jansen[1] and Haiko Müller[2]

[1] FB IV - Mathematik und Informatik, Universität Trier, 54 317 Trier, Germany.
[2] Fakultät für Mathematik und Informatik, Universität Jena, 07 740 Jena, Germany.

Abstract. Broadcasting is the information dissemination process in a communication network. A subset of processors $V_0 \subset V$ called originators knows an unique message which has to be transferred by calls between adjacent processors. Each call requires one time unit and each processor can participate in at most one call per time unit. The problem is to find a schedule such that the time needed to inform all processors is less than or equal to a deadline $k \in IN$. We present NP-completeness results for this problem restricted to several communication networks (bipartite planar graphs, grid graphs, complete grid graphs, split graphs and chordal graphs) with constant deadline $k = 2$ or one originator $V_0 = \{v\}$.

1 Introduction

Broadcasting is the information dissemination process in a processor network where all processors become informed of a message by calls over lines in the network. A communication network is modelled by an undirected graph $G = (V, E)$ consisting of a set V of vertices (processors) and a set E of edges (network connections). The process of information dissemination is described by the following constraints:

(i) Each call involves two adjacent processors.
(ii) Each call requires one time unit.
(iii) Each processor can participate in at most one call per time unit.

More formally, the *broadcasting* can be defined as a sequence of sets $V_0 \subseteq \ldots \subseteq V_k = V$ where each set V_i represents the processors informed after time unit i, $0 \leq i \leq k$. The vertices in V_0 are called *originators*. For each vertex $v \in V_i \setminus V_{i-1}$, there exists an adjacent vertex $\bar{v} \in V_{i-1}$ who has informed v. Moreover, for each pair $v, w \in V_i \setminus V_{i-1}$ with $v \neq w$ we have $\bar{v} \neq \bar{w}$.

The minimum value of k for a given network G and originator set V_0 is defined as the *broadcast time* of G with respect to V_0. The minimum number of time units required to broadcast a message on a network with n processors and one originator is $\lceil log(n) \rceil$. A *minimum broadcast network* with n processors is a network in which each processor can broadcast in $\lceil log(n) \rceil$ time units. Constructing minimum broadcast networks has been studied extensively, see for example [1, 2, 3, 5, 8, 12]. For a survey on the broadcast time problem we refer to [9].

Slater, Cockayne and Hedetniemi [15] have described an algorithm to compute the broadcast time for an arbitrary tree network with $|V_0| = 1$. An algorithm for trees with a general number of originators is given in [6]. The broadcast time problem restricted to complete grid graphs and $|V_0| = 1$ is also solvable in polynomial time [4].

The general problem of determining the broadcast time for a given network has been shown algorithmic hard (NP-complete) by Garey and Johnson [7]. A proof of this result for an arbitrary graph with deadline $k = 4$ and unbounded degree is contained in [15]. Using a reduction from the three dimensional matching problem, Jakoby, Reischuk and Schindelhauer [10] proved the NP-completeness for graphs with constant maximum degree 5 and constant deadline $k \geq 3$. The complexity of the problem for deadline $k = 2$ and an arbitrary network was unknown. Using a complicated reduction, they showed that the problem remains NP-complete for planar graphs and $|V_0| = 1$.

In this paper, we improve their results by simpler reduction from the planar satisfiability problem. We investigate on several processor networks as bipartite planar graphs, grid graphs, complete grid graphs, chordal graphs and split graphs. For grid graphs with maximum degree 3, complete grid graphs and split graphs, we show that broadcast time problem is NP-complete even for deadline $k = 2$. For $|V_0| = 1$, the problem restricted to grid graphs and chordal graphs remains NP-complete.

2 Preliminaries

Planar 3-SAT. Our NP-completeness proofs are based on the planar 3-SAT problem which is NP-complete shown by Lichtenstein [11].

INPUT: A set of unnegated variables $X = \{x_1, \ldots, x_n\}$ and negated variables $\overline{X} = \{\overline{x_1}, \ldots, \overline{x_n}\}$, a collection of clauses C over $X \cup \overline{X}$ such that

(i) The graph $G = (X \cup C, E)$ with edge set $E = \{\{x, c\} | x \in c \vee \overline{x} \in c\}$ is planar.

(ii) Each clause $c \in C$ contains two or three literals $y \in X \cup \overline{X}$.

QUESTION: Does there exist a truth mapping for the variables such that each clause is satisfied?

We say x_i appears in c if $x_i \in c$ or $\overline{x_i} \in c$. In the first case x_i appears unnegated, and in the second case negated. In the reductions with constant deadline, we use a modification of the planar 3-SAT problem. In this problem, called planar 3,4-SAT we assume that each clause $c \in C$ contains exactly three literals and that each variable x_i appears in at most four clauses. The problem 3,4-SAT without the planarity constraint is NP-complete shown by Tovey [16].

Fig. 1. Clauses for planar 3,4-SAT

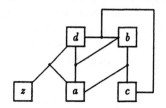

Theorem 1. *Planar 3,4-SAT is NP-complete.*

Proof. We give a reduction from planar 3-SAT. First, for each variable x which appears in $n > 3$ clauses, we define a cycle $(y_1 \vee \overline{y_2}) \wedge (y_2 \vee \overline{y_3}) \wedge \ldots \wedge (y_n \vee \overline{y_1})$ and replace the i.th occurrence of x with y_i, $1 \leq i \leq n$. After this step, each variable appears in at most three clauses and each clause contains at most three literals.

In the second step, we replace each clause $(x \vee y)$ containing two literals in the following way. We introduce a new variable z such that the clause becomes $(x \vee y \vee \overline{z})$ and force that z must be true. To achieve this, we add the following clauses:

(i) $(\overline{a_i} \vee b_i \vee d_i)$, $(\overline{a_i} \vee \overline{b_i} \vee c_i)$, $(\overline{b_i} \vee \overline{c_i} \vee d_i)$ for each $i = 1, 2, 3$.
(ii) $(z \vee a_i \vee d_i)$ for each $i = 1, 2, 3$.
(iii) $(\overline{d_1} \vee \overline{d_2} \vee \overline{d_3})$.

Assume that z is false. Then, $(a_i \vee d_i)$ must be true for $i = 1, 2, 3$. Using (iii), at least one variable d_j must be false and the corresponding variable a_j must be true. By (i), the clauses (b_j), $(\overline{b_j} \vee c_j)$ and $(\overline{b_j} \vee \overline{c_j})$ are true. This gives a contradiction. Therefore, z is forced to be true.

It is possible to insert these clauses so that the corresponding graph $G = (X \cup C, E)$ remains planar (see also Figure 1). In total, we obtain a planar 3-SAT instance where each variable appears in at most four clauses. □

Fig. 2. The separability constraint

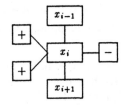

For reductions with $|V_0| = 1$, we use another modification of planar 3-SAT. In this problem, called planar separable 3-SAT, we assume that the graph $G = (X \cup C, E \cup E')$ with edge set $E = \{\{x, c\} | x \in c \vee \overline{x} \in c\}$ and $E' = \{\{x_i, x_{i+1}\} | 1 \leq i \leq n-1\} \cup \{x_n, x_1\}$ is planar. Moreover, each variable x_i appears in at most three clauses and each unnegated or negated variable at least once. Furthermore, the following separability constraint holds. For each variable x_i, the edges $\{x_{i-1}, x_i\}$ and $\{x_i, x_{i+1}\}$ separate the edges $\{x_i, c\} \in E$ such that all edges representing unnegated variables are incident to one side and all edges representing negated variables are incident to the other side. An illustration of this separability constraint is given in Figure 2.

Theorem 2. *Planar separable 3-SAT is NP-complete.*

Proof. See Lichtenstein [11]. □

Embedding of a planar graph. A complete grid graph is a graph consisting of n^2 points labeled by all ordered pairs (i, j) of integers between 1 and n with an edge between each pair of points (i, j) and $(i, j+1)$ and each pair (i, j) and $(i+1, j)$. A grid graph $G = (V, E)$ is a connected subgraph of a complete grid graph $G' = (V', E')$ with $V \subset V'$, $E \subset E'$.

For the NP-completeness results of grid graphs, we compute for a planar graph G a rectilinear planar layout. This layout maps vertices of G to horizontal line segments and edges to vertical line segments with all endpoints at positive integer coordinates. Two horizontal line segments are connected by a vertical one if and only if the corresponding vertices are adjacent in G. An example for this transformation is given in Figure 3.

Fig. 3. A planar graph and its rectilinear planar layout

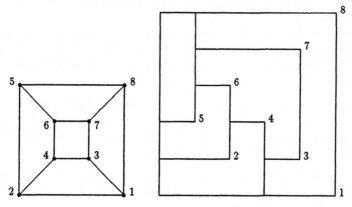

Fig. 4. The channel layout

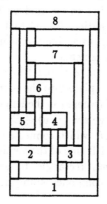

For a planar graph with n vertices a rectilinear planar layout can be computed in $O(n)$ time. The height of the layout is at most n and the width is at most $2n - 4$. For details regarding the algorithm used to obtain a rectilinear planar layout, we refer to [14].

For our reduction we use a modified layout which can be obtained directly from the planar rectilinear layout. Let d be a positive even integer.

(i) Vertices are mapped to disjoint axes parallel rectangles of height d with integer endpoints. If ℓ is the length of the horizontal line, then the width of the rectangle is $d \cdot \ell + \frac{d}{2}$.

(ii) Edges are mapped to disjoint axes parallel rectangles of width $\frac{d}{2}$ with integer endpoints. If ℓ is the length of the original line, then the height of the rectangle is $d \cdot (2 \cdot \ell - 1)$.

(iii) Two horizontal rectangles are connected by a vertical rectangle if and only if the corresponding vertices are adjacent. In this case the vertical rectangle touches the upper horizontal rectangle at the lower side and the lower horizontal rectangle at the upper side.

(iv) If ℓ is the distance between two vertical lines, the distance between the left sides of the corresponding rectangles is $\ell \cdot d$. If ℓ is the distance between two horizontal lines, the distance between the upper sides of the corresponding rectangles is $2 \cdot \ell \cdot d$.

We obtain the modified layout, which we call channel layout, by stretching the horizontal and vertical lines to rectangles with height d and width $\frac{d}{2}$, respectively. Clearly, this transformation can be done in polynomial time. For our example in Figure 3 we get the channel layout illustrated in Figure 4.

Lemma 3. *Let $G = (V, E)$ be a planar graph with $\Delta(G) \leq 3$. Then, there is a positive integer $\ell = O(|V|^2)$ such that the graph $G' = (V \cup V', E')$ with $V' = \{a_{e,j} | e \in E, 1 \leq j \leq \ell - 1\}$ and*

$$E' = \bigcup_{e=\{u,v\} \in E} \{\{u, a_{e,1}\}, \{a_{e,1}, a_{e,2}\}, \ldots, \{a_{e,\ell-1}, v\}\}$$

is a grid graph.

Proof. We assume that G contains no isolated vertices. Clearly, isolated vertices can be added after the transformation. In the first step we compute a rectilinear planar layout for G where w.l.o.g. each line segment has even length. Let ℓ_{max} be the maximum length of the line segments.

In the second step, we generate a channel layout with stretching factor $d = 4 \cdot \ell_{max}$. For each vertex $v \in V$, we place into the corresponding horizontal rectangle a horizontal path P_v of length $d \cdot \ell$ (where ℓ is the length of the line segment) and with vertices of distance 1. For each edge $e = \{u, v\} \in E$, we insert into the corresponding vertical rectangle a vertical path P_e which connects the horizontal paths P_u and P_v. We place the vertical path at the left side of the vertical rectangle and the horizontal path in the middle of the horizontal rectangle. In dependence on the degree $\delta(v)$ of the vertices, we obtain different components and define an unique point on P_v as follows (see also Figure 5).

Fig. 5. The grid graphs in dependence on the degrees of vertices $v \in V$

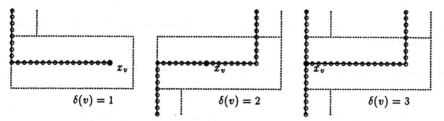

$\delta(v) = 1$ $\delta(v) = 2$ $\delta(v) = 3$

Fig. 6. A stretching of a vertical path

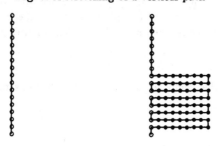

(1) If $\delta(v) = 1$, then x_v is the rightmost point on P_v.
(2) If $\delta(v) = 2$, then x_v is the middle point on P_v.
(3) If $\delta(v) = 3$, then x_v is the point on P_v adjacent to the middle vertical path.

Since each horizontal path has even length, a middle point always exists. For each edge $e = \{u, v\}$ in G, we obtain a path between x_u and x_v consisting of at most two horizontal subpaths of P_u and P_v and of the vertical path P_e. With exeception of the endpoints, these paths are pairwise disjoint. The total length of a path between points x_u and x_v is $(a_1 + 2 \cdot a_2 + a_3) \cdot d$ with $a_1, a_2, a_3 \leq \ell_{max}$ and $a_2 \geq 2$ (note that each line segment has even length). The maximum possible length is $4 \cdot \ell_{max} \cdot d = 16 \cdot \ell_{max}^2$.

In the third step, we replace each vertical path P_e, for $e = \{u, v\} \in E$, by a path P_e' such that the path between x_u and x_v has the length $16 \cdot \ell_{max}^2$. To stretch a path with length $(a_1 + 2 \cdot a_2 + a_3) \cdot d$, we insert

$$[(\ell_{max} - a_1) + 2 \cdot (\ell_{max} - a_2) + (\ell_{max} - a_3)] \cdot d < 16 \cdot \ell_{max}^2$$

new points. An original vertical path with white colored points and a stretched path with new black colored points are given in Figure 6. Since the length of a vertical path P_e is at least $4 \cdot d = 16 \cdot \ell_{max}$ and since the width of a vertical rectangle is $2 \cdot \ell_{max}$, the construction is always possible. The grid graph constructed in this way has the desired properties; where each edge is replaced by a path of length $16 \cdot \ell_{max}^2 = O(|V|^2)$.   □

3 Planar bipartite graphs

In this section we analyse the complexity of the broadcast time problem for bipartite planar graphs.

Theorem 4. The broadcast time problem is NP-complete for bipartite planar graphs, deadline $k = 2$ and maximum degree at most three.

Proof. By reduction from planar 3,4SAT. Given an instance with planar graph $(X \cup C, E)$ we construct the reduction graph as follows. Each vertex $x \in X$ is replaced by a variable component and each vertex $c \in C$ is replaced by a clause component. We describe two variants of the variable component. Both are working here, but later on we use the small one in case of complete grids and the big one in case of split and chordal graphs. The components are depicted in Figures 7, 8 and 9.

Fig. 7. Small variable component   **Fig. 8.** Big variable component

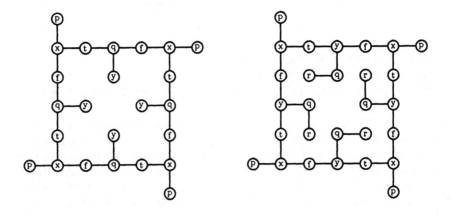

Fig. 9. Clause component

The construction of the reduction graph is complete by identifying corresponding b and t or f vertices. For each edge $\{x, c\}$ in $(X \cup C, E)$ we identify a t-vertex of

the variable component of x with a b-vertex of the clause component of c if $x \in c$. Otherwise ($\bar{x} \in c$) we identify a f-vertex of the variable component of x with a b-vertex of the clause component of c. This identification is made with respect to the cyclic ordering of the neighbourhood of each vertex in $(X \cup C, E)$ such that the resulting reduction graph is planar. It is easy to see that the reduction graph is bipartite too. We define the set V_0 of originators to be the set of all x and all y-vertices.

Suppose there is a truth mapping $b : X \to \{t, f\}$ satisfying all clauses of C. We define $a(x) = t$ if $b(x) = f$ and $a(x) = f$ if $b(x) = t$ for all $x \in X$. Then there is a broadcasting scheme of our reduction graph with deadline $k = 2$. In the first step the vertices $x \in X$ inform their neighbours $b(x)$. In the second step the vertices $c \in C$ are informed by one of its neigbours informed in the first step, i.e. if $x \in c$ and $b(x) = t$ then the common neighbour t of x and c informs c, otherwise ($\bar{x} \in c$ and $b(x) = f$) the common neighbour f informs c. Also in the second step the p-vertices are informed by their neighboured x-vertices. The remaining vertices (q, $a(x)$ and in case of the big variable component r) receive information originating in y-vertices. This takes two steps.

On the other hand, suppose there is a broadcast scheme of our reduction graph with deadline $k = 2$. Observe that in this case all x-vertices in a variable component inform in the first step their neighbours t, or all of them inform their neighbours f. Otherwise it is impossible to inform the remaining vertices of the component in a second step. We define a truth mapping $b : X \to \{t, f\}$ such that the x-vertices of the variable component of x inform in the first step their neighbours $b(x)$. We consider a clause $c \in C$. The corresponding neighbours of the vertex c in our reduction graph are t or f-vertices of variable components. Hence c is informed in the second step by a vertex $b(x)$ informed in the first step and the variable x or $\bar{x}$ satisfies the clause c. □

A graph $G = (V, E)$ is called a *split graph* if the vertex set V can be patitioned into a clique C of G and an independent set I of G, i.e. $V = C \cup I$, $C \cap I = \emptyset$ and $\binom{C}{2} \subseteq E \subseteq \binom{V}{2} \setminus \binom{I}{2}$. A *chord* of a cycle in a graph G is an edge of G connecting two nonconsecutive vertices of the cycle. G is a *chordal graph* if each cycle of length at least four admits a chord. Clearly, split graphs are chordal graphs.

Corollary 5. *The broadcast time problem with deadline $k = 2$ is NP-complete for split graphs.*

Proof. If we use the big variable components in the above proof, then all the c, r, x and y-vertices belong to one color class of the bipartite reduction graph. We obtain a split graph by completing the other color class. Clearly, added edges can not be used in a broadcast scheme with deadline $k = 2$ to inform c-vertices. □

Theorem 6. *The broadcast time problem with one originator is NP-complete for chordal graphs.*

Proof. We reduce from MTB for split graphs and $k = 2$. The complete construction is given in the full paper. □

Fig. 10. The graph G_i

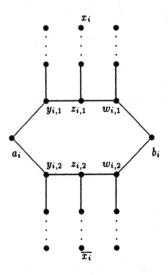

In the following we consider the problem for planar graphs with the constraint $|V_0| = 1$.

Theorem 7. *The broadcast time problem is NP-complete even if the graph G is planar with maximum degree $\Delta(G) \leq 3$ and $|V_0| = 1$.*

Proof. By reduction from planar separable 3-SAT. Given an instance with planar graph $G = (X \cup C, E \cup E')$, we construct a planar graph G^* as follows.

First, we define a graph G_i given in Figure 10, for each variable $x_i \in X$. The vertices $y_{i,1}, y_{i,2}, z_{i,1}, z_{i,2}, w_{i,1}$ and $w_{i,2}$ are adjacent to disjoint paths. The number of vertices in the pending paths is $5 \cdot (|X| - i + 1) - 3$ for $y_{i,1}$ and $y_{i,2}$ and $5 \cdot (|X| - i) - 1$ for the other. The last vertices in the pending paths for $z_{i,1}$ and $z_{i,2}$ are the variable x_i and the negated variable $\overline{x_i}$, respectively.

For each edge $\{x_i, x_{i+1}\}$ in G with $1 \leq i \leq n - 1$, we insert an edge $\{b_i, a_{i+1}\}$ which connects the graphs G_i and G_{i+1}. If x_i appears as unnegated variable in the clause c, we add an edge $\{x_i, c\}$, and if x_i appears negated, we add $\{\overline{x_i}, c\}$. The set $V_0 = \{a_1\}$ and the deadline k is $5 \cdot |X| - 1$. Since G is planar and separable, the constructed graph G^* is planar, too.

For a feasible broadcast of G^* within k time steps, we get

(i) $a_i \in V_{5(i-1)}$ and $b_i \in V_{5i-1}$ and

(ii) either $y_{i,1} \in V_{5(i-1)+1}$, $z_{i,1} \in V_{5(i-1)+2}$, $w_{i,1} \in V_{5(i-1)+3}$, $x_i \in V_{5|X|-3}$
     or $y_{i,2} \in V_{5(i-1)+1}$, $z_{i,2} \in V_{5(i-1)+2}$, $w_{i,2} \in V_{5(i-1)+3}$, $\overline{x_i} \in V_{5|X|-3}$.

Notice, if $y_{i,1} \in V_{5(i-1)+2}$ then $y_{i,1}$ must inform in time step $5(i-1)+3$ the first vertex in his pending path (otherwise the last vertex there gets the information too late). Therefore, if $y_{i,1} \in V_{5(i-1)+2}$ then $z_{i,1} \in V_{5(i-1)+4}$. Again, using the number of pending vertices of $w_{i,1}$, $w_{i,1}$ must lie in $V_{5(i-1)+5}$. Since the first vertex in the pending path of $z_{i,1}$ lies in $V_{5(i-1)+6}$, x_i is informed in the last step $k = 5|X| - 1$. Therefore, using the number of vertices in the pending paths, only the following two cases are possible:

(i) $y_{i,2} \in V_{5(i-1)+2}$, $z_{i,2} \in V_{5(i-1)+4}$, $w_{i,2} \in V_{5(i-1)+5}$ and $\overline{x_i}$ is informed in the last step k.

(ii) $y_{i,1} \in V_{5(i-1)+2}$, $z_{i,1} \in V_{5(i-1)+4}$, $w_{i,1} \in V_{5(i-1)+5}$ and x_i is informed in the last step k.

Therefore, the instance of planar separable 3-SAT is satisfiable if and only if there is a broadcast for the graph G^* and $V_0 = \{a_1\}$ within $k = 5|X| - 1$ time steps. $\quad\Box$

Using a similar transformation, we can prove the following result.

Theorem 8. *The broadcast time problem is NP-complete even if the graph G is bipartite and planar with maximum degree $\Delta(G) \leq 3$ and $|V_0| = 1$.*

4 Grid graphs

In this section we extend our results to grid graphs.

Theorem 9. *The broadcast time problem is NP-complete even for a grid graph with deadline $k = 2$ and maximum degree at most three.*

Fig. 11. The variable setting

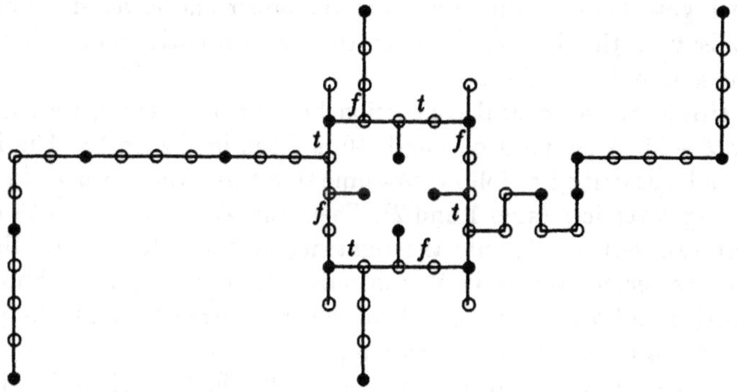

Proof. By reduction from planar 3,4-SAT. Given an instance, a planar bipartite graph $G = (X \cup C, E)$, we construct a channel layout with stretching factor $d = 10$. We place into the horizontal and vertical rectangles a set of points (at integer positions) and a set of edges such that we obtain a grid graph. In the following, we give the designs for the connection between a variable and a clause, for the variable setting and for the clause function.

Fig. 12. A path of length 10

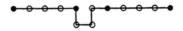

Fig. 13. The clause function

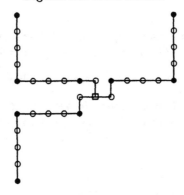

Variable setting. In each horizontal rectangle corresponding to a variable we place a component as illustrated in Figure 11. This component contains the same gadget as for the bipartite planar graph. The vertices colored black belong to V_0 and the other vertices to $V \setminus V_0$. For each occurrence of a variable in a clause we construct a path connected with a t-vertex for an unnegated variable and connected with a f-vertex for a negated variable. The height of each variable component is 10. We observe (see Figure 11) that the vertices at the upper and lower side belong to V_0. For synchronisation, the distance between two vertical paths connecting a variable with a clause is a multiple of 10.

Connections. For a vertical connection between a variable, we use a path of length $10 \cdot (2\ell - 1)$. A subpath of length 10 is given in Figure 12. The idea of the signal flow can be described as follows. Assume that the leftmost vertex in V_0 informs the next two right vertices (step 1 and 2). Then, the second vertex in V_0 can inform also the next two vertices. If only the right neighbor is informed by the leftmost vertex in V_0, the second vertex in V_0 can only inform his right neighbor. In other words, a positive and a negative signal can be transferred through the path and a negative signal can not be converted into a positive.

Clause component. A clause component of height 10 is given in Figure 13. It contains three paths corresponding to the three literals of the clause connected at a square. This square is informed in two time units, only if at least one positive signal arrives in a connecting path. Clearly, this simulates the function of the clause.

In total, we derive that the instance of planar 3,4-SAT is satisfiable if and only if there is a broadcast for G within two time steps. $\Box$

Theorem 10. *The broadcast time problem remains NP-complete for a grid graph with $|V_0| = 1$.*

Proof. By modification of the graph G^* constructed in Theorem 7. The complete construction is given in the full paper.   □

5   Complete grid graphs

In this section we analyse the complexity of the broadcast time problem for complete grid graphs.

Fig. 14. The bend

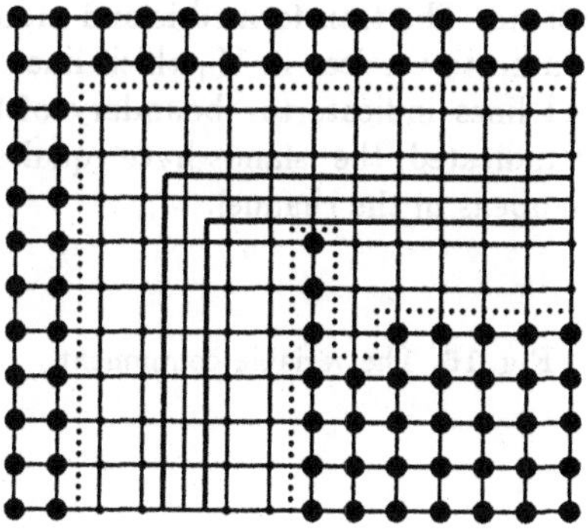

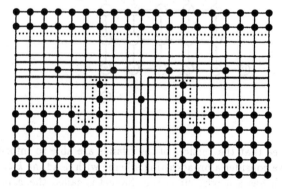

Fig. 15. The clause component

Theorem 11. *The broadcast time problem with deadline $k = 2$ is NP-complete for complete grid graphs.*

In what follows we describe the main idea. We use the reduction graph for grid graphs and stretch all paths outside any variable component such that there is an embedding in a complete grid with each such path of length 8 is embedded with at most one bend. We fix such an embedding.

All the vertices of the embedded graph form the *channels*. The complete host grid is choosen such that the vertices on its boundary have distance at least 3 from the channels. The vertices of distance 3 or more from the channels form *islands*. In particular, all the vertices of degree 2 or 3 of the complete grid belong to one island. The vertices between the islands and the channel form strips of breadth two. These vertices form the *banks*. We add the islands to V_0. The vertices in the islands inform the vertices in the neighboured bank. Vertices in a channel are informed by other vertices in the channel as before. This does not work correctly around bends and components. Therefore we have to transform the islands in these areas.

First we consider the bend. The transformed island around a bend is shown in Figure 14. Thick points indicate vertices in V_0, thick lines indicate the boundary of the channel and dotted lines indicate the boundary of the islands. The channel vertices in V_0 are not indicated, the islands have equal shape around the bend independent from what happens in the channel.

Fig. 16. The variable component

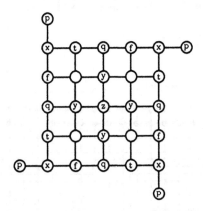

It is easy to see that the vertices of the islands are able to inform the banks in two steps. Now we consider a broadcasting scheme with deadline $k = 2$. That vertex in an island which has a neighbour in the channel is called the *point*. The point is the only vertex in an island able to inform vertices in the channel. In this case, for each vertex in the channel, receiving information originating in the point, there is a vertex in the bank receiving information originating in the channel. Therefore we may assume that vertices in the island inform vertices in the bank only.

Now it easy to see that the clause component shown in Figure 15 with the islands around works as in the prior cases. Our variable component is based on the small variable component for bipartite planar graphs. First we add four further vertices to

Fig. 17. The variable component for a variable appearing four times unnegated

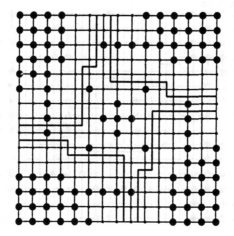

Fig. 18. The variable component for a variable appearing twice unnegated and twice negated in alternate order

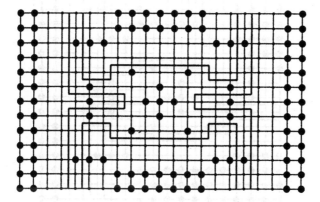

be informed by y-vertices in the second step and a z-vertex in V_0. This enlarged component is shown in Figure 16 and works as before. The shape of the islands around the variable component depends on sequence of negated and unnegated appearences. Some examples are given in Figures 17, 18, 19 and 20.

If one is not convinced that the variable components and clause components are connectable as before, then consider the following construction. For each component, either variable component or clause component, we create a sufficiently large box, say 99×99, in the host grid. Then we put the component in the box at central position. For each channel leaving the component we append pieces of ordinary channel and up to three bends, such that the channels leave the box at a midpoint of a side of the box. Clearly, the remaining part of the box is filled with vertices in V_0. Outside these boxes the channels use the grid on rows and columns divisible by 100 only.

Fig. 19. The variable component for a variable appearing twice unnegated and twice negated in consequtive order

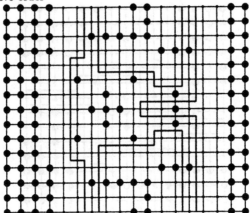

Fig. 20. The variable component for a variable appearing once unnegated and three times negated

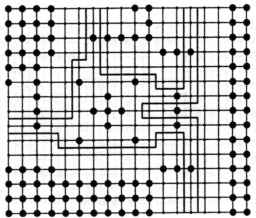

6 Conclusion

In this paper, we have given an overview about the complexity of the broadcast time problem restricted to several communication networks. Independently, Middendorf [13] proved that the minimum broadcast time problem is NP-complete for 3-regular planar graphs with constant deadline $k \geq 2$. We note that the complexity is unknown for split graphs with $|V_0| = 1$.

References

1. J.C. Bermond, P. Hell, A.L. Liestman and J.G. Peters: Broadcasting in bounded degree graphs, SIAM Journal on Discrete Mathematics **5** (1992), 10–24.
2. J.C. Bermond and C. Peyrat: Broadcasting in de Bruijn networks, Proc. 19. S-E Conference on Combinatorics, Graph Theory, and Computing (1978), 283–292.
3. R.M. Capocelli, L. Gargano and U. Vaccaro: Time bounds for broadcasting in bounded degree graphs, Workshop Graphtheoretical Concepts in Computer Science (1991), 19–33.
4. A.M. Farley and S.T. Hedetniemi: Broadcasting in grid graphs, Proc. 9. S-E Conference on Combinatorics, Graph Theory, and Computing (1978), 275–288.
5. A.M. Farley: Minimum broadcast networks, Networks **9** (1979), 313–332.
6. A.M. Farley and A. Proskurowski: Broadcasting in trees with multiple originators, SIAM Journal on Algebraic and Discrete Methods **2** (1981), 381–386.
7. M.R. Garey and D.S. Johnson: Computers and Intractability: A Guide to the Theory of NP-Completeness, Freeman, San Francisco, 1979.
8. M. Grigni and D. Peleg: Tight bounds on minimum broadcast networks, SIAM Journal on Discrete Mathematics **4** (1991), 207–222.
9. S.M. Hedetniemi, S.T. Hedetniemi and A.L. Liestman: A survey of gossiping and broadcasting in communication networks, Networks **18** (1988), 319–349.
10. A. Jakoby, R. Reischuk and C. Schindelhauer: Minimum Broadcast Time auf Graphen mit konstantem Grad, 18. Workshop über Komplexitätstheorie und effiziente Algorithmen, Forschungsbericht 92-17, Universität Trier (1992).
11. D. Lichtenstein: Planar formulae and their uses, SIAM Journal on Computing **11** (1982), 320–343.
12. A.L. Liestman and J.G. Peters: Broadcast Networks of bounded degree, SIAM Journal on Discrete Mathematics **1** (1988), 531–540.
13. M. Middendorf: Minimum broadcast time is NP-complete for 3-regular planar graphs and deadline 2, Information Processing Letters **46** (1993), 281–287.
14. P. Rosenstiehl and R.E. Tarjan: Rectilinear planar layouts of planar graphs and bipolar orientations, Discrete Computational Geometry **1** (1986), 434 – 353.
15. P.J. Slater, E.J. Cockayne and S.T. Hedetniemi: Information dissemination in trees, SIAM Journal on Computing **10** (1981), 692–701.
16. C.A. Tovey: A simplified NP-complete satisfiability problem, Discrete Applied Mathematics **8** (1984), 85–89.

The Complexity of Systolic Dissemination of Information in Interconnection Networks *

Extended Abstract

Juraj Hromkovič** , Ralf Klasing, Walter Unger, Hubert Wagener***
Department of Mathematics and Computer Science
University of Paderborn
33095 Paderborn, Germany

Dana Pardubská†
Faculty of Mathematics and Physics
Comenius University
84215 Bratislava, Slovakia

Keywords: communication algorithms, parallel computations.

Abstract. A concept of systolic dissemination of information in interconnection networks is introduced, and the complexity of systolic gossip in one-way (telegraph) and two-way (telephone) communication mode is investigated. Optimal systolic algorithms in both communication modes are constructed for paths and complete k-ary trees for any $k \geq 2$.

1 Introduction

One of the most intensively investigated areas of computation theory is the study and comparison of the computational power of distinct interconnection networks as candidates for the use as parallel architectures for existing parallel computers. There are several approaches enabling to compare the efficiency and the "suitability" of different parallel architectures from distinct point of views. One extensively used approach deals with the possibility to simulate one network by another without any

* This work was partially supported by grants Mo 285/4-1, Mo 285/9-1 and Me 872/6-1 (Leibniz Award) of the German Research Association (DFG), and by the ESPRIT Basic Research Action No. 7141 (ALCOM II).

** This author was partially supported by SAV Grant No. 88 and by EC Cooperation Action IC 1000 Algorithms for Future Technologies.

*** This author was supported by the Ministerium für Wissenschaft und Forschung des Landes Nordrhein-Westfalen.

† This author was partially supported by MSV SR and by EC Cooperative Action IC 1000 Algorithms for Future Technologies (ALTEC). Most of the work of this author was done when she was visiting the University of Paderborn.

essential increase of computational complexity (parallel time, number of processors). Such an effective simulation of a network A by a network B surely exists if the network A can be embedded into B (more details and an overview about this research direction can be found in [MS90]).

Another approach to measure the computational power of interconnection networks is to investigate which class of computing problems can be computed by a given class of networks. Obviously, this question is reasonable only by additional restrictions on the networks because each class of networks of unbounded number of processors (like paths, grids, complete binary trees, hypercubes, etc.) can recognize all recursive sets. These additional restrictions mostly restrict the time of computations (for example, to $\log_2 n$ by complete binary trees or to real time by paths) and/or the kind of computation assuring a regular flow of data in the given network. A nice concept for the study of the power of networks from this point of view has been introduced by Culik II et al. [CGS84], and investigated in [IK84, CC84, IPK85, CSW84, CGS83, IKM85]. This concept considers classes of languages recognized only by systolic computations on the given parallel architecture (network) in the shortest possible time for a given network. The notion "systolic computation" has been introduced by Kung [Ku79], and it means that the computation consists only of the repetition of simple computation and communication steps in a periodic way. The reason to prefer systolic computations is based on the fact that each processor of a network executing a systolic algorithm works very regularly repeating only a short sequence of simple instructions during the whole computation. Thus, the hardware and/or software realization of systolic algorithms is essentially cheaper than the realization of parallel algorithms containing many irregularities in the data flow or in the behaviour of the processors.

The last of the approaches mentioned here helping to search for the best (most effective) structures of interconnection networks is the study of the complexity of information dissemination in networks (for an overview see [HHL88, HKMP93]). This approach is based on the observation that the realization of the communication (data flow between the processes) of several parallel algorithms on networks requires at least as much (or sometimes even more) time as the computation time of the processors. This means that the time spent with communication is an important parameter of the quality of interconnection networks. To get a comparison of networks from the communication point of view, the complexity of the realization of some basic communication tasks like broadcast (one processor wants to tell something to all other processors), accumulation (one processor wants to get some pieces of information from all other processors) or gossip (each processor wants to tell something to each other) is investigated for different networks.

The aim of this paper is to combine the ideas of the last two approaches mentioned above to get a concept of systolic communication algorithms enabling to study the communicational effectivity of networks when a very regular behaviour of each processor of the network is required.

The first step in this direction was made by Liestman and Richards [LR93a] who introduced a very regular form of communication based on graph coloring. (This kind of communication has later been called "periodic gossiping" by Labahn et al. [LHL93].) This "periodic" communication was introduced in order to solve a special gossip problem introduced by Liestman and Richards [LR93b] and called "perpetual gossiping", where each processor may get a new piece of information at any time from the outside of the network and the never halting communication algorithm has to broadcast it to all other processors as soon as possible.

The concept of Liestman and Richards [LR93a] includes some restrictions which bound the possibility of systolic communication in a non-necessary way (more about this in the next section). Another drawback is that the complexity considered in [LR93a, LHL93] is the number of systolic periods and not the number of communication rounds, i.e. only rough approximations on the precise number of rounds executed are achieved in [LR93a, LHL93]. Here, we introduce a more general concept of systolic (periodic) dissemination of information in order to evaluate the quality of interconnection networks from this point of view. Our main aim of the investigation of systolic communication is not only to establish the complexity of systolic realisation of basic communication tasks in distinct networks, but also to learn how much must be paid for the change from arbitrary "irregular" communication to nice, regular systolic one.

This paper is organized as follows. Section 2 contains the formal description of the concept of systolic communication and some basic observations comparing general communication algorithms with systolic ones and systolic gossip with systolic broadcast. Section 3 is devoted to the complexity of systolic gossip in paths, and Section 4 to systolic gossip in trees. In the Conclusion, the results achieved are discussed and some open problems are formulated. More information about the contents of Sections 3 and 4 is given in Section 2 after the introduction of the concept of systolic communication.

2 The Concept of Systolic Communication

The aim of this section is to introduce systolic communication algorithms and to give some fundamental observations about systolic broadcast and gossip. Before doing this, we give a more precise description of the communication problems investigated and of the communication modes considered.

An (interconnection) network is viewed as a connected undirected graph $G = (V, E)$ where the nodes of V correspond to the processors and the edges correspond to the communication links of the network. An infinite sequence $\{G_i\}_{i=1}^{\infty}$ with $G_i = (V_i, E_i), |V_i| > |V_j|$ for $i > j$, is called a class of interconnection networks. Examples of classes of interconnection networks are paths - $\{P_n\}_{n=1}^{\infty}$ (P_n is the path of n nodes), cycles - $\{C_n\}_{n=1}^{\infty}$ (C_n is the cycle of n nodes), complete balanced k-ary trees - $\{T_k^h\}_{h=1}^{\infty}$ (T_k^h is the complete balanced k-ary tree of depth h), cube-connected cycles - $\{CCC_k\}_{k=1}^{\infty}$ (CCC_k is the cube-connected cycles network of dimension k), and two-dimensional square grids - $\{Gr_m^2\}_{m=1}^{\infty}$ (Gr_m^2 is the $m \times m$ grid).

In what follows, we shall investigate the following three fundamental communication problems in networks:

1. Broadcast problem for a network G and a node v of G.

   Let $G = (V, E)$ be a network and let $v \in V$ be a node of G. Let v know a piece of information $I(v)$ which is unknown to all nodes in $V - \{v\}$. The problem is to find a communication strategy such that all nodes in G learn the piece of information $I(v)$.

2. Accumulation problem for a network G and a node v of G.

   Let $G = (V, E)$ be a network, and let $v \in V$ be a node of G. Let each node $u \in V$ know a piece of information $I(u)$ which is independent of all other pieces of information distributed in other nodes (i.e. $I(u)$ cannot be derived from $\bigcup_{v \in V - \{u\}} \{I(v)\}$). The set $I(G) = \{I(w) | w \in V\}$ is called the cumulative message of G. The problem is to find a communication strategy such that the node v learns the cumulative message of G.

3. Gossip problem for a network G.

   Let $G = (V, E)$ be a network, and let, for all $v \in V$, $I(v)$ be a piece of information residing in v. The problem is to find a communication strategy such that each node from V learns $I(G)$.

Now, it remains to explain what the notion "communication strategy" means. The communication strategy is meant to be a communication algorithm from an allowed set of synchronized communication algorithms. Each communication algorithm is a sequence of simple communication steps called communication rounds (or simply rounds). To specify the set of allowed communication algorithms one defines a so-called communication mode which precisely defines what may happen in one communication step (round). Here, we consider the following two basic communication modes:

a, one-way mode (also called telegraph mode)

   In this mode, in a single round, each node may be active only via one of its adjacent edges either as a sender or as a receiver. This means that if one edge (u, v) is active as a communication link, then the information is flowing only in one direction. Formally, let $G = (V, E)$ be a network, $\vec{E} = \{(v \rightarrow u), (u \rightarrow v) | (u, v) \in E\}$. A one-way communication algorithm for G is a sequence of rounds $A_1, A_2, \ldots, A_k$, where $A_i \subseteq \vec{E}$ for every $i \in \{1, \ldots, k\}$, and if $(x_1 \rightarrow y_1), (x_2 \rightarrow y_2) \in A_i$ and $(x_1, y_1) \neq (x_2, y_2)$ for some $i \in \{1, \ldots, k\}$, then $x_1 \neq x_2 \wedge x_1 \neq y_2 \wedge y_1 \neq x_2 \wedge y_1 \neq y_2$ (i.e. each A_i is a matching in the directed graph $(V, \vec{E})$). If $(u \rightarrow v) \in A_i$ for some $i \in \{1, \ldots, k\}$, then it is assumed that the whole current knowledge of the node u is known to the node v after the execution of the i-th round A_i.

b, two-way mode (also called telephone mode)

   In two-way mode, in a single round, each node may be active only via one of its adjacent edges and if it is active then it simultaneously sends a message and receives a message through the given, active edge. Formally, let G be a network. A two-way communication algorithm for G is a sequence of rounds $B_1, B_2, \ldots, B_r$, where each round $B_j \subseteq E$, and for each $i \in \{1, \ldots, r\}, \forall (x_1, y_1), (x_2, y_2) \in B_i :$ $(x_1, y_1) \neq (x_2, y_2)$ implies $x_1 \neq x_2 \wedge x_1 \neq y_2 \wedge y_1 \neq x_2 \wedge y_1 \neq y_2$ (i.e. B_i is a matching in G). If $(u,v) \in B_i$ for some i, then it is assumed that the whole current knowledge of u is submitted to v, and the whole current knowledge of v is submitted to u in the i-th round.

Another possibility to describe a communication algorithm for $G = (V,E)$ is to say what happens on which edge of E in which time unit. For every two-way communication algorithm $A = A_1, A_2, \ldots, A_m$ for G, $\underline{\text{edge (A)}} = \{(e, S_e^A) \mid e \in E, S_e^A = \{j \mid e \in A_j\}$ for any $e \in E\}$.

Analogously, for every one-way communication algorithm $B = B_1, \ldots, B_m$ for G, $\underline{\text{edge (B)}} = \{(u \to v, S_{u \to v}^B) \mid (u,v) \in E$, and $S_{u \to v}^B = \{j \mid (u \to v) \in B_j\}$ for any $u \to v$ such that $(u,v) \in E\}$. We define $\underline{\text{edge (C)}} \leq \underline{\text{edge (D)}}$ for two two-way (one-way) communication algorithms C, D for $G = (V,E)$, if for every edge $e \in E$, $S_e^C \subseteq S_e^D$ (if for every $(u \to v)$ such that $(u,v) \in E$, $S_{u \to v}^C \subseteq S_{u \to v}^D$). Obviously, if C is a broadcast (accumulation, gossip) algorithm for some G, and edge$(C) \leq$ edge (D) for some communication algorithm D for G, then D is a broadcast (accumulation, gossip) algorithm for G, too.

Now, we are prepared to introduce systolic communication algorithms.

Definition 1. Any one-way (two-way) communication algorithm $A = A_1, A_2, \ldots, A_m$ for some $m \in \mathbb{N}$ is called $\underline{k\text{-systolic}}$ for some positive integer k, if there exist some $r \in \{1, \ldots, m\}$ and some $j \in \{1, \ldots, k\}$ such that

$$A = (A_1, A_2, \ldots, A_k)^r, A_1, A_2, \ldots, A_j.$$

$P = A_1, \ldots, A_k$ is called the $\underline{\text{period/cycle}}$ of A, k is called the $\underline{\text{length}}$ of P. □

In what follows the complexity of communication algorithms is considered as the number of rounds they consist of.

Definition 2. Let $G = (V,E)$ be a network. Let $A = A_1, A_2, \ldots, A_m$ be a one-way (two-way) communication algorithm on G. The $\underline{\text{complexity of } A}$ is $c(A) = m$ (the number of rounds of A). Let v be a node of V. The $\underline{\text{one-way complexity of}}$ $\underline{\text{the broadcast problem for } G \text{ and } v}$ is

$$\underline{b_v(G)} = \min \{c(A) | A \text{ is a one-way communication algorithm solving the broadcast problem for } G \text{ and } v\}.$$

The one-way complexity of the accumulation problem for G and v is

$$\underline{a_v(G)} = \min \{c(A) | A \text{ is a one-way communication algorithm solving the accumulation problem for } G \text{ and } v\}.$$

We define

$b(G) = \max \{b_v(G)|v \in V\}$ as the broadcast complexity of G,

$\text{minb}(G) = \min \{b_v(G)|v \in V\}$ as the min-broadcast complexity of G,

$a(G) = \max \{a_v(G)|v \in V\}$ as the accumulation complexity of G, and

$\text{mina}(G) = \min \{a_v(G)|v \in V\}$ as the min-accumulation complexity of G.

The one-way gossip complexity of G is

$r(G) = \min \{c(A)|A$ is a one-way communication algorithm solving
the gossip problem for $G\}$,

and the two-way gossip complexity of G is

$r_2(G) = \min \{c(A)|A$ is a two-way communication algorithm solving
the gossip problem for $G\}$. □

Note that we do not define the broadcast (accumulation) complexity for two-way mode because each two-way broadcast (accumulation) algorithm for a network G can be transformed into a one-way broadcast (accumulation) algorithm consisting of the same number of rounds. Observe also that $a_v(G) = b_v(G)$ (and consequently $a(G) = b(G)$, $\text{mina}(G) = \text{minb}(G)$) for any network G and any node v of G (cf. [HKMP93]).

Now, we give the notation for the complexity of systolic broadcast, accumulation, and gossip.

Definition 3. Let $G = (V, E)$ be a network, and let k be a positive integer. Let v be a node of V. The k-systolic complexity of the broadcast problem for G and v is

$[k] - sb_v(G) = \min \{c(A)|A$ is a one-way k-systolic communication
algorithm solving the broadcast problem for G and $v\}$,

and the k-systolic complexity of the accumulation problem for G and v is

$[k] - sa_v(G) = \min \{c(A)|A$ is a one-way k-systolic communication
algorithm solving the accumulation problem for G and $v\}$.

We define

$[k] - sb(G) = \max \{[k] - sb_v(G)|v \in V\}$

as the k-systolic broadcast complexity of G, and

$[k] - \text{minsb}(G) = \min \{[k] - sb_v(G)|v \in V\}$

as the k-systolic min-broadcast complexity of G.
The one-way k-systolic gossip complexity of G is

$$[k] - sr(G) = \min \{c(A)|A \text{ is a one-way } k\text{-systolic communication}$$
$$\text{algorithm solving the gossip problem for } G\},$$

and the two-way k-systolic gossip complexity of G is

$$[k] - sr_2(G) = \min \{c(A)|A \text{ is a two-way } k\text{-systolic communication}$$
$$\text{algorithm solving the gossip problem for } G\}. \qquad \square$$

Obviously, each communication algorithm is k-systolic for some sufficiently large k. But we want to consider k-systolic communication algorithms for fixed k for some classes of networks. In this approach, k is a constant independent of the sizes of the networks of the class. This means that our k-systolic algorithms are simply realized by the repetition of a cycle of k simple instructions by any processor of the network.

We observe that each k-systolic algorithm uses (activates) at most k adjacent edges of every node of the network during the whole work of the algorithm. Thus, there is no reason to consider classes of networks like hypercubes and complete graphs, because a k-systolic algorithm can use only a subgraph of these graphs with the degree bounded by k. For this reason, we shall investigate systolic complexity of broadcast and gossip for constant-degree bounded classes of networks only. Our aim is not only to get some lower and upper bounds on the systolic broadcast and gossip complexity of some concrete networks, but also to compare the general complexities of unrestricted communication algorithms with the systolic ones. In this way, we can learn which is the price for our systolization, i.e. how many additional rounds are needed to go from an optimal broadcast (gossip) algorithm to an optimal systolic one.

Our first result shows that, in some sense, the broadcast complexity is the same as the systolic broadcast complexity for any network.

Lemma 4. *For every class of networks* $\{G_i\}_{i=1}^{\infty}$ *and every positive integer* d *such that* G_i *has the degree at most* d *for any* $i \in I\!N$,

$$[d] - sb_v(G_i) = b_v(G_i)$$

for any $i \in I\!N$ *and any node* v *of* G_i.

Proof. Let $B = B_1, B_2, \ldots, B_m$ be a broadcast algorithm from v in $G_i = (V, E)$ for some positive integer i. W.l.o.g. we may assume $B_i \bigcap B_j = \emptyset$ for $i \neq j$. One can reconstruct B to get another broadcast algorithm $A = A_1, A_2, \ldots, A_m$ for v and G_i with the following three properties:

(i) $\displaystyle\bigcup_{i=1}^{m} A_i \subseteq \bigcup_{j=1}^{m} B_j$, $c(B) = c(A)$,

(ii) $T = (V, \displaystyle\bigcup_{i=1}^{m} A_i)$ is a directed tree with the root v (all edges are directed from the root to the leaves), and

(iii) for every node $w \in V$, if w gets $I(v)$ in time t and the degree (indegree plus outdegree) of w in T is k, then w submits $I(v)$ in the rounds $t+1, t+2, \ldots, t+k-1$ to all its $k-1$ descendants in T.

The construction of broadcast algorithm A' from B with the properties (i) and (ii) is simple and can be found in Observation 1.2.5 of [HKMP93]. How to get the property (iii) (i.e. to construct A from A') is obvious.

To define a d-systolic broadcast algorithm $C = (C_1, C_2, \ldots, C_d)^r, C_1, C_2, \ldots, C_j$ for some $r \in \mathbb{N}$, $j \in \{0, \ldots, d-1\}$, $m = r \cdot d + j$, it is sufficient to specify C_i for $i = 1, \ldots, d$. For $1 \leq s \leq d$, let $In(s) = \{n | 1 \leq n \leq m, n \bmod d = s - 1\}$. Then,

$$C_s = \bigcup_{\ell \in In(s)} A_\ell \text{ for } s \in \{1, \ldots, d\}.$$

Since $\text{edge}(A) \leq \text{edge}(C)$, it is obvious that C is a broadcast algorithm for G and v.
$\square$

Clearly, the same consideration as in the proof of Lemma 4 leads to the following result.

Lemma 5. *Let $\{G_i\}_{i=1}^{\infty}$ be a class of networks, where for every $i \in \mathbb{N}$, G_i has degree bounded by some constant $d \in \mathbb{N}$. Then, for every $i \in \mathbb{N}$ and every node v of G_i*

$$[d] - sa_v(G_i) = a_v(G_i).$$
$\square$

The next important question is which relation holds between gossip complexity and systolic gossip complexity. The following sections show that, as opposed to broadcast (accumulation), there are already essential differences between the complexities of general gossip and systolic gossip. Here we shall still deal with the relation between gossip and broadcast. It is well-known (see, for instance [HKMP93]) that $r(G) \leq \text{mina}(G) + \text{minb}(G) = 2\,\text{minb}(G)$ and $r_2(G) \leq 2\,\text{minb}(G) - 1$ for any graph G. Note that for trees [BHMS90] and some cyclic graphs with "weak connectivity" [HJM93] the equalities $r(G) = 2\,\text{minb}(G)$ and $r_2(G) = 2\,\text{minb}(G) - 1$ hold. The idea of the proof of $r(G) \leq \text{mina}(G) + \text{minb}(G)$ is very simple: One node of G first accumulates $I(G)$, and then it broadcasts $I(G)$ to all other nodes. Unfortunately, we cannot use this scheme to get systolic gossip from systolic broadcast and systolic accumulation, because we have to use every edge of an optimal broadcast (accumulation) scheme in both directions in each repetition of the cycle of a systolic gossip algorithm which already increases the time for the broadcast phase twice. Thus, using this straightforward idea we only obtain the following.

Theorem 6. *Let G be a communication network of degree bounded by some positive integer k. Then*

$$[2k]\text{-}sr(G) \leq 4 \cdot [k]\text{-}minsb(G) + 2k.$$

Proof. Let $B = B_1, B_2, \ldots, B_m$, $m = \text{minb}(G) = [k]\text{-minsb}(G)$, be an optimal broadcast algorithm for G and some node v of G with the properties (i), (ii), (iii) as in Lemma 4. Let $T_B = (V, \bigcup\limits_{i=1}^{m} B_i)$. Obviously, $A = A_1, A_2, \ldots, A_m$, where $A_i = \bar{B}_{m-i+1} = \{(x \to y)|(y \to x) \in B_{m-i+1}\}$ is an optimal accumulation algorithm for G and v. Moreover, $T_A = (V, \bigcup\limits_{i=1}^{m} A_i)$ is the same scheme as T_B, only the edges are directed in opposite directions. Building

$$C_s = \bigcup_{j \in In(s)} B_j \text{ for } s \in \{1, \ldots, k\} \text{ (where } In(s) \text{ as defined in Lemma 4)},$$

$$D_s = \bigcup_{j \in In(s)} A_j \text{ for } s \in \{1, \ldots, k\},$$

one obtains optimal k-systolic broadcast and accumulation algorithms $C = (C_1, C_2, \ldots, C_k)^r, C_1, C_2, \ldots, C_j$ and $D = (D_1, \ldots, D_k)^r, D_1, D_2, \ldots, D_j$ resp. for some positive integer r, and $j \in \{0, \ldots, k-1\}$. Now, we consider the $2k$-systolic communication algorithm

$$F = (D_1, \ldots, D_k, C_1, \ldots, C_k)^{2r+1}.$$

The initial part of F, $(D_1, \ldots, D_k, C_1, \ldots, C_k)^r D_1, \ldots, D_k$, is a one-way $2k$-systolic accumulation algorithm for G and the node v. The rest, $C_1, \ldots, C_k$ $(D_1, \ldots, D_k, C_1, \ldots, C_k)^r$, is a one-way $2k$-systolic broadcast algorithm for G and the node v. Thus, F is a one-way $2k$-systolic gossip algorithm with $c(F) = 2k \cdot (2r+1) = 4r \cdot k + 2k \leq 4 \cdot [k]\text{-minsb}(G) + 2k$. □

The next sections deal with the systolic gossip problem in concrete networks. The next Section 3 is devoted to gossiping in paths P_n. For systolic algorithms in two-way (telephone) communication mode, the optimal gossip algorithm for paths is in fact a 2-systolic communication algorithm. On the other hand, we show that for the one-way communication mode one can systolically gossip faster in P_n with a longer period k. More precisely, for growing period k, the function $r(P_n)$ of n can be approached more and more but never achieved (namely $[k]\text{-}sr(P_n) \approx (1+\varepsilon_k) \cdot r(P_n)$ and $\lim_{k \to \infty} \varepsilon_k = 0$).

Section 4 is devoted to gossiping in complete k-ary trees T_k^h for $k \geq 2$. Surprisingly we show for sufficiently large periods d (d independent of the depth h of the tree and depending only on the degree k) that $[d]\text{-}sr_2(T_k^h) = r_2(T_k^h) = 2\text{minb}(T_k^h) - 1$ and $[d]\text{-}sr(T_k^h) = r(T_k^h) = 2\text{minb}(T_k^h)$ for any $h \in \mathbb{N}$, i.e. we can systolically gossip in complete trees in optimal gossip time in both modes. We also show for the minimal possible period length $d = k + 1$ of any two-way systolic communication algorithm for k-ary trees that $[k+1]\text{-}sr_2(T_k^h) \leq r_2(T_k^h) + 1$.

3 Systolic Gossiping in Paths

In this section we consider systolic gossiping in the path P_n of n nodes.

For the two-way mode, we can find systolic gossip algorithms for P_n and C_n with an optimal period length that work as efficiently as the algorithms in the general gossip mode.

Theorem 7.

(i) $[2]$-$sr_2(P_n) = n - 1 = r_2(P_n)$ *for even $n \geq 2$,*

(ii) $[2]$-$sr_2(P_n) = n = r_2(P_n)$ *for odd $n \geq 3$.*

Proof. The lower bounds for the general gossip mode are presented in [HKMP93]. The upper bounds are variations of the algorithms described in [HKMP93].

Algorithm A for P_n (where $V(P_n) = \{x_1, x_2, \ldots, x_n\}$, $E(P_n) = \{(x_1, x_2), (x_2, x_3), \ldots, (x_{n-1}, x_n)\}$) has the following systolic period:

$$A_1 = \{(x_1, x_2), (x_3, x_4), (x_5, x_6), \ldots\},$$
$$A_2 = \{(x_2, x_3), (x_4, x_5), (x_6, x_7), \ldots\}.$$

A simple analysis shows that Algorithm A takes $n - 1$ rounds if n is even and n rounds if n is odd. □

Note that for an optimal gossip in P_n, the period lengths in Theorem 7 are the best possible. Any systolic algorithm for P_n must have at least period length 2.

As a consequence of Theorem 7, we have that for every even period length, optimal gossip time is achievable. The next theorem states that all algorithms of odd period length are less effective. We omit the proof here.

Theorem 8. *For any $n \in \mathbb{N}$ and any odd $k \geq 3$:*

$$[k]\text{-}sr_2(P_n) = \frac{k}{k-1} \cdot n \pm O(k). \qquad\qquad □$$

This means that if one has an odd period length k for gossip in P_n, the best what can be done is to take one empty round in the period and $k - 1$ rounds alternatively using rounds A_1 and A_2 from the proof of Theorem 7.

Let us turn to the one-way mode of communication now. For the complexity of systolic gossiping in the path P_n of n nodes, we obtain upper and lower bounds which are tight up to a constant if the period length k is even. An important observation contrasting to the two-way case is that there is no constant d such that $[d]$-$sr(P_n) = r(P_n)$ for every $n \in \mathbb{N}$.

Theorem 9. *For any $n \geq 2$:*

(i) $\frac{k}{k-2} \cdot (n-1) - 1 \leq [k]\text{-}sr(P_n) \leq k \cdot \left\lceil \frac{n}{k-2} \right\rceil - 2$ *for k even,*

(ii) $\frac{k+1}{k-1} \cdot (n-1) - 1 \leq [k]\text{-}sr(P_n) \leq k \cdot \left\lceil \frac{n-1}{k-2} \right\rceil$ *for k odd.*

Proof. Because of the restricted space of this extended abstract and the length of this proof, we omit it completely here. □

Theorem 9 shows that one can essentially gossip faster in P_n with a longer period k. For growing k, $r(P_n)$ can be approached.

4 Systolic Gossiping in Complete k–ary Trees

In this section we investigate the systolic gossip complexity of complete, balanced k–ary trees. The main result of this section is that there exist systolic gossip algorithms with a period of constant length whose complexity matches the lower bound for even non–systolic algorithms.

Let us first state the lower bound for gossiping in complete, balanced k–ary trees. It is shown in [BHMS90] that the gossip complexity in two–way mode $r_2(T)$ for any tree T is exactly $2 \cdot \text{minb}(T) - 1$ and that for one–way mode $r(T) = 2 \cdot \text{minb}(T)$ holds. For a complete, balanced k–ary tree T_k^h of height h it is not hard to see that $\text{minb}(T_k^h)$ is given by $k \cdot h$ (for a proof consult [HKMP93]). This implies the following proposition.

Proposition 10. *For a complete, balanced k–ary tree T_k^h of height h and period p*

(i) $[p] - sr(T_k^h) \geq r(T_k^h) = 2 \cdot k \cdot h,$

(ii) $[p] - sr_2(T_k^h) \geq r_2(T_k^h) = 2 \cdot k \cdot h - 1.$ □

Next we will provide upper bounds on the systolic gossip complexity. We consider the two–way mode of communication first, and give a nearly optimal gossip scheme with minimal period.

Theorem 11. *For $k \geq 2$ and $h \geq 0$*

$$[k+1]\text{-}sr_2(T_k^h) \leq 2 \cdot k \cdot h.$$ □

Note that there exists no systolic algorithm of period $\leq k$, if $h > 1$, because in this case there are vertices of degree $k + 1$. Any algorithm with period $\leq k$ would ignore some edge and no information between the components of the tree connected by this edge can be exchanged.

Outline of the Proof. We specify a communication pattern for the systolic algorithm of period $k + 1$. We say that vertex v performs pattern $(x_0, x_1, \cdots x_k)$ with $x_i \in \{c_1, \ldots, c_k, p\}$, where $x_i = c_j$ if v communicates with its j-th child in round i, and $x_i = p$ describes a communication with its parent in round i.

The actual pattern performed is as follows: Each vertex performs a cyclic shift of the pattern $(p, c_1, c_2, \cdots, c_k)$. The root of a tree of height h uses the pattern obtained by shifting the above pattern h positions to the left. The i-th child of vertex v uses the pattern of vertex v shifted i positions to the right.

Note that the pattern of vertices are compatible in the sense, that the i-th child communicates with its parent in the same round, as the parent is communicating with its i-th child. Note also that each vertex is exchanging its current information with its father vertex each $k + 1$ rounds. This immediately implies that after $(k + 1) \cdot (h + 1)$ rounds, the root has accumulated all messages, and that after additional $(k + 1) \cdot (h + 1)$ rounds, this accumulated messages is broadcast to all vertices of the tree. Thus the algorithm performs a gossiping.

A tighter analysis now yields the result. Easy inductive arguments show:

1. In a tree T_k^h, with a communication pattern obtained from the one given above by a shift to the right by i positions for each vertex, the accumulated message is available at the root after $k \cdot h + i$ rounds for $0 \leq i \leq k$. Moreover at the same time the accumulated message is available in the rightmost child of the root, too.
2. A message stored in the root of T_k^h is broadcast to all vertices in the tree after $k \cdot h$ rounds, if in the starting round (first delivery of this message) the root delivers the message to its first child c_1. Otherwise one additional round is required.

Now the time for gossiping can be bounded as follows. After $k \cdot h$ rounds the accumulated message is known to the root and its rightmost child. After k additional rounds all children have obtained this message, and $k \cdot (h - 1)$ rounds now suffice to broadcast the accumulated message in each of the subtrees of height h. Thus overall $2 \cdot k \cdot h$ rounds have to be performed. □

Next we argue that we can obtain an optimal algorithm in two–way mode when using a larger period.

Theorem 12. *For $k \geq 3$ and $h \geq 0$*

$$[3(k + 1)] - sr_2(T_k^h) = 2 \cdot k \cdot h - 1 = r_2(T_k^h).$$

Outline of the Proof. Given a tree T_k^h of height $h \geq 3$, we cut off the top three levels, i.e. a tree of height 2. For T_k^2 we design a systolic algorithm A of period $3(k + 1)$ that has the following properties:

1. A runs in time $4k - 1$ (i.e. in time $2hk - 1$).

2. A does not require all information in the beginning of its execution. More precisely, for any j, $1 \leq j \leq 2k - 2$, there exists at least one leaf that accesses its information in round j for the first time.

3. Any leaf requesting its information in round j holds the cumulative message after round $j + 2k + 1$

4. Each leaf performs only two communications (with its parent), namely in round j and $j + 2k + 1$.

This algorithm can be specified by the action of level–1–vertices, since such a vertex is involved in any communication. The i-th level-1-vertex starts its communication in round k. First, messages with all children are exchanged followed by a parent communication, $k - 1$ no-ops, again a parent communication and finally a sequence of k child communications. The period of activity in this algorithm is bounded for each vertex by $3k + 1$. Thus it is easy to formulate this algorithm as a systolic one of period $3(k + 1)$. It is now easy to check the properties given above.

The algorithm is now composed from algorithm A and the one given in the previous theorem called algorithm B for the rest of this proof. Recall that algorithm B has period $k + 1$ and thus is also systolic for period $3(k + 1)$. Algorithm B performs the accumulation in hk rounds and the root can deliver the cumulative message in round $hk + 1$ to the environment. Moreover any message given to the root in a round dedicated to parent communication is broadcast in hk rounds. Note also that if we perform a cyclic right shift of i positions for the communication pattern of each vertex, we obtain an algorithm B_i that requires $hk + i$ rounds for accumulation, and hk rounds for broadcasting. The actual composition is as follows:

Algorithm A is adjusted by a cyclic shift such that round 1 of A is performed as round $(h - 2)k + 1$ in the composed one. The vertices of the top three levels perform this algorithm. The level-3-vertices have to perform some additional communications serving as an interface between the two algorithms. Consider some level-3-vertex ℓ. Being a leaf of the top tree, ℓ should receive the cumulative message of its subtree before round $(h-2)k+j$ for some $j \leq 2k-2$. This guarantees that after $h(k-2)+4k-1$ rounds the entire cumulative message is known to all vertices of the top tree. To achieve this behaviour, we let algorithm $B_{j+i-2 \mod (k+1)}$ run on the i-th subtree of ℓ, if $j < 2k - 2$, and algorithm $B_{j+i-3 \mod (k+1)}$, if $j = 2k - 2$. Thus the cumulative message is available just in time, i.e after round $(h - 2)k + j - 1$ for $j < 2k - 2$ and after round $(h - 2)k + j - 2$ for $j = 2k - 2$. Note that in no case information from any subtree is requested before round $(h - 3)k + 1$.

A crucial point is to check that ℓ is available for child communication at the postulated rounds. Indeed, all these requests fit in the communication pattern. After an appropriate shift the communication pattern for $j < 2k - 2$ is given by $(c_1, c_2, \cdots, c_k, p, noop, \cdot, noop, p, noop)$ and by $(c_1, c_2, \cdots, c_k, noop, p, noop, \cdot, noop, p)$ for $j = 2k - 2$. Note that the cumulative message arrives in the second occurrence of the parent communication. Thus the message has to wait in ℓ for at most one round, if $j < 2k-2$, i.e. if the message arrives before round $(h-2)k+4k-1$. Otherwise, the message is directly forwarded to its children. Thus after additional k rounds all subtrees have received the cumulative message, and according to the broadcast property of algorithm B the gossip is completed after round $(h-2)k+4k-1+k+(h-3)k = 2hk-1$.

□

Using similar arguments we have shown similar upper bounds for the one–way mode of communication. We state in this abstract the result without proof.

Theorem 13. *For $k \geq 2$ and $h \geq 0$*

$$\left[\left(3 + \left\lceil \frac{4}{k-1} \right\rceil \right)(k+1) \right] - sr(T_k^h) = 2 \cdot k \cdot h = r(T_k^h).$$

□

Thus, we have shown in this section that the communication complexity of systolic gossiping for reasonable, constant periods matches the complexity of unrestricted gossiping.

5 Conclusion

Here we discuss the results achieved and formulate some of the main resulting open problems.

In this paper we have introduced the concept of systolic communication. In Section 2 we have shown that the complexity of systolic gossip is at most four times the complexity of systolic min-broadcast. This contrasts to the general relation $r(G) \leq 2\text{minb}(G)$ for any G [BHMS90].

Open problem 1. Can the multiplicative constant 4 in the result $[2k]-sr(G) \leq 4 \cdot [k]-minsb(G)+2k$ of Theorem 6 be improved? Note that 2 does not suffice because due to Theorem 9, $[k]-sr(P_n) \geq d_k \cdot 2\text{minb}(P_n)$, where $d_k > 1$ for any $k \in \mathbb{N}$, holds. On the opposite, trees are the hardest graphs for the relation between general gossip and min-broadcast ($r(T) = 2\text{minb}(T)$ for any tree T), and we can prove $[d]-sr(T_k^h) = 2\text{minb}(T_k^h)$ for some suitable constant d. This gives us hope for a much better relation between systolic gossip and broadcast than the relation given in Theorem 6.

Section 4 shows that we can systolically gossip in T_k^h in the optimal gossip time $r(T_k^h)$ $[r_2(T_k^h)]$. We only have to pay for this with a systolic period longer than the minimal possible period length $k + 1$ $[2k + 2]$ for one–way [two–way] systolic communication algorithms for T_k^h.

Open problem 2. What is the minimal period length for a time-optimal gossip? Which time can be achieved by a $[2k + 2]$-systolic one-way gossip algorithm?

References

[BHMS90] A. Bagchi, S. L. Hakimi, J. Mitchem, E. Schmeichel: *Parallel algorithms for gossiping by mail.* Information Process. Letters 34 (1990), 197-202.

[CC84] C. Choffrut, K. Culik II: *On real-time cellular automata and trellis automata.* Acta Informatica 21 (1984), 393-407.

[CGS83] K. Culik II, J. Gruska, A. Salomaa: *Systolic automata for VLSI on balanced trees.* Acta Informatica 18 (1983), 335-344.

[CGS84] K. Culik II, I. Gruska, A. Salomaa: *Systolic trellis automata. Part I.* Intern. J. Comput. Math. 15 (1984), 195-212.

[CSW84] K. Culik II, A. Salomaa, D. Wood: *Systolic tree acceptors.* R.A.I.R.O. Theoretical Informatics 18 (1984), 53-69.

[HHL88] S. M. Hedetniemi, S. T. Hedetniemi, A. L. Liestmann: *A survey of gossiping and broadcasting in communication networks.* Networks 18 (1988), 319-349.

[HJM93] J. Hromkovič, C.-D. Jeschke, B. Monien: *Note an optimal gossiping in some weak-connected graphs.* Theoretical Computer Science, to appear.

[HKMP93] J. Hromkovič, R. Klasing, B. Monien, R. Peine: *Dissemination of information in interconnection networks (broadcasting & gossiping).* In: Combinatorial Network Theory (Frank Hsu, Ding-Zhu Du, Eds.), Science Press, AMS 1993, to appear.

[IK84] O. H. Ibarra, S. M. Kim: *Characterizations and computational complexity of systolic trellis automata.* Theoretical Computer Science 29 (1984), 123-153.

[IKM85] O. H. Ibarra, S. M. Kim, S. Moran: *Sequential machine characterizations of trellis and cellular automata and applications.* SIAM J. Comput. 14 (1985), 426-447.

[IPK85] O. H. Ibarra, M. A. Palis, S. M. Kim: *Fast parallel language recognition by cellular automata.* Theoretical Computer Science 41 (1985), 231-246.

[Ku79] H.T. Kung: *Let's design algorithms for VLSI systems.* In: Proc. of the Caltech Conference of VLSI (CL.L. Seifz Ed.), Pasadena, California 1979, pp. 65-90.

[LHL93] R. Labahn, S.T. Hedetniemi, R. Laskar, *Periodic gossiping on trees.* to appear in: Discrete Applied Mathematics.

[LR93a] A.L. Liestman, D. Richards: *Network communication in edge-colored graphs: Gossiping.* IEEE Trans. Par. Distr. Syst. 4 (1993), 438-445.

[LR93b] A.L. Liestman, D. Richards: *Perpetual gossiping.* Technical Report, Simon Fraser University, Vancouver, Canada, 1993.

[MS90] B. Monien, I. H. Sudborough: *Embedding one interconnection network in another.* Computing Suppl. 7 (1990), 257-282.

Representations of Gossip Schemes

David W. Krumme[1]

Tufts University, Medford MA 02155, USA. krumme@cs.tufts.edu

Abstract. Formalisms for representing gossip problems are surveyed. A new method "calling schemes" is presented which generalizes existing methods. This survey is intended to serve primarily as a basis for future work.

1 Introduction

In the gossip problem [6], each of a number of people (or processing elements) has an item of gossip to share with the others through a sequence of communication events called calls. In general we can have full-duplex or two-way calls ("gossiping by telephone"), half-duplex or one-way calls ("gossiping by telegraph"), multiway or conference calls, and we may have limits on the number of calls a person can participate in simultaneously (usually just one). One can study gossiping in terms of the time it takes, the number of calls, the "cost," the structure of communication patterns, and so on. This paper presents and compares several formalisms that can represent these problems, and it introduces a new formalism "calling schemes" which is seen to be a generalization of existing methods.

Two kinds of graphs are useful. First is what we shall call a *communication graph*. Its vertices represent people gossiping and its edges denote actual or potential communication events. None of our graphs will have loops (self-edges); in some cases we allow multigraphs where multiple edges between a pair of vertices represent multiple calls, while in other cases we consider only simple graphs with at most one edge between any pair of vertices which may be used repeatedly. We deal with both directed and undirected graphs; most of the definitions apply at once to both cases.

The other kind of graph is what we shall call an *information flow diagram*, or simply a *flow diagram*. Its vertices denote an accumulation of information at some specific place and time, and its edges denote the transfer of that information to other places and times.

2 Preliminaries

A poset $(P, \prec)$ is an antisymmetric, transitive relation $\prec$ defined on a set P. By $\stackrel{\cdot}{\prec}$ we denote the immediate successor relation: $p \stackrel{\cdot}{\prec} q$ means $p \prec q$ and if $p \prec r \prec q$, then $r = p$ or $r = q$. If we form the symmetric transitive closure of $\prec$, we obtain an equivalence relation on P; each of the resulting equivalence classes forms a poset under $\prec$ that we shall call a *component* of $\prec$. An *upper* bound r of p and q satisfies $p \preceq r$ and $q \preceq r$; it is a *minimal upper bound* if it has no predecessor that is an upper bound of p and q. A *lower bound* r of p and q satisfies $r \preceq p$ and $r \preceq q$; it is a *maximal lower bound* if it has no successor that is a lower bound of p and q. If $\prec$ and $\prec_1$ are relations such that $p \prec_1 q$ implies $p \prec q$, then $\prec$ is a refinement of $\prec_1$. Posets are ordered by refinement.

A poset can be represented by its *Hasse diagram* [3, 10, 11] which is a digraph whose vertex set is P and whose edges correspond to $\prec\!\!\cdot$. Conventionally, Hasse diagrams are illustrated using undirected links with the understanding that the links are oriented in the upward direction on the page. The Hasse diagram representation provides a 1:1 correspondence between posets and directed acyclic graphs. A weakly connected component of the Hasse diagram, in the usual graph-theoretic sense, corresponds to a component, as defined above, of the corresponding poset.

Given a partition Π of P, the contraction $\prec_\Pi$ is the relation on Π defined by: $P_1 \prec_\Pi P_2$ iff $p_1 \prec p_2$ for some $p_1 \in P_1$ and $p_2 \in P_2$. A contraction of a poset may fail to be a poset because either the antisymmetry or the transitive property may be lost (as in Fig. 2—see §4). Contracting a relation corresponds to contracting its Hasse diagram.

Similarly, given a partition Π of the vertices of a graph (V, E), the contraction under Π is the simple graph whose vertex set is Π and where (V_1, V_2) is an edge iff $(v_1, v_2) \in E$ for some $v_1 \in V_1$ and $v_2 \in V_2$. In a similar vein, we define the multigraph contraction of a multigraph to consist of an edge (V_1, V_2) for every $(v_1, v_2) \in E$ with $v_1 \in V_1$ and $v_2 \in V_2$.

An undirected multigraph is a T^*-*graph* [2, 3] if it consists of just a single vertex or if it has two edge-disjoint spanning trees. Given an undirected multigraph $G = (V, E)$, let V^* be the collection of vertex sets of the maximal T^*-subgraphs of G. V^* is a partition of V [3, 9]. The T^*-*contraction* of (V, E) is the contraction under V^*. Because of the maximality of the sets in V^*, the T^*-contraction is necessarily a simple graph. For any $v_1 \in V_1 \in V^*$, we denote V_1 as v_1^* (nonuniquely). If $e = (v_1, v_2)$ where $v_1 \in V_1 \in V^*$ and $v_2 \in V_2 \in V^*$, we denote the edge between V_1 and V_2 as e^* (uniquely, because of the maximality of the sets in V^*).

3 Communication graphs

A *weak information flow* (G, φ) (not to be confused with a flow diagram) is a kind of communication graph $G = (V, E)$ whose edges denote allowable communication events and where each edge is labeled with the times at which communication takes place through a function $\varphi : E \to 2^{\mathbb{Z}}$. A call is a pair (e, t) with edge $e \in E$ and time $t \in \varphi(e)$. A *regular information flow* satisfies the additional condition $\varphi(e_1) \cap \varphi(e_2) = \emptyset$ for all adjacent edges e_1 and e_2. This condition prohibits a vertex from participating in concurrent calls. A *strong information flow* is one where $\varphi(e_1) \cap \varphi(e_2) = \emptyset$ for all e_1 and e_2. This condition prohibits concurrent calls entirely, so that each call is performed at a distinct time.

If we take G to be an undirected graph in the above definitions, we obtain weak, regular, and strong *two-way information flows*. A regular two-way information flow is also known as simply an *information flow* [10]. The calls of an information flow naturally define a poset called the *minimal order* [1]: it is the transitive closure of the relation $(e_1, t_1) \prec (e_2, t_2)$ if e_1 and e_2 have a vertex in common and $t_1 < t_2$. Information flows and minimal orders appear in Fig. 4.

A *weak labeling* of a multigraph is a communication graph whose edges denote actual communication events, where each edge is labeled with exactly one integer denoting the time at which the communication event takes place, through a function $\psi \colon E \to \mathbf{Z}$. A call is a pair $(e, \psi(e))$ with edge e and time $\psi(e)$. A *regular labeling* is a weak labeling with the additional condition that $\psi(e_1) \neq \psi(e_2)$ if e_1 and e_2 are adjacent edges. This condition prohibits concurrent calls with a common vertex. A *strong labeling* specifies that $\psi(e_1) \neq \psi(e_2)$ for every $e_1 \neq e_2$. In a strong labeling, each call occurs at a distinct time.

A (weak, regular, strong) information flow naturally determines a (weak, regular, strong) labeling of the multigraph that has one instance of each edge used in the information flow. Conversely, a (weak, regular, strong) labeling naturally determines a (weak, regular, strong) information flow on the simple graph obtained by collapsing multiple edges in the multigraph. A weak or regular information flow or multigraph labeling can always be converted to a regular or strong one by serializing concurrent calls.

A regular labeling of a multigraph amounts to a solution to the problem of gossiping without reusing any edge.

4 Flow diagrams

A *calling scheme* is a set P and relations $\prec$ and $\prec_1$ such that:

(S1) $(P, \prec)$ and $(P, \prec_1)$ are posets;
(S2) $\prec$ is a refinement of $\prec_1$; and
(S3) if $p \overset{\cdot}{\prec} q$ and $p \not\prec_1 q$, then p and q belong to different components of $\prec_1$.

We say $p \prec_2 q$ if $p \overset{\cdot}{\prec} q$ and $p \not\prec_1 q$. Note that $\prec_2$ is not transitive and that $\overset{\cdot}{\prec}_2 = \prec_2$.

Since $\prec$ is a refinement of $\prec_1$, we can show the Hasse diagrams of both relations together. In this paper we depict elements of $\overset{\cdot}{\prec}_1$ with light lines and elements of $\prec_2$ with bold lines. In these depictions, deleting the bold lines yields the Hasse diagram of $\prec_1$, whereas the Hasse diagram of $\prec$ is obtained by ignoring the difference between light and bold and deleting light lines implied by transitivity.

Typically each component of $\prec_1$ corresponds to a vertex in a communication graph, and $\prec_2$ denotes the transfer of information by calls. We refer to the elements of P as the *events* of the calling scheme, and in particular, p is a *sending event* (*receiving event*) if $p \prec_2 q$ ($q \prec_2 p$) for some event q. An event is an *active event* if it is a sending or receiving event, otherwise it is a *passive event*.

Fig. 1. From left to right: one-way, two-way, three-way, broadcast, and accumulation calls. Bold links denote transfer of information (in the upward direction). Light links denote retention of information at a vertex of the communication graph.

A calling scheme is *reduced* if it has no passive events. One can reduce a scheme without loss of essential information by simply omitting its passive events, since the relationships among the active events determined by $\prec$ and $\prec_1$ will be unchanged. The scheme in the middle of Fig. 5 has two passive events.

A calling scheme is *ordered* if the relation $\prec_1$ is a disjoint union of chains (totally ordered subsets). This corresponds to an information flow or multigraph labeling being regular.

4.1 Calls

A one-way call is a pair of events $\{p, q\}$ such that $p \prec_2 q$ and there is no event r such that $p \prec_2 r$, $q \prec_2 r$, $r \prec_2 p$, or $r \prec_2 q$. An n-way call for $n \geq 2$ is a set of $2n$ events $\{p_1, q_1, \ldots, p_n, q_n\}$ such that for all $1 \leq i \leq n$:

(C1) the immediate successors of p_i under $\prec$ are precisely the $\{q_j\}$;
(C2) the immediate predecessors of q_j under $\prec$ are precisely the $\{p_i\}$;
(C3) the unique immediate successor of p_i under $\prec_1$ is q_i;
(C4) the unique immediate predecessor of q_j under $\prec_1$ is p_j;
(C5) p_i has no predecessor under $\prec_2$;
(C6) q_j has no successor under $\prec_2$.

Figure 1 illustrates some calls, including accumulation and broadcast calls not covered by the above definitions. Other types of calls are certainly possible, according to the type of information transfer one wishes to model.

The events in a call are said to be its participants, and we can define an equivalence relation among events by making two events equivalent if they participate in the same call. (Conditions (C5) and (C6) ensure that an event cannot participate in more than one call.) The contraction of $\prec$ under the partition induced by such an equivalence relation is called a *contraction by calls*. We treat this kind of contraction as an ordinary poset rather than a calling scheme, in other words we do not pay attention to what happens to $\prec_1$ under this contraction. For any given n, we say a calling scheme is an n-way scheme iff every event is a participant in an n-way call and if the contraction by n-way calls is a

Fig. 2. The contraction by one-way calls may lack the antisymmetric (left) or the transitive property (right). The contraction by one-way calls amounts to collapsing each bold link to a point.

poset. Note that an n-way calling scheme is necessarily a reduced scheme. Note also that when every event is a participant in an n-way call, the contraction eliminates $\prec_2$; this is why it makes sense to ignore the distinction between $\prec_1$ and $\prec$. Conditions (C1) and (C2) ensure that if every event is a participant in an n-way call for $n \geq 2$, then the contraction by n-way calls is necessarily a poset. That this is not the case for $n = 1$ is illustrated in Fig. 2, where the contraction by one-way calls lacks antisymmetry or transitivity.

4.2 Example

Consider the problem of gossiping in minimal time using one-way calls, using the complete graph on n points as the communication graph. The objective is to perform a complete gossiping in as few rounds of calls as possible, where completeness (§7) means that $p \preceq q$ for every event p minimal under $\prec_1$ and every q maximal under $\prec_1$. The optimal algorithm [4, 8] uses seven rounds and is presented as a calling scheme in Fig. 3, for a communication graph of 16 points. This is the smallest graph on which this approach is better than the simple method of accumulating to one location and then broadcasting.

5 Labeled schemes

A *labeled calling scheme* is a calling scheme $(P, \prec, \prec_1)$, a directed simple graph $G = (V, E)$, and a function $\theta \colon P \to V$ such that

(L1) if $p \prec_1 q$, then $\theta(p) = \theta(q)$; and
(L2) if $p \prec_2 q$, then $(\theta(p), \theta(q)) \in E$.

If $p \prec_1 q$, then we may write $p \prec_v q$, where $v = \theta(p) = \theta(q)$.

A labeled calling scheme naturally denotes a flow of information in the following way. Each event p represents the information that is present at location $\theta(p)$ at a particular time. The relation $\prec_2$ denotes the transfer of information by calls, while $\prec_1$ denotes the retention of information at vertices of G. If the scheme is not ordered, then information at a vertex of the communication graph

Fig. 3. The fastest way to gossip among 16 points with one-way calls, represented as a calling scheme. Each vertex in the communication graph is represented as a vertical line. Diagonal lines are calls. For clarity, calls that would cross from right to left are depicted going off the right side and wrapping around to the left. If each pair of vertical lines joined in the first call is considered a column, then every round of concurrent calls involves every column, and the horizontal displacement of each call in round $(n+2)$, measured as a number of columns in the diagram, is the n-th Fibonacci number. The scheme provides a path from every spot at the bottom to every spot at the top.

might be completely or partially forgotten at times, or the vertices might have internal structure limiting the availability of information at different times.

A calling scheme can always be converted to a labeled one: each component of $\prec_1$ can be labeled with a distinct value, and a communication graph can be formed whose vertices are these values and whose edges are determined by the $\prec_2$ relation.

We shall use the term *2-contraction* to refer to the contraction by two-way calls of a two-way calling scheme. If a two-way scheme is labeled from communication graph G, then we define function γ from the calls into the edges of an undirected version of G in this way: for call $Q = \{p_1, q_1, p_2, q_2\}$ with $\theta(p_1) \neq \theta(p_2)$, $\gamma(Q)$ is the undirected edge $(\theta(p_1), \theta(p_2))$. A 2-contraction with such a function γ will be called a *labeled 2-contraction*. The labeled 2-contraction is the representational method used in [9].

In a labeled 2-contraction, we define relations $\prec_v$ among the calls for all $v \in V(G)$ by: $P_1 \prec_v Q_1$ iff $P_1 \stackrel{\cdot}{\prec} Q_1$, $v \in \gamma(P_1)$, and $v \in \gamma(Q_1)$. This is equivalent to the condition $P_1 \prec_v Q_1$ iff $p_1 \prec_v q_1$ for some $p_1 \in P_1$ and $q_1 \in Q_1$.

Theorem 1. *Let $(P, \prec, \prec_1, \theta)$ be a two-way calling scheme labeled from $G = (V, E)$, and $(\Pi, \prec_\Pi, \gamma)$ its labeled 2-contraction. Then:*

(A1) *if $P_1 \stackrel{\cdot}{\prec}_\Pi Q_1$, then $\gamma(P_1)$ and $\gamma(Q_1)$ have a vertex in common.*

If the calling scheme is ordered, then

(A2) *for any $v \in V$ and $Q_1 \in \Pi$, there is at most one $P_1 \in \Pi$ such that $P_1 \stackrel{\cdot}{\prec}_v Q_1$, and at most one $P_2 \in \Pi$ such that $Q_1 \stackrel{\cdot}{\prec}_v P_2$.*

Proof. (A1) There must be $p_1 \in P_1$, $q_1 \in Q_1$ with $p_1 \stackrel{\cdot}{\prec}_1 q_1$. Then by (L1), $\theta(p_1) = \theta(q_1)$ is in both $\gamma(P_1)$ and $\gamma(Q_1)$. (A2) Suppose $P_0 \stackrel{\cdot}{\prec}_v Q_1$ and $P_1 \stackrel{\cdot}{\prec}_v Q_1$, $P_0 \neq P_1$. By (C2) and (S3), we can say there must be $q_1 \in Q_1$, $p_0 \in P_0$, and $p_1 \in P_1$ such that $p_0 \prec_v q_1$ and $p_1 \prec_v q_1$. If the calling scheme is ordered, we can assume without loss of generality that $p_0 \prec_v p_1$, and thus $P_0 \prec_v P_1$, contradicting $P_0 \stackrel{\cdot}{\prec}_v Q_1$. A similar argument applies in the case of P_2. $\square$

The reverse also holds:

Theorem 2. *Let $(\Pi, \prec_\Pi)$ be a poset and γ a map from Π into the edge set of an undirected graph G satisfying (A1); then there is a unique two-way labeled calling scheme whose labeled 2-contraction is $(\Pi, \prec_\Pi, \gamma)$. If (A2) also holds, then the calling scheme is ordered.*

Proof. For each $P_1 \in \Pi$, where $\gamma(P_1) = (u, v)$, we create active events denoted P_1^u, P_1^v, P_{1u}, P_{1v} such that $P_{1u} \stackrel{\cdot}{\prec}_1 P_1^u$, $P_{1v} \stackrel{\cdot}{\prec}_1 P_1^v$, $P_{1u} \prec_2 P_1^v$, $P_{1v} \prec_2 P_1^u$, $\theta(P_{1u}) = \theta(P_1^u) = u$, and $\theta(P_{1v}) = \theta(P_1^v) = v$. If $P_1 \stackrel{\cdot}{\prec}_\Pi Q_1$, let u be common between $\gamma(P_1)$ and $\gamma(Q_1)$. Then we make $P_1^u \stackrel{\cdot}{\prec}_1 Q_{1u}$. This is clearly a two-way calling scheme. Clearly its 2-contraction is $(\Pi, \prec_\Pi, \gamma)$. It is uniquely determined because any such scheme must contain the calls and relationships among calls that this one does, and yet if it contained additional events or relationships among events, its labeled 2-contraction would differ. If condition (A2) holds, then each P_1^u has at most one immediate successor under $\prec_1$ and each P_{1u} has at most one immediate predecessor. By construction, each P_{1u} has only P_1^u as immediate successor and each P_1^u has only P_{1u} as immediate predecessor under $\prec_1$. Thus if (A2) holds, the calling scheme is ordered. □

While a two-way labeled calling scheme can be completely represented by its labeled 2-contraction, the contraction by one-way calls of a one-way calling scheme necessarily loses information that distinguishes information at senders from information at receivers. Thus the general calling scheme formalism must usually be used with one-way schemes and schemes with a mixture of one-way and two-way calls.

6 Types of contractions

We recognize several kinds of contractions. The first is the contraction by calls of an n-way calling scheme, as discussed above. It eliminates the $\prec_2$ relation and produces a poset on the calls of the scheme. The contraction by calls of an ordered two-way calling scheme produces the minimal order of the scheme [1].

Second, we observe that the components of $\prec_1$ represent a partition of the events of a calling scheme, so we can contract using that partition to obtain the *contraction by time*. This yields the communication graph of the calling scheme, while discarding information about the order of calls. If the links in the uncontracted Hasse diagram are labeled with integers in a manner consistent with $\prec$, the labels that survive the contraction describe a function φ that defines an information flow.

Since these two contractions omit complementary information, one wonders whether a scheme is uniquely determined by its unlabeled contractions by calls and by time. Unfortunately, this is not the case, as Fig. 4 shows. Two essentially different (non-isomorphic) calling schemes both have the same unlabeled 2-contraction and the same communication graph. (Either of the labeled 2-contraction or the information flow alone contains enough information to reconstruct a two-way calling scheme.) To see that the calling schemes are essentially different, note that the edge in the communication graph that is used twice appears in different places in the two calling schemes.

The third type of contraction is induced by the T^*-contraction of a two-way calling scheme's communication graph. Suppose a contraction of the communication graph joins vertices u and v. Then the participants of each call between u and v can be joined in a contraction of the calling scheme. That is, we contract calls as in a contraction by calls, but we contract only those calls whose participants correspond to the same vertex in the contracted communication graph. We say the T^*-*contraction* of a two-way calling scheme is the one induced in this way by the T^*-contraction of the scheme's communication graph.

These three kinds of contractions are illustrated in Fig. 5, where examples are organized in three columns. The middle column contains labeled calling schemes. The left column shows their labeled 2-contractions. The right column shows

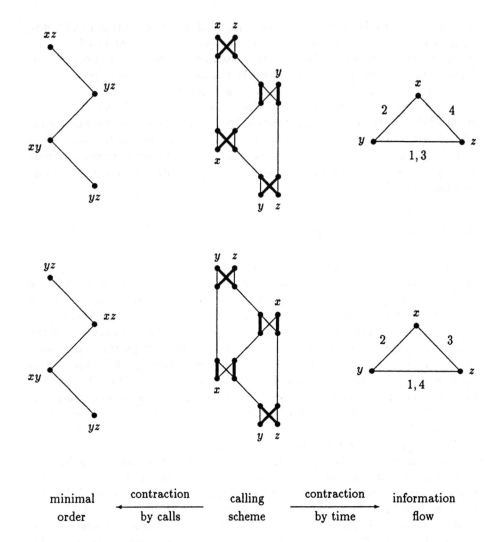

minimal   contraction   calling   contraction   information
order   by calls   scheme   by time   flow

Fig. 4. Two different calling schemes whose unlabeled contractions by calls and by time are the same. This diagram also serves to illustrate three formalisms for representing gossip schemes. Unlabeled versions of all three may be obtained by erasing letters and numbers. The multigraph form of the communication graph is obtained by duplicating the edge between y and z in the information flow.

their communication graphs, given as labeled multigraphs. The calling scheme at the top is given. Its T^*-contraction is below. There is only one nontrivial T^*-subgraph, containing u and v, and for simplicity the singleton T^*-subgraphs are denoted w, x, y, and z rather than w^*, x^*, y^*, and z^*. It can be seen that the contracted scheme is not ordered, and, indeed, that it connotes concurrent calls at vertex u^* and is thus not a two-way calling scheme. At the bottom is a modified scheme in which the two concurrent calls are performed sequentially, thus producing an ordered two-way calling scheme.

Yet another kind of contraction yields the "instances" used by Wolfson and Segall to study atomic commitment algorithms in distributed databases [13]. In an ordered, one-way calling scheme, say $p \approx q$ if $p \stackrel{.}{\prec}_1 q$ and p is a receiving event or q is a sending event. If we contract the calling scheme under the transitive closure of $\approx$, we obtain one of their "instances." Their "participant, event, message, order arc, commit instance, C-event, V-event, and boundary C-event" correspond to our "vertex of G, event, pair of events related by $\prec_2$, pair of events related by $\stackrel{.}{\prec}_1$, complete calling scheme, final event, non-final event, and convergence event," respectively. (See the next section for the definitions of the last four of these.)

7 Completeness

The above formalisms characterize the flow of information generally. In gossiping, we want information to flow from every location of the communication graph to every other.

In a one-way or two-way information flow on G, a *monotone sequence of calls* is one whose edges define a path in G and whose times are strictly increasing. The information flow is *complete* iff for every $u \in V$ and $v \in V$ there is a monotone sequence of calls defining a path from u to v. All the information flows depicted in this paper are complete. A complete regular information flow is also called a *complete gossiping*.

In a labeling of a directed or undirected multigraph, an *increasing path* is one whose edges are labeled with strictly increasing numbers. The labeling is *complete* if there is an increasing path from u to v for every $u \in V$ and $v \in V$. An *admissible labeling* [3, 5] is a complete strong labeling of an undirected graph. A *label-connected graph* [3, 5] is an undirected multigraph for which there exists an admissible labeling. The natural correspondence between information flows and multigraph labelings preserves completeness.

In a calling scheme, an event that is minimal (maximal) under $\prec_1$ will be known as a *first (last)* event. Note that these are not necessarily minimal (maximal) under $\prec$, but that in a labeled scheme they are minimal (maximal) points with a particular label. An event q such that $p_1 \preceq q$ (i.e. a common upper bound) for all first events p_1 is called a *final* event. A minimal final event is a *convergence event*. An event p such that $p \preceq q_1$ (i.e. a common lower bound) for all last events q_1 is called an *initial event*. A maximal initial event is a *divergence event*. A calling scheme need not have any final, convergence, initial, or divergence events.

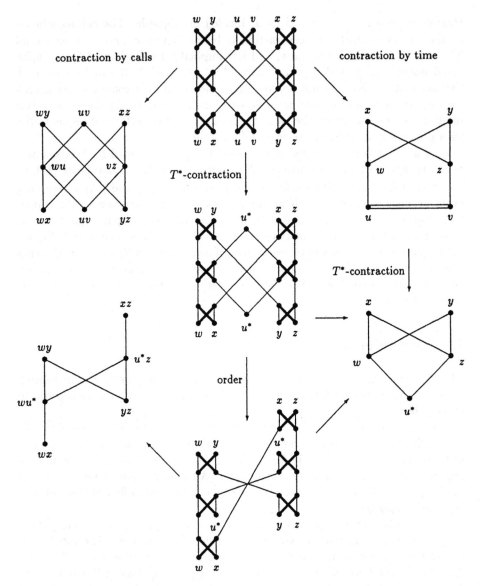

Fig. 5. Various contractions of a calling scheme. See text for details.

A calling scheme is *complete* iff $p \preceq q$ for every first event p and last event q. It is necessary and sufficient for a scheme to be complete that for each last event q there is a convergence event $p \preceq q$; similarly it is necessary and sufficient that for each first event p there is a divergence event q such that $p \preceq q$. A calling scheme is complete iff its corresponding information flow or multigraph labeling is complete.

If a call contains a (final, initial, convergence, divergence) event, then we say it is a (final, initial, convergence, divergence) call. In a 2-contraction, final (initial) calls are upper (lower) bounds of the set of minimal (maximal) calls, and convergence (divergence) calls are minimal upper bounds (maximal lower bounds) of the set of minimal (maximal) calls. A two-way calling scheme is complete iff its 2-contraction has the property that every maximal element is an upper bound of the set of all minimal elements, or, equivalently, that every minimal element is a lower bound of the set of all maximal elements. If a scheme is complete, then its T^*-contraction will be complete; the converse does not hold.

From top to bottom in Fig. 5, the number of convergence (and divergence) events in the three calling schemes are six, five, and four, respectively; the two 2-contractions have three and two convergence (and divergence) calls, respectively.

A compact representation of the 2-contraction of an unlabeled calling scheme can be obtained by deleting all final calls that are not convergence calls and all initial calls that are not divergence calls. This is called the *kernel of an information flow* [11]. This representation omits information, both because there is no labeling and because some calls are omitted. It depicts essential aspects of the structure of a two-way calling scheme. A calling scheme is complete iff the kernel of its 2-contraction has the property that every maximal element is an upper bound for all minimal elements, or, equivalently, that every minimal element is a lower bound for all maximal elements.

8 Redundancy

If a call can be deleted from a complete scheme without losing the completeness property, then the call is said to be *redundant*. Deleting a call corresponds to deleting one or more elements of $\prec_2$, depending on the type of call, and it is natural to consider elements of $\overset{\cdot}{\prec}_1$ in a similar way. In the Hasse diagram of a complete n-way calling scheme, we can say an edge $p \overset{\cdot}{\prec}_1 q$ is redundant if it can be deleted without losing the completeness property among the events that were first and last events in the given scheme. If we eliminate all such redundant elements of $p \overset{\cdot}{\prec}_1 q$ from the contraction by calls of an n-way calling scheme, we obtain the *reduced minimal order* [10].

Yet another type of redundancy arises when there are multiple paths connecting a pair of events in the Hasse diagram. A scheme with no such multiple paths is said to satisfy the NODUP condition, in that no call transmits information that duplicates information already know to the recipient. A weaker condition is the NOHO condition, meaning that no vertex hears its own information, which

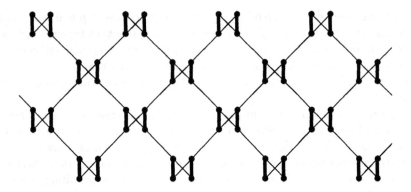

Fig. 6. Calling scheme depiction of a NOHO gossiping ("no one hears own" information) where every pair of calls possible violates the NODUP condition (no one hears information that duplicates what is already known). Links are assumed to wrap around left and right boundaries. The NOHO condition is easy to verify in this depiction since all paths based on $\overset{.}{\prec}_1$ are straight, diagonal lines.

says there is no pair of events joined by one path composed entirely from $\prec_1$ and another path of any sort.

Figure 6 shows a calling scheme on eight vertices which satisfies the NOHO property but otherwise fails the NODUP property in the worst possible way: for every pair of events connected by a path of length two or more, if the NOHO property does not prohibit multiple paths, then there are multiple paths. No edge is redundant. Every call not a first or last call is redundant. Indeed, four calls can be removed from this scheme without losing the completeness property, as shown in Fig. 7. The communication graph of that scheme (not shown), known as D_8, is used in an existence proof regarding NOHO gossiping in a minimal number of calls [12].

9 Discussion

The calling scheme formalism presented here provides several benefits. It completely characterizes the flow of information without using any type of labeling. It can be readily rendered in illustrations that show all aspects of the flow of information in a compact and aesthetically pleasing form. Finally, other major schemes can be obtained as contractions.

The minimal order, kernel of an information flow, labeled 2-contraction, etc., have been useful in proofs [1, 3, 7, 9, 10, 11] for two main reasons. The first is that calls with critical properties can be identified as lower or upper bounds of given calls, and/or as being maximal or minimal with respect to some property. This characteristic derives from the existence of the partial order, and is thus shared by calling schemes. The second utility, not so much shared by calling

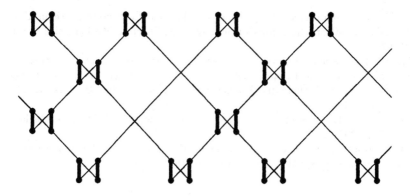

Fig. 7. The result of deleting four calls from the above scheme. This scheme is still complete and has a minimal number of calls for gossiping.

schemes, is that connectedness arguments regarding the Hasse diagram can be used. When minimal schemes are involved, one can then make graph-theoretic and counting arguments about spanning trees or forests in the Hasse diagram.

Communication graphs, the structures on which the original problems are most naturally defined, seem not to play such important roles in proofs. A notable exception is the T^*-contraction technique, which enables questions about general schemes to be reduced to questions about minimal schemes [2, 3, 9].

Of course, communication graphs necessarily figure into problems that explicitly involve its structure [2, 3, 5], and in such cases the interplay between other representations and the communication graph provides leverage.

References

1. R.T. Bumby, A problem with telephones, SIAM J. Algebraic Discrete Methods 2 (1981) 13–18.
2. G. Burosch, V.K. Leont'ev, and A.S. Markosyan, On the possibility of information propagation on graphs, Soviet Math. Doklady 37 (1988) 52–55.
3. G. Burosch and J.M. Laborde, Label-connected graphs – an overview, Technical Report 898, Institut Imag, Grenoble, France, 1992.
4. R.C. Entringer and P.J. Slater, Gossips and telegraphs, J. Franklin Institute 307 (1979) 353–360.
5. F. Göbel, J.O. Cerdeira, and H.J. Veldman, Label-connected graphs and the gossip problem, Discrete Mathematics 87 (1991) 9–40.
6. S.M. Hedetniemi, S.T. Hedetniemi, and A.L. Liestman, A survey of gossiping and broadcasting in communication networks, Networks 18 (1988) 319–349.
7. D.J. Kleitman and J.B. Shearer, Further gossip problems, Discrete Mathematics 30 (1980) 151–156.
8. D.W. Krumme, K.N. Venkataraman, and G. Cybenko. Gossiping in minimal time, SIAM J. Computing 21 (1992) 111–139.

9. D. Krumme, Reordered gossip schemes, submitted to Discrete Mathematics.

10. R. Labahn, Information flows on hypergraphs, Discrete Mathematics 113 (1993) 71–97.

11. R. Labahn, Kernels of minimum size gossip schemes, to appear in Discrete Mathematics. Also Technical Reports 92774 (Part I) and 92785 (Part II), Forschungsinstitut für Diskrete Mathematik, Rheinische Friedrich-Wilhelms-Universität, Nassestr 2, D-5300 Bonn, Germany, 1992.

12. D. West, A class of solutions to the gossip problem, Part I, Discrete Mathematics 39 (1982) 307–326.

13. O. Wolfson and A. Segall, The communication complexity of atomic commitment and of gossiping, SIAM J. Computing 20 (1991) 423–450.

ON THE MULTIPLY-TWISTED HYPERCUBE

Priyalal Kulasinghe
Department of Elect. & Comp. Engineering
Louisiana State University, Baton Rouge, LA 70803

and

Said Bettayeb
Department of Computer Science
Louisiana Sate University, Baton Rouge, LA 70803

Abstract In this paper, we prove that the multiply-twisted hypercube is a Cayley graph and hence it possesses the desirable properties such as vertex symmetry, optimal fault tolerance, and small node degree. We also prove the conjecture that the $2^n - 1$ node complete binary tree is a subgraph of the 2^n node multiply twisted hypercube.

1 Introduction

Recently, a new version of the hypercube, called multiply-twisted hypercube, was introduced [4]. The Multiply-twisted hypercube and the ordinary hypercube have the same set of vertices but the set of edges are different. The twisted hypercube introduced originally in [5] is obtained from the ordinary hypercube by interchanging the end points of only two edges. The multiply-twisted hypercube presented in [4] is a more generalized version of the twisted hypercube. The multiply-twisted hypercube has some superior properties over the conventional hypercube. Its diameter is half the diameter of the ordinary hypercube. Certain *SIMD* algorithms, such as matrix multiplication and sorting, require half the number of communication steps compared to the ordinary hypercube [4]. Efficient routing algorithms were presented for the multiply-twisted hypercube in [4, 8].

Interconnection networks belonging to the class of graphs, called Cayley graphs, are very popular for multiprocessor systems [2]. This popularity is due to many desirable properties such as symmetry, optimal fault tolerance, and small node degree. Many of the interconnection networks such as hypercube, cube connected cycles, star, and pancake belong to this class. Whether the multiply-twisted hypercube belongs to this class or not has not been established before. In this paper, we show that the multiply-twisted hypercube also belong to this class. This result will increase the usage of the multiply twisted hypercube in multiprocessor systems. Furthermore, Cayley graph representation gives us an easier method to represent adjacencies among vertices in the multiply-twisted hypercube. The multiply-twisted hypercube of dimension n will be denoted by TQ_n.

A binary tree is a very important data structure encountered in many applications. As a consequence, a great deal of research has been done on embedding binary trees into various interconnection networks [1, 3]. A complete binary tree with $2^n - 1$ vertices cannot be embedded into Q_n with dilation 1; but it can be embedded into Q_n with dilation 2 [3]. Whether the complete binary tree with $2^n - 1$ can be embedded into TQ_n with dilation 1 has stayed as an open problem. In [4], it was shown that TQ_n can be reconfigured such that it can embed the complete binary tree with dilation 1. In [1], a method was described to embed the complete binary tree into TQ_n with dilation 2. In [8], it was conjectured that the complete binary tree on $2^n - 1$ vertices can be embedded into TQ_n with dilation 1. In this paper, we prove that the conjecture is true.

The rest of the paper is organized as follows. Section 2 provides preliminaries and the conventional definition of the multiply-twisted hypercube. Section 3 proves that multiply-twisted hypercube TQ_n is the Cayley graph associated with a certain finite group and one of its generating sets. Section 4 proves that the complete binary tree with $2^n - 1$ vertices can be embedded into TQ_n with dilation 1. We conclude the paper in Section 5.

2 Preliminaries

The n-dimensional hypercube Q_n and the n-dimensional twisted hypercube TQ_n have the same set of vertices. We will represent the address of each vertex in TQ_n as in Q_n, as a binary string of length n. (Later, we will also represent vertices as binary vectors of length n.) We would not distinguish between vertices and their binary addresses. In Q_n, two vertices are adjacent if and only if their binary addresses differ only in one bit position. For the multiply-twisted hypercube, adjacency requirement is little more involved.

Definition 1: Two binary strings $x = x_1x_0$ and $y = y_1y_0$ of length two are said to be *pair related* (denoted by $x \sim y$) if and only if $(x, y) \in \{(00, 00), (10, 10), (01, 11), (11, 01)\}$. If x and y are not pair related, we write $x \nsim y$.

The n dimensional multiply-twisted hypercube TQ_n is recursively defined as follows. TQ_1 is the complete (undirected) graph on two vertices whose addresses are 0 and 1. TQ_n consists of two subcubes TQ_{n-1}^0 and TQ_{n-1}^1. The most significant bit of the addresses of the vertices in TQ_{n-1}^0 (TQ_{n-1}^1) is 0 (1). The vertices $u = u_{n-1}u_{n-2}...u_1u_0 \in TQ_{n-1}^0$ with $u_{n-1} = 0$, and $v = v_{n-1}v_{n-2}...v_1v_0 \in TQ_{n-1}^1$ with $v_{n-1} = 1$, are joined by an edge in TQ_n if and only if
(1) $u_{n-2} = v_{n-2}$ if n is even, and
(2) $u_{2i+1}u_{2i} \sim v_{2i+1}v_{2i}$, for $0 \le i < \lfloor (n - 1)/2 \rfloor$.

An edge joining a vertex in TQ_{n-1}^0 and a vertex in TQ_{n-1}^1 will be said to belong to *dimension* $n - 1$. It is clear from the definition, that each vertex is incident with n edges belonging to dimensions 0 through $n - 1$. Therefore, TQ_n is an n-regular graph. In Section 3, we prove that TQ_n is indeed a Cayley graph. Therefore, TQ_n is vertex symmetric. Next, we extend the definition of the term 'pair related' to involve binary strings of length more than two.

Definition 2: Binary strings $x = x_{k-1}x_{k-2}...x_1x_0$ and $y = y_{k-1}y_{k-2}...y_1y_0$ $(k > 0)$ of the same length are said to be *pair related relative to dimension* τ if there exists a constant τ, for $0 \le \tau < k$, such that

(a) $y_i = x_i$ for $\tau < i < k$.

(b) $y_\tau = \overline{x}_\tau$

(c) $y_{\tau-1} = x_{\tau-1}$ if τ is even.

(d) $y_{2i+1}y_{2i} \sim x_{2i+1}x_{2i}$, for $0 \le i \le \lfloor (\tau - 2)/2 \rfloor$.

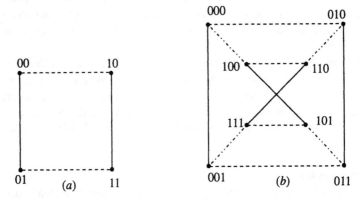

Fig. 1. (a) TQ_2, (b) TQ_3.

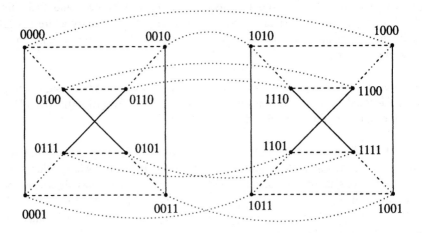

Fig. 2. TQ_4

From the definition, it is clear that two vertices in TQ_n are adjacent along dimension τ if and only if their binary addresses are pair related relative to dimension τ. Figure 1 shows TQ_2 and TQ_3. Figure 2 shows TQ_4. Different line styles are used to represent edges belonging to different dimensions. Notice that the multiply-twisted hypercube with two or less dimensions is isomorphic with its hypercube counterpart. But when the number of dimensions is 3 or more, they are no longer isomorphic. For example, TQ_3 shown in Figure 1(b) is not isomorphic with Q_3. In the next section we prove that TQ_n is the Cayley graph associated with a certain group and one of its generating sets.

3 Representation of TQ_n as a Cayley graph

Let Γ be a group and let Δ be a generating set for Γ. Then we associate a Cayley graph, denoted by $G_\Delta(\Gamma)$, with Γ and Δ. Vertices of $G_\Delta(\Gamma)$ are the group elements. Two vertices γ_1 and γ_2 in $G_\Delta(\Gamma)$ are adjacent if and only if $\gamma_2 = \gamma_1\delta$ (or $\gamma_1 = \gamma_2\delta$), for some $\delta \in \Delta$. Hypercubes, Cube Connected Cycles, Pancakes, and Stars are some popular interconnection networks which are Cayley graphs. Usually, groups associated with the above Cayley graphs are expressed as permutation groups. In a permutation group, each element is a permutation on a given set of symbols. In fact, by Cayley's theorem [7], any group is isomorphic with a permutation group. In this paper, in order to prove that the multiply-twisted hypercube is a Cayley graph, we use a different representation. We consider group elements as binary vectors. However, for notational convenience, we will write a binary vector as a binary string.

Let T_n be the set of all binary vectors of length n. Clearly, there are 2^n elements in T_n. We will use upper case letters to represent vectors in T_n and lower case letters to represent the elements (bits) of a vector. For example, if A is a vector in T_n, then $A = a_{n-1}a_{n-2}...a_1a_0$. Element a_i can take only two values, 0 or 1. A *unit vector* is a vector which has 0 in all bit positions except in one. The all zero vector in T_n will be denoted by **0**.

Definition 3: The bit position of the leading 1 in vector A will be denoted by $\eta(A)$.

Definition 4: Let A be any vector in T_n. Then $B = \zeta(A, j)$ is defined as follows.
(1) $b_i = a_i$, for $n > i > j$
(2) $b_j = \bar{a}_j$
(3)   if j is odd, then $b_{j-1} = a_{j-1}$
(4) $b_{2i+1}b_{2i} \sim a_{2i+1}a_{2i}$, for $0 \le i \le \lfloor (j - 2)/2 \rfloor$.

For example, $\zeta(11011, 2) = 11101$. The following observation can be made as a consequence of previous definitions. Let A and B be two vectors in T_n. Then the following three statements are equivalent.
(a)   There is an integer $j < n$ such that $B = \zeta(A, j)$.
(b)   The binary strings corresponding to the two vectors A and B are pair related.
(c)   The vertices in TQ_n corresponding to the vectors A and B are adjacent.

Now, we define the binary operator $\otimes$ for the vectors in T_n in algorithmic form as follows.

Definition 5: Let A and B be two vectors in T_n. Then the following procedure computes $A \otimes B$.

```
procedure composition;
        input: vectors A and B;
        output: A ⊗ B;
```

```
temporary variables: X and Y;
    begin
    X = A; Y = B;
    while (Y ≠ 0) do
        begin
        push η(Y) into the stack;
        Y = ζ(Y, η(Y));
        end;
    while (not the top of the stack) do
        begin
        pop the stack into j;
        X = ζ(X, j);
        end;
    output A ⊗ B = X;
    end.
```

Due to the limitation of space, we state the following theorem without the proof. The interested reader can refer to [6] for a proof.

Theorem 1: T_n forms a group under the composition $\otimes$. Furthermore, the set of unit vectors in T_n is a generating set for the group T_n.

Notice that, if U is a unit vector which has 1 at bit position j, then $A \otimes U = \zeta(A, j)$, for any vector A. Also notice that the group T_n is not commutative in general. Furthermore, each unit vector is its self inverse. We denote the set of unit vectors in T_n by Δ_n. The following theorem is a direct result of the previous observations (a), (b), and (c).

Theorem 2: The multiply-twisted hypercube TQ_n is the Cayley graph associated with the group T_n and its generating set Δ_n.

Therefore, TQ_n possesses all the desirable properties of Cayley graphs. In particular, TQ_n is vertex symmetric, that is, for every pair of vertices x and y, there exists an automorphism of TQ_n, which maps x to y. However, unlike the case for the conventional hypercube Q_n, TQ_n is not edge symmetric in general. It is edge symmetric for $n < 3$. In the next section, we prove that the complete binary tree with $2^n - 1$ vertices can be embedded into TQ_n with dilation 1. We denote the complete binary tree with $2^n - 1$ vertices by B_n.

4 Embedding of B_n into TQ_n

In the rest of the paper, unless otherwise stated, embedding implies dilation 1 embedding. Since $|TQ_n| = |B_n| + 1$, in any one-to-one embedding of B_n into TQ_n, there will be a vertex in TQ_n which is not the image of any vertex of B_n. This vertex will be called the *extra vertex* of the embedding. We will first obtain some structural properties of the multiply-twisted hypercube.

Definition 6: Let P and Q be two vertex-induced subgraphs of a graph G. Then the subgraph of G induced by $V(P) \cup V(Q)$ will be denoted by $P \cup Q$.

Definition 7: $TQ_n(s)$ is the subgraph of TQ_n induced by the set of vertices whose addresses have the same prefix s. The following lemma is straightforward from the definition.

Lemma 1: $TQ_n(s)$ is isomorphic with $TQ_{n-|s|}$.

Theorem 3: Let s and t be two binary strings such that $|s| = |t| = l < n$ and $n - l$ is even. If s and t are pair related, then $TQ_n(s) \cup TQ_n(t)$ is isomorphic with TQ_{n-l+1}.
Proof: Since $n > l$ and $n - l$ is even, $n - l \geq 2$. For a vertex $u = su_{n-l-1}u_{n-l-2}...u_1u_0$ in $TQ_n(s)$, define the mapping ψ_s: $u \rightarrow 0u_{n-l-1}u_{n-l-2}...u_1u_0$. Also, for a vertex $v = tv_{n-l-1}v_{n-l-2}$ $...v_1v_0$ in $TQ_n(t)$, define the mapping ψ_t: $v \rightarrow 1v_{n-l-1}v_{n-l-2}...v_1v_0$. It is clear that, both $\psi_s(u)$ and $\psi_t(v)$ are in TQ_{n-l+1}. Let u^1 and u^2 be two adjacent vertices in $TQ_n(s)$. Then the two addresses u^1 and u^2 are pair related. Therefore, since $n - l$ is even by the hypothesis, $\psi_s(u^1)$ and $\psi_s(u^2)$ are also pair related. Hence $\psi_s(u^1)$ and $\psi_s(u^2)$ are adjacent in TQ_{n-l+1}. Similarly, if v^1 and v^2 are two adjacent vertices in $TQ_n(t)$, then $\psi_t(v^1)$ and $\psi_t(v^2)$ are adjacent in TQ_{n-l+1}. Now suppose that the two vertices $u \in TQ_n(s)$ and $v \in TQ_n(t)$ are adjacent in TQ_n, that is, they are adjacent in $TQ_n(s) \cup TQ_n(t)$. Then the strings u and v are pair related. Since u and v belong to different subgraphs ($TQ_n(s)$ and $TQ_n(t)$, respectively), adjacency must be along a dimension $j \geq n - l$. Therefore, $u_{2i+1}u_{2i} \sim v_{2i+1}v_{2i}$, for $0 \leq i \leq (n - l - 2)/2^\dagger$. Therefore, $0u_{n-l-1}u_{n-l-2}...u_1u_0 = \psi_s(u)$ and $1v_{n-l-1}v_{n-l-2}$ $...v_1v_0 = \psi_t(v)$ are pair related. That is, $\psi_s(u)$ and $\psi_t(v)$ are adjacent in TQ_{n-l+1}. Thus the mappings ψ_s and ψ_t preserve adjacency, and therefore, $TQ_n(s) \cup TQ_n(t)$ and TQ_{n-l+1} are isomorphic. □

Lemma 2 : When $n = 4k + 1$, for any non negative integer k, B_n can be embedded into TQ_n such that the root and the extra vertex are adjacent along the highest dimension.
Proof: We use induction on k. The base case ($k = 0$) corresponds to the embedding of B_1 into TQ_1. It is trivially true. Now suppose that the lemma is true for $k = 0$ through τ. We prove that the lemma is true for $k = \tau + 1$. That is, we prove that $B_{4\tau+5}$ can be embedded into $TQ_{4\tau+5}$ such that the root and the extra vertex are adjacent along the highest dimension. The multiply-twisted hypercube $TQ_{4\tau+5}$ can be considered as consisting of 32 ($= 2^5$) supernodes such that each supernode is a subgraph isomorphic with $TQ_{4\tau}$ (Lemma 1). Vertices in $TQ_{4\tau+5}$ belonging to the same supernode have the same 5-bit prefix. Let S_x be the supernode consisting of vertices whose 5-bit prefix is the binary representation of x, for $0 \leq x \leq 31$. Define the *origin* of a supernode as the vertex in the supernode whose last 4τ bits are all zeros. For example, the address of the origin of the supernode S_5 is 0010100...00. Figure 3 shows the supernodes (represented as circles) of $TQ_{4\tau+5}$ in four rows; the first

---

†Note that, by the hypothesis in the theorem, $n - l - 2$ is even and non negative.

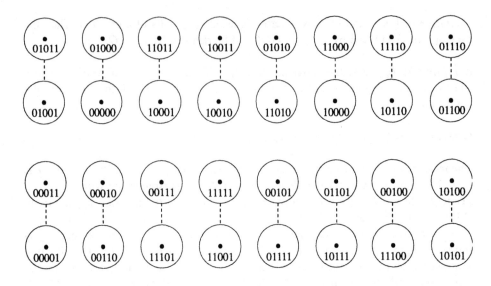

Fig. 3. Supernodes of $TQ_{4\tau+5}$. Label of each supernode is the 5-bit prefix of the vertices belonging to the supernode.

row is at the top and the fourth row is at the bottom. Each supernode is labeled with the corresponding 5-bit prefix. The vertex shown in each supernode is its origin. Two supernodes joined by a dotted line in Figure 3, are said to form a *couple*. Careful observation of the labels of the supernodes in Figure 3 reveals that labels of a couple are pair related. Let S_x and S_y be a couple such that S_x is in row 1 (4) and S_y is in row 2 (3). Then, according to Theorem 3, since the 5-bit binary strings x and y are pair related, $S_x \cup S_y$ is isomorphic with $TQ_{4\tau+1}$. Furthermore, according to the induction hypothesis, $B_{4\tau+1}$ can be embedded into $TQ_{4\tau+1}$ such that the root and the extra vertex are adjacent along the highest dimension. By symmetry, the root can be mapped to any vertex in $TQ_{4\tau+1}$. Moreover, by construction, the origins of S_x and S_y are adjacent in $S_x \cup S_y$ along the highest dimension. Therefore, $B_{4\tau+1}$ can be embedded into $S_x \cup S_y$ such that the root is at the origin of S_x and the extra vertex is at the origin of S_y. Suppose we embed $B_{4\tau+1}$ into each such S_x - S_y couple in Figure 3. Then, as shown in Figure 4, we can connect the origins of the supernodes in order to form the complete binary tree $B_{4\tau+5}$. A triangle is attached to the origin of each of the supernodes in rows 1 and 4 to indicate that it is the root of the corresponding subtree $B_{4\tau+1}$. The root of $B_{4\tau+5}$ in Figure 4 is the origin of the supernode S_0; that is, the vertex whose label consists of all zeros. Also, the extra vertex is the origin of the supernode S_{16}; that is the vertex whose label consists of all zeros except at the most significant bit position. Thus, we have found an embedding of $B_{4\tau+5}$ into $TQ_{4\tau+5}$ such that the root and the extra vertex are adjacent along the highest dimension. $\square$

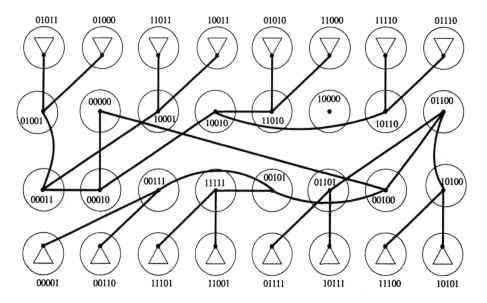

Fig. 4. Embedding of $B_{4\tau+5}$ into $TQ_{4\tau+5}$. Each triangle represents a subtree $B_{4\tau+1}$ which spans the corresponding supernode couple.

Lemma 3: When $n = 4k + 7$, for any non negative integer k, B_n can be embedded into TQ_n such that the root and the extra vertex are adjacent along the highest dimension.
Proof: Again, we use induction on k. The induction step is similar to the one used in the previous lemma. We need only to prove the base case ($k = 0$). That is, we should prove that B_7 can be embedded in TQ_7 such that the root and the extra vertex are adjacent along the highest dimension. We provide such an embedding. Figures 5a, 5b, 5c, and 5d represent vertices of TQ_7 with prefixes 00, 01, 11, and 10 respectively. In Figure 5a, a complete subtree is constructed with root 00\01010 (for readability, we use '\' to separate 2-bit prefix from the rest) such that one leaf of the vertex 00\10000 (marked as x) is missing, and the vertices 00\00000 and 00\01000 are unused. In Figure 5b, a complete subtree is constructed such that the root is 01\10000 and the extra vertex is 01\00000. Similarly, in Figure 5c, a complete subtree is constructed such that the root is 11\00000 and the extra vertex is 11\10000. Also, in Figure 5d, a complete subtree is constructed such that the root is 10\01000 and the extra vertex is 10\00000. Figure 6 shows how those subtrees can be connected together to obtain the complete binary tree B_7, where the missing leaf of 'x' in Figure 5a is the extra vertex of Figure 5c, namely 11\10000. Notice that the root is 00\00000 and the extra vertex is 10\00000. □

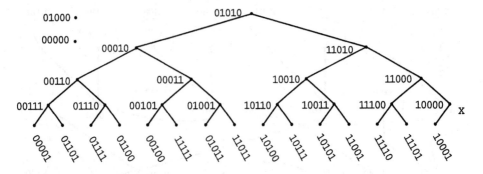

Fig. 5a. Subtree with prefix 00. The root is 01010. Vertices 00000 and 01000
are unused and the leaf of the vertex marked 'x' is missing.

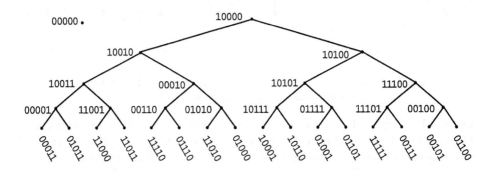

Fig. 5b. Complete with prefix 01. The root and
the extra vertex are 10000 and 00000, respectively.

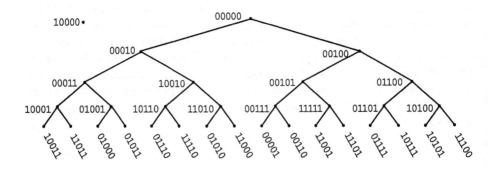

Fig. 5c. The complete subtree with prefix 11. The root and
the extra vertex are 00000 and 10000, respectively.

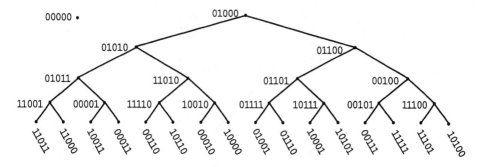

Fig. 5d. The complete subtree with prefix 10. The root and the extra vertex are 01000 and 00000, respectively.

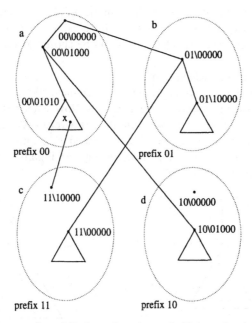

Fig. 6. Construction of B_7 from the subtrees in Figures 5.a through 5.d. The root and the extra vertex are 00\00000 and 10\00000, respectively.

Lemma 4: For any odd n, B_n can be embedded into TQ_n such that the root is adjacent to the extra vertex.

Proof: Lemmas 2 and 3 provide the proofs for every odd n except for $n = 3$. It remains to prove that B_3 can be embedded into TQ_3 such that the root and the extra vertex are adjacent. Figure 7 shows such an embedding. The root and the extra vertex are 000 and 010, respectively. □

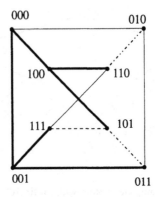

Fig. 7. Embedding of B_3 into TQ_3. The root and
the extra vertex are 000 and 010, respectively.

Notice that, in each of the embeddings in the proofs of Lemmas 2, 3, and 4, the root is the vertex whose label consists of all zeros. By vertex symmetry of the multiply-twisted hypercube, any other vertex can be chosen as the root.

Theorem 4: B_n can be embedded into TQ_n.
Proof: When n is odd, the proof is provided by Lemma 4. When n is even, let TQ_{n-1}^0 and TQ_{n-1}^1 be the subcubes of dimension $n - 1$. Now embed the subtree B_{n-1}^0 in TQ_{n-1}^0 such that the root R_0 is adjacent with the extra vertex X_0. Let R_1 be the vertex in TQ_{n-1}^1 which is adjacent with X_0 in TQ_{n-1}^0. Embed B_{n-1}^1 into TQ_{n-1}^1 such the root is at R_1 (position of the extra vertex X_1 is irrelevant). As Figure 8 shows, now we can obtain the embedding of B_n into TQ_n. □

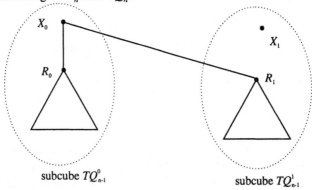

Fig. 8. Construction of B_n from two subtrees B_{n-1}.

5. Conclusion

The recently introduced interconnection network multiply-twisted hypercube has some properties which are better than the ordinary hypercube. We prove that TQ_n is the Cayley graph associated with group T_n and its generating set Δ_n. Due to inherent desirable properties of a Cayley graph, this result makes the multiply-twisted hypercube more attractive for use as an interconnection network. We also prove that the complete binary tree with $2^n - 1$ can be embedded into TQ_n with dilation 1. This provides an attractive property of the multiply-twisted hypercube which is not found in the ordinary hypercube.

References

1. E. Abuelrub and S. Bettayeb, "Embedding of Complete Binary trees into Twisted Hypercubes", Proceedings of the International Conference on Computer Applications in Design, Simulation and Analysis, 1993, pp 1 - 4.

2. S. B. Akers and B. Krishnamurthy, "A Group-Theoretic Model for Symmetric Interconnection Networks", IEEE Trans. on Computers, Vol. 38, No. 4, April 1989, pp 555 - 566.

3. S. Bhatt and I. Ipsen, "How to embed trees in Hypercubes", Research Report 443, Dept. of Computer Science, Yale University, 1985.

4. K. Efe, "The Crossed Cube Architecture for Parallel Computing", IEEE Trans. on Parallel and Distributed Systems, Vol. 3, No. 5, September 1992, pp 513 - 524.

5. A. H. Esfahanian, Lional M. Ni, and Bruce E. Sagan, "The Twisted N-Cube with Application to Multiprocessing", IEEE Trans. on Computers, Vol. 40, No. 1, January 1991, pp 88 - 93.

6. D. P. Kulasinghe and S. Bettayeb, "Cayley Graph Representation of the Multiply-Twisted Hypercube", Tech. Rept. No. TR# 93 - 006, Department of Computer Science, Louisiana State University, Baton Rouge.

7. C. C. Pinter, "A Book of Abstract Algebra", McGraw-Hill Publishing Company, 1990.

8. S. Q. Zheng, "SIMD Data Communication Algorithms for Multiply Twisted Hypercube", Proceedings of the 5th IEEE International Parallel Proc. Symposium, 1991, pp 120 - 125.

Springer-Verlag
and the Environment

We at Springer-Verlag firmly believe that an international science publisher has a special obligation to the environment, and our corporate policies consistently reflect this conviction.

We also expect our business partners – paper mills, printers, packaging manufacturers, etc. – to commit themselves to using environmentally friendly materials and production processes.

The paper in this book is made from low- or no-chlorine pulp and is acid free, in conformance with international standards for paper permanency.

Lecture Notes in Computer Science

For information about Vols. 1–724
please contact your bookseller or Springer-Verlag

Vol. 762: K. W. Ng, P. Raghavan, N. V. Balasubramanian, F. Y. L. Chin (Eds.), Algorithms and Computation. Proceedings, 1993. XIII, 542 pages. 1993.

Vol. 763: F. Pichler, R. Moreno Díaz (Eds.), Computer Aided Systems Theory – EUROCAST '93. Proceedings, 1993. IX, 451 pages. 1994.

Vol. 764: G. Wagner, Vivid Logic. XII, 148 pages. 1994. (Subseries LNAI).

Vol. 765: T. Helleseth (Ed.), Advances in Cryptology – EUROCRYPT '93. Proceedings, 1993. X, 467 pages. 1994.

Vol. 766: P. R. Van Loocke, The Dynamics of Concepts. XI, 340 pages. 1994. (Subseries LNAI).

Vol. 767: M. Gogolla, An Extended Entity-Relationship Model. X, 136 pages. 1994.

Vol. 768: U. Banerjee, D. Gelernter, A. Nicolau, D. Padua (Eds.), Languages and Compilers for Parallel Computing. Proceedings, 1993. XI, 655 pages. 1994.

Vol. 769: J. L. Nazareth, The Newton-Cauchy Framework. XII, 101 pages. 1994.

Vol. 770: P. Haddawy (Representing Plans Under Uncertainty. X, 129 pages. 1994. (Subseries LNAI).

Vol. 771: G. Tomas, C. W. Ueberhuber, Visualization of Scientific Parallel Programs. XI, 310 pages. 1994.

Vol. 772: B. C. Warboys (Ed.),Software Process Technology. Proceedings, 1994. IX, 275 pages. 1994.

Vol. 773: D. R. Stinson (Ed.), Advances in Cryptology – CRYPTO '93. Proceedings, 1993. X, 492 pages. 1994.

Vol. 774: M. Banâtre, P. A. Lee (Eds.), Hardware and Software Architectures for Fault Tolerance. XIII, 311 pages. 1994.

Vol. 775: P. Enjalbert, E. W. Mayr, K. W. Wagner (Eds.), STACS 94. Proceedings, 1994. XIV, 782 pages. 1994.

Vol. 776: H. J. Schneider, H. Ehrig (Eds.), Graph Transformations in Computer Science. Proceedings, 1993. VIII, 395 pages. 1994.

Vol. 777: K. von Luck, H. Marburger (Eds.), Management and Processing of Complex Data Structures. Proceedings, 1994. VII, 220 pages. 1994.

Vol. 778: M. Bonuccelli, P. Crescenzi, R. Petreschi (Eds.), Algorithms and Complexity. Proceedings, 1994. VIII, 222 pages. 1994.

Vol. 779: M. Jarke, J. Bubenko, K. Jeffery (Eds.), Advances in Database Technology — EDBT '94. Proceedings, 1994. XII, 406 pages. 1994.

Vol. 780: J. J. Joyce, C.-J. H. Seger (Eds.), Higher Order Logic Theorem Proving and Its Applications. Proceedings, 1993. X, 518 pages. 1994.

Vol. 781: G. Cohen, S. Litsyn, A. Lobstein, G. Zémor (Eds.), Algebraic Coding. Proceedings, 1993. XII, 326 pages. 1994.

Vol. 782: J. Gutknecht (Ed.), Programming Languages and System Architectures. Proceedings, 1994. X, 344 pages. 1994.

Vol. 783: C. G. Günther (Ed.), Mobile Communications. Proceedings, 1994. XVI, 564 pages. 1994.

Vol. 784: F. Bergadano, L. De Raedt (Eds.), Machine Learning: ECML-94. Proceedings, 1994. XI, 439 pages. 1994. (Subseries LNAI).

Vol. 785: H. Ehrig, F. Orejas (Eds.), Recent Trends in Data Type Specification. Proceedings, 1992. VIII, 350 pages. 1994.

Vol. 786: P. A. Fritzson (Ed.), Compiler Construction. Proceedings, 1994. XI, 451 pages. 1994.

Vol. 787: S. Tison (Ed.), Trees in Algebra and Programming – CAAP '94. Proceedings, 1994. X, 351 pages. 1994.

Vol. 788: D. Sannella (Ed.), Programming Languages and Systems – ESOP '94. Proceedings, 1994. VIII, 516 pages. 1994.

Vol. 789: M. Hagiya, J. C. Mitchell (Eds.), Theoretical Aspects of Computer Software. Proceedings, 1994. XI, 887 pages. 1994.

Vol. 790: J. van Leeuwen (Ed.), Graph-Theoretic Concepts in Computer Science. Proceedings, 1993. IX, 431 pages. 1994.

Vol. 791: R. Guerraoui, O. Nierstrasz, M. Riveill (Eds.), Object-Based Distributed Programming. Proceedings, 1993. VII, 262 pages. 1994.

Vol. 792: N. D. Jones, M. Hagiya, M. Sato (Eds.), Logic, Language and Computation. XII, 269 pages. 1994.

Vol. 793: T. A. Gulliver, N. P. Secord (Eds.), Information Theory and Applications. Proceedings, 1993. XI, 394 pages. 1994.

Vol. 794: G. Haring, G. Kotsis (Eds.), Computer Performance Evaluation. Proceedings, 1994. X, 464 pages. 1994.

Vol. 795: W. A. Hunt, Jr., FM8501: A Verified Microprocessor. XIII, 333 pages. 1994.

Vol. 796: W. Gentzsch, U. Harms (Eds.), High-Performance Computing and Networking. Proceedings, 1994, Vol. I. XXI, 453 pages. 1994.

Vol. 797: W. Gentzsch, U. Harms (Eds.), High-Performance Computing and Networking. Proceedings, 1994, Vol. II. XXII, 519 pages. 1994.

Vol. 798: R. Dyckhoff (Ed.), Extensions of Logic Programming. Proceedings, 1993. VIII, 362 pages. 1994.

Vol. 799: M. P. Singh, Multiagent Systems. XXIII, 168 pages. 1994. (Subseries LNAI).

Vol. 800: J.-O. Eklundh (Ed.), Computer Vision – ECCV '94. Proceedings 1994, Vol. I. XVIII, 603 pages. 1994.

Vol. 801: J.-O. Eklundh (Ed.), Computer Vision – ECCV '94. Proceedings 1994, Vol. II. XV, 485 pages. 1994.

Vol. 802: S. Brookes, M. Main, A. Melton, M. Mislove, D. Schmidt (Eds.), Mathematical Foundations of Programming Semantics. Proceedings, 1993. IX, 647 pages. 1994.

Vol. 803: J. W. de Bakker, W.-P. de Roever, G. Rozenberg (Eds.), A Decade of Concurrency. Proceedings, 1993. VII, 683 pages. 1994.

Vol. 804: D. Hernández, Qualitative Representation of Spatial Knowledge. IX, 202 pages. 1994. (Subseries LNAI).

Vol. 805: M. Cosnard, A. Ferreira, J. Peters (Eds.), Parallel and Distributed Computing. Proceedings, 1994. X, 280 pages. 1994.

Vol. 806: H. Barendregt, T. Nipkow (Eds.), Types for Proofs and Programs. VIII, 383 pages. 1994.